GENERAL BIOLOGY I

Molecules, Cells, and Genes

DENNIS HOLLEY

Illustrated by George Barile Accurate Art Inc.
Production Manager Louis C. Bruno, Jr.
Interior Design by John Arbutante

First published by Dog Ear Publishing
4011 Vincennes Road
Indianapolis, IN 46268
www.dogearpublishing.net

ISBN: 978-145755-274-8

CONTENTS

UNIT TWO: STRUCTURE OF LIFE

Chapter 6 -   Cell Processes: Transport and Metabolism 93

Chapter 7 -   Cell Processes: Respiration 109

UNIT FOUR: EVOLUTION

Chapter 17 - The Human Condition: Rise of the Cultural Ape....................................299

PREFACE

Greetings biology student and welcome to the always astonishing, sometimes strange, and occasionally even bizarre realm of life and living things.

> *Biology gives you a brain. Life turns it into a mind.*

> —Jeffery Eugenides

Biology or any scientific endeavor should be thought of as consisting of two phases: the first being the *Investigation* and *Exploration* phase and the final is the *Accumulation* phase. Biologists attempt to answer questions about life and living things by actively investigating organisms through experimentation and by discovering new organisms through exploration. Investigation and exploration lead to the accumulation of facts and information. These accumulated facts and information lead to even more questions that, in turn, lead to more investigation resulting in even more facts and information being accumulated. And the cycle continues.

In this course, you will confront the facts and concepts of biology in your textbook (*Accumulation*). However, you will also be challenged to think, act, and work like a biologist (*Investigation*) at certain points in your textbook, and especially in the laboratory segment of this course. As you investigate, you will use the same information, develop the same scientific skills, and employ the same scientific processes as do professional biologists.

Science Process Skills

Organizing Information

- Classify
- Sequence
- Describe
- Summarize
- Explain
- Definition and proper use of terminology
- Accessing and using reference materials
- Reading comprehension

Critical Thinking

- Critical and creative thinking
- Observe
- Infer
- Compare and contrast
- Recognize cause and effect
- Formulate and use models

Experimentation

- Experimental design
- Formulate hypothesis/prediction
- Establish variables and controls
- Collect and organize data
- Accurate measurement
- Analyze data
- Draw reasonable conclusions

Graphics and Numbers

- Make and interpret graphs
- Construct and interpret tables
- Interpret scientific illustrations
- Calculate and compute

Communication

- Brainstorming
- Collaboration
- Communicating

Developing and using these skills effectively is very important if you are biology major, but even if you are not majoring in a scientific field, mastering these skills will help you function as a clear-thinking and scientifically literate citizen of a society that grows ever more science-based and technologically oriented.

Approach and Organization

Approach Biology textbooks and related curricular materials at all levels have come under harsh but justified criticism by various scientific and educational groups in the past decade. From the inception of this text, it has been the goal to write a general biology program that acts on the criticisms and recommendations of those authorities and is based on current educational research. This textbook has been designed and written to be:

> ➢ **Readable and Interesting**. The goal has been to write a textbook in which the chapters read more like an interesting magazine or newspaper article and less like a dry and detailed technical

entry from an encyclopedia. Increasing reader interest increases readability and to aid in that goal, I include out-of-the-ordinary things in each chapter that would not normally be found in general biology texts. I have also taken a different approach than other zoology books in that while I firmly believe that evolution is driving force and cornerstone of all things biological, I did not make the theoretical and often speculative aspects of origins and patterns of evolution the focal point of each chapter. Instead, I opted for a more concrete "here-and-now" approach in which our focus is mainly on animal systematics, phyla and class characteristics, and ecology. Hopefully, less emphasis on the theoretical translates into a work that is more relevant to you the student.

➢ **Understandable.** As I wrote this textbook, I tried to avoid the "Huh? Factor" as much as possible. That is; you as a student should not be obliged to reread a passage several times all the while armed with a biological dictionary to understand what you just read. The chapters of this textbook are centered on concepts and ideas. Specific facts, terms and terminology, and scientific names are used only when necessary and appropriate to illustrate and explain the concepts and ideas inherent in a particular chapter. This textbook is concept (idea) driven, not terminology (definitions) driven.

➢ **Connected.** Living things are all around us, on us and possibly in us, and they affect our daily lives directly and indirectly in ways we are continuing to uncover. In an attempt to connect you the reader directly to the living things around you, each chapter concludes with a discussion on how the organisms encountered in that chapter connect to humans economically, environmentally, medically, and even culturally.

➢ **Personable** Many textbooks are written by teams of writers, some of which are anonymous. As a result, the reader (student) lacks a personal connection with the author(s). Again, this text is different. First, this text was written in entirety only by the name you see stamped on the front of this book—Dennis Holley. Secondly, I have attempted to write each chapter in the tone of enthusiastic and passionate, but caring and concerned teacher speaking directly to you the student. Hopefully, I have succeeded.

Organization A quick glance at the table of contents reveals that what the science of biology is all about and how it works is detailed in Chapter 1. With this foundation in place, in *Unit One* you will examine the chemistry of life providing you the background needed to understand the properties of life and the chemistry of living things. In *Unit Two* you will delve into the cellular structure of living things, the processes cells undergo, and discover how cells replicate. In *Unit Three* you will explore genetics and meet Mendel and his garden peas, learn about gene and chromosome structure, and appreciate the application of genetics through biotechnology. Lastly, in *Unit Four* you will investigate modern evolutionary history, and ponder the history of life on Earth and the rise of the human species."

At the end of each chapter, you will find both a set of *Review and Reflect* questions that will test your critical thinking skills while reviewing the main concepts of that particular chapter and a set of *Create and Connect* challenges that will help you develop and use important science process skills. Some or all of these questions and challenges may be assigned by the instructor as part of the assessment package for this course. In these assignments, you will be asked to write everything from formal scientific reports to essays to position papers to short stories. The exact format and details will be given with each assignment. Consult the appendix on scientific writing for guidelines and suggestions for correct scientific writing.

I believe this textbook represents a major paradigm shift in the way college biology textbooks are written and presented because it was written by a teacher (not a research scientist) for students. I have labored to make this textbook accurate, understandable, and interesting so that you can and will read it. And if you do indeed bother to read it, I guarantee that you will gather not only a wealth of information but also a never-ending respect for those amazing creatures with which we share this planet.

A Personal Note from the Author

I am a biologist to the core, always have been, and always will be. My interest in all things living is broad and generic. If it's a living creature—plant, animal, or microbe—I find it fascinating. How did I get this way? Understanding parents and a nurturing habitat are to blame. My mother was constantly contending with tadpoles in jars, aquariums of fish, mice in cages, and occasionally rewashing the clothes she had just hung out to dry because my flock of pigeons flew too low overhead. She pretended to make a fuss but encouraged my every adventure. My father helped me build cages and traps and was quite adept at capturing and helping me rear the many kinds of small animals that constantly caught my attention and interest.

I was blessed with growing up in a very small rural village where my family's acreage was only several blocks from a meandering stream aptly known by the locals as "Muddy Creek." This brook was shaded by many huge overhanging trees and was full of snails, fish, frogs, turtles, and even beavers and muskrats. Many inquisitive hours were spent around and in that stream.

Two events sealed my fate and set me on my course. In my early high school years, my parents finally gave in to my pestering and bought me a small, simple microscope (which they couldn't afford even though it cost only around $30). This amazing black beauty came complete with a wooden box of slides and a few dissecting instruments. Once I dove into the microscopic world, I was hooked on all things biological. Later, I stumbled on Paul de Kruif's 1926 book, *Microbe Hunters* and was inspired to get the education that would allow me to become a professional biologist. At that point, I didn't know exactly what I wanted to do professionally, but I did know my future would have something to do with biology.

I eagerly devoured every biology course I could take in college, and while I flirted for a time with the idea of becoming a marine biologist, I eventually became an educator. For nearly forty years, high schools and universities have actually paid me for merely doing what I love—teaching biology and teaching others how to teach biology and science. I am a very inquiry-oriented, hands-on type of teacher whose philosophy as an educator is best and most simply articulated in the words of Louis Agassiz:

Study nature, not books.

My love of all things biological continues unabated to this day. As such, I would consider the day poorly spent were I not to stumble upon at least several biological "WOW! Moments" (*WoMos*) during the course of that day. Such moments are not hard to find for they are everywhere. You just have to be receptive to them. Stop, look, and appreciate the natural world around you

I would like to dedicate this book to my parents for their nurturing and understanding, my wife and family for their patience and support, and to my students—past and present—who have taught me more than they will ever know.

Dennis Holley

As a teacher there is nothing more I enjoy professionally than "talking shop," so please feel free to contact me if we might be of any assistance in your biological endeavors either in the classroom or laboratory. Also, I certainly want to address any complaints or problems you have with the book, and I am anxious for feedback and input from you regarding any suggestions you might have for future editions. Please contact me through the book website at www.generalbiologytextbook.com.

THE SCIENCE OF BIOLOGY

Back to the sea again! Down there today I watched the goings-on of the sea snails, patellas (mussels with a single shell), and the pocket crabs, and it gave me a glow of pleasure observing them. No, really, how delightful and magnificent a living thing is! How exactly matched to its condition, how true, how intensely being! And how much I'm helped by the small amount of study I've done and how I look forward to taking it further.

—Johann Wolfgang von Goethe

Introduction

This planet pulsates with life. Interlaced within the larger physical world of the **atmosphere** (air), the **hydrosphere** (water), and the **lithosphere** (soil) are countless smaller biological worlds teeming with creatures of such variety of size, shape, and behavior as to boggle the mind. The study of these many and varied forms of life is the science of biology.

Rise of Biology

Biology (Gr. *bios*, living + *logos*, to study) is a natural science concerned with the study of life and living organisms, including their structure, function, growth, distribution, taxonomy, and evolution. In broadest scientific and historical terms, the first investigations of the natural world made by humankind were undoubtedly biological in nature. Using trial-and-error methods to fulfill the basic need for food, humans quickly learned which animals and plants they could eat and which would eat or poison them.

Although the concept of biology as a single coherent scientific field arose in the 19th century, the biological sciences emerged from the traditions of medicine and natural history reaching back to ancient Egyptian medicine and the works of philosopher scientists such as Aristotle (classification and comparative anatomy of animals (**Figure 1.1**), Galen (human anatomy), Theophrastus (botany), and Pliny the Elder (natural history compilations) in the ancient Greco-Roman world. The work of these ancients was further developed during the Middle Ages by Muslim physicians.

Figure 1.1 Aristotle (384-322 B.C.) is regarded by many as the father of biology.

During the European Renaissance (1300s-1500s) biological thought reemerged and was revolutionized by the rise of empiricism, a philosophy of science emphasizing evidence over traditional beliefs, especially as discovered through experimentation (scientific method). Modern science was born.

1600-1700

The early modern period (1600s-1700s) saw the rise of prominent biologists such as Harvey who used experimentation and careful observation to investigate human physiology, Linnaeus and Ray who labored to incorporate the flood of newly discovered organisms coming in from global expeditions into a coherent taxonomy, Buffon who laid the groundwork for evolutionary thought, and the microscopists Leuwenhoek and Hooke who perfected the microscope and were the first to observe microscopic creatures and cells. Also appearing during this time were explorer-naturalists such as Humboldt who laid the foundations for biogeography, ecology, and ethology. Our understanding of the true nature of life took a giant leap forward during this period when first Redi (**Figure 1.2**) with large organisms and then Spallanzani with microorganisms disproved the theory of spontaneous generation.

Figure 1.2 For his work on disproving spontaneous generation, Redi (1626-1697) is regarded as the father of experimental biology.

1800-1899

The 1800s began with the first appearance of the term *biology* used in its modern sense in 1802. Schleiden (1838) and Schwann (1837) proposed that cells (plant or animal) are the elementary particles of life and when Virchow postulated that cells can only arise from preexisting cells (1858), **Cell Theory** was born. In 1858 Charles Darwin and Alfred Wallace independently proposed a theory of biological evolution by means of natural selection. Louis Pasteur would earn the title "Father of Bacteriology" for his work on the germ theory of disease and for producing vaccines against diseases such as anthrax and rabies. The Austrian monk Gregor Mendel would lay the foundation for modern genetics when he proposed his Principles of Inheritance based on his work with pea plants. The 19th century would end with Beijerinck using filtering experiments to show that tobacco mosaic disease is caused by something smaller than a bacterium when he named a "virus."

1900-1949

The dawn of a new century began with the rediscovery of Mendel's work in 1900. That quickly allowed Sutton and Boveri to propose that chromosomes carry the hereditary information. A short time later Morgan postulated that genes are arranged in a linear fashion on chromosomes. These discoveries would form the basis of the modern science of genetics. The period from 1900-1949 saw our understanding of animal behavior greatly advanced when Ivan Pavlov demonstrated conditioned response in salivating dogs and Konrad Lorenz described the imprinting behavior of young birds (**Figure 1.3**). And while Wendall Stanley was crystallizing the tobacco mosaic virus, Oswald Avery was demonstrating that deoxyribonucleic acid (DNA) carries hereditary information.

Figure 1.3 Konrad Lorenz being trailed by greylag goslings that had imprinted on him as their parent.

1950-1990

Based on the work of Franklin, Henry, Chase, and Chargaff, James Watson and Francis Crick proposed the double-helix structure of DNA in 1953. They further suggested a method by which the molecule can replicate itself and serve to transmit genetic information. This Nobel Prize-winning work is the foundation upon which today's molecular biology and biotechnology is built. During this period we increased our understanding of the structure of biological molecules such as vitamin B_{12}, hemoglobin, and insulin as well as the development of some synthetic biologic molecules. Molecular biology begins to come of age with a number of major discoveries during this period:

- 1955—Arthur Kornberg discovers DNA polymerase enzymes.
- 1958—John Gurdon uses nuclear transplantation to clone the first vertebrate using the nucleus from a fully differentiated adult cell of an African clawed frog (family Pipidae).
- 1970—Hamilton Smith and Daniel Nathans discover DNA restriction enzymes.
- 1977—Working independently, the teams of Walter Gilbert and Alan Maxam and Frederick Sanger and Alan Coulson develop techniques for rapid gene sequencing using gel electrophoresis.
- 1978—Frederick Sanger presents the 5,386 base sequences for the virus PhiX 174, the first sequencing of an entire genome.
- 1983—Kary Mullis invents "PCR" (polymerase chain reaction), an automated method for rapidly copying sequences of DNA.
- 1984—Alec Jeffreys devises a method of genetic fingerprinting.

1990-Present

Our understanding of the structure and function of DNA and our ability to construct DNA from its basic components gives rise to biotechnology during this period:

- 1990—The first approved gene therapy is performed on a human patient.
- 1996—A sheep named Dolly is the first clone of an adult mammal.
- 1998—Publication of the first complete genome of a free-living animal, the soil nematode *Caenorhabditis elegans*.
- 2000—Publication of the first complete genome of a plant, thale cress (*Arabidopsis thaliana*).

- 2001—The Human Genome Project (HUGO) and Celera Genomics publish the complete genome of humans.
- 2010—Researchers at the J. Craig Venter Institute create the first synthetic cell by transplanting a synthetic genome capable of self-replication into a recipient bacterial cell.

This timeline represents only the major highpoints in the history of biology. For a more complete listing search "History of Biology" timeline, "History of Biotechnology" timeline and/or "History of Genetics" timeline.

Branches of Biology

The sheer magnitude of the number of living things and the varying complexity of those organisms' bodies, internal processes, heredity, and evolutionary origins necessitates that the science of biology be divided into many separate branches and that some of those branches be divided further into subdisciplines. The main branches of biology are:

- *Agriculture*—the practical application of raising crops and producing livestock
- *Anatomy*—the study of the internal structure of organisms
  - *Histology*—the study of the microscopic anatomy of cells and tissues
- *Biogeography*—the study of the spatial and temporal distribution of organisms
- *Biotechnology*—the manipulation (as through genetic engineering) of living organisms or their components to produce useful commercial products or medical compounds
  - *Synthetic biology*—the design and construction of new biological parts, structures or systems and/or the re-design of existing natural biological systems for useful purposes
- *Botany*—the study of plants
- *Cytology* (*Cell biology*)—the study of the structural components of a cell and the molecular and chemical interactions that occur within that cell.
- *Conservation biology*—the study of the preservation, protection, or restoration of the natural environment, natural ecosystems, vegetation, and wildlife
- *Developmental biology*—the study of the preservation, protection, or restoration of the natural environment, natural ecosystems, vegetation, and wildlife
  - *Embryology*—the study of the development of the embryo
- *Ecology*—the study of how organisms interact with each other and the physical environment
- *Evolutionary biology*—the study of the origin and descent of species over time
- *Genetics*—the study of genes and heredity
- *Marine biology*—the study of ocean ecosystems and the living things that inhabit those ecosystems
- *Microbiology*—the study of microscopic organisms
  - *Bacteriology*—the study of bacteria
  - *Mycology*—the study of fungi
  - *Protozoology*—the study of protists
  - *Virology*—the study of viruses
- *Molecular biology*—the study of biological functions at the molecular level

- *Neurobiology*—the study of nervous systems
- *Parasitology*—the study of parasites and parasitism
- *Population ecology*—the study of the dynamics of populations
- *Population genetics*—the study of changes in the gene frequencies within populations
- *Paleontology*—the study of fossils
- *Physiology*—the study of the normal functions of organisms and their parts
- *Zoology*—the study of all aspects of animals
  - *Entomology*—the study of insects
  - *Ethology*—the study of animal behavior
  - *Ichthyology*—the study of fish
  - *Herpetology*—the study of amphibians and reptiles
  - *Ornithology*—the study of birds
  - *Mammalogy*—the study of mammals

Nature of Science

The task of biology is to understand the living world at all levels of the biological hierarchy—from the molecular level of the cell and the mechanisms of inheritance to the complex ecological interactions that shape the biosphere

Science cannot and should not address the realms of the mythical, the imaginary, the metaphysical, or the spiritual. For too long people accepted the musings of authority figures as truth. Often the more bizarre the speculation, the more eager people were (and, unfortunately, many still are) to believe it. The sole aim of science is to classify, understand, and unify the objects and phenomena of the material world. By using a combination of accurate observation and experimentation, logic and intuition, and the occasional fortunate happenstance of **serendipity**, biologists seek to divine and understand the rules that govern all levels of the natural universe

To understand and fully appreciate the art and craft of biology, one must understand the nature of science, how scientists think, and how they develop those amazing tests known as experiments.

What is Science?

Simply put, *science is the search for natural truths by exacting individuals (scientists) using precise and reliable methods*. Although we find this a practical and workable definition for student use in our work-a-day educational world, there is more to it than that.

The answer to this question has two parts or components. The first component is *investigation*, and the second is *accumulation*.

Investigation Science is an on-going and never-ending search (investigation) for the truth. However, this search must always be tempered by the realization that (1) we might not recognize the truth when we see it and that (2) what we regard as the truth is always subject to change. Investigative science must be grounded and guided by the understanding that there are not and never can be any absolute scientific "truths."

Accumulation Science is also a body of knowledge obtained (accumulation) by exacting individuals (scientists) using precise and reliable methods. The knowledge (truths) accumulated through investigation are assimilated into our biological knowledge base. The sheer amount of biological knowledge is staggering, and humankind continues to accumulate ever more of this knowledge at breath-taking speed. (Visualize the volume our knowledge of biology as an upside down pyramid growing upward and outward at an incredible pace.)

Unfortunately, science is often taught and learned solely as accumulation with prodigious amounts of information and terminology to be memorized and often quickly forgotten. Presenting such a myopic view as the entirety of science is a detriment to student understanding and appreciation of the true nature of science. As our society becomes ever more science-driven, a scientifically literate citizenry becomes ever more critical.

Characteristics of Science

Investigative science rests on certain cornerstones:

1. **Materialism** Scientific explanations must be grounded in material causes and cannot violate natural law. Magic, myth and mysticism have no place in science and only hinder the search for truths and obscure such truths as we might discover.

2. **Testability** Science forms hypotheses that can and must be tested experimentally against the material world.

3. **Repeatability** Results obtained through experimental testing must, for the most part, hold true time and again. If an experiment yields significantly different results each time it is conducted, none of the data sets collected from the experiment can be regarded as having any probability of being the truth. Regardless of who conducts the experiment or how many times it is repeated, the results must be substantially the same time and again. Nonreproducible results cannot be accorded any reliability.

4. **Self-Correcting** A scientific theory makes a statement and draws some conclusion about the material world. If later observations or experiments contradict this conclusion, the theory must be revised or rejected. Ideally, this should make science self-correcting thus validating the accuracy of any knowledge (truth) gained. However, such is not always the reality of the scientific endeavor as science historians, philosophers, and working scientists will attest.

How Do Scientists Think?

The single most important tool in a scientist's arsenal of discovery is clear and logical thought. Ancient Greeks, like Aristotle, attempted to explain the natural world through reason alone applying what is known as the **classical approach** to science. They felt that if the reasoning is sound, then the conclusion is trustworthy. This method works well in much of mathematics and logic, but in science, it is not enough, because what appears true in the mind and what actually exists in nature are often not the same. **Empiricism**, the notion that reason alone is not sufficient and that ideas must be tested, appeared in Western culture at the time of the Renaissance. As the empirical approach—accurate observation, measurement, and experimentation—became the central aspect of science, modern science as we know it came to be.

Scientists apply empiricism to solve problems and answer questions in two ways:

1. **Induction** or **Inductive Reasoning** With induction, the researcher measures or carefully observes aspects of the phenomena being studied. Then those measurements and observations are ana-

lyzed, and generalizations (theories) are formed from the analysis. Using the induction approach results in reasoning progressing from the specific results to a general principle. For example, if finches, ostriches, and penguins possess feathers (specific results), one might logically reason that all birds have feathers (general principle).

Inductive reasoning as an important tool of science blossomed in Europe during the 1500s and1600s when scientists like Isaac Newton, Galileo Galilei, William Gilbert, Robert Boyle and others began to construct models of how the biological and physical world operates from the results of their experiments.

2. **Deduction** or **Deductive Reasoning** In deduction the reasoning progresses in a fashion opposite to that of induction, moving from general to specific. For example, livestock and pet animals need minerals, vitamins, and other trace elements to grow properly (general principle). With this general knowledge, we can formulate foods that contain the exact components needed by particular types of animals (specific results). Both deductive and inductive reasoning can lead to specific hypotheses that can then be tested.

How Does Science Work?

Although the working of science (or the **scientific method** as it is often known) is often taught and learned as a linear series of steps, it is more accurate to visualize science as a circular enterprise—the science cycle (**Figure 1.4**).

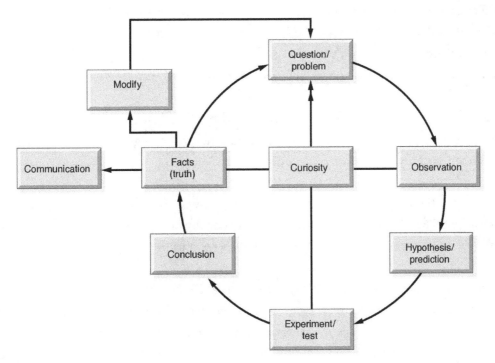

Figure 1.4 The Science Cycle. Science is a curiosity-driven cycle that usually generates more questions than answers.

Curiosity Science begins and ends with curiosity. Curiosity is the fuel that drives the science machine and the central hub around which the science cycle spins. Without curiosity, humankind would not be driven to divine the truths of the universe. Albert Einstein once said, "I have no special gifts; I am only passionately curious."

Questions Curiosity raises questions and questions present problems to be solved. Science is the enterprise humans have developed to discern the answers to our questions about nature and the universe around us.

Observation Humans are not born accurate observers. People have certain prejudices and preconceptions about how they perceive the world around them, especially as they grow older. Because of these prejudices, scientists, and even the common person must strive to separate *reality* (the way things actually are) from *inference* (the way we think things are or the way we want or expect things to be) in any situation in which observation plays a role. In the words of Albert Szent-Gyorgi: "Discovery consists of seeing what everybody else has seen but then thinking what nobody else has thought."

Hypothesis and Prediction Curiosity raises questions. Accurate observations reveal information. We then use this information to suggest explanations and answers through hypotheses. A prediction is a statement made in advance that states the results that will be obtained from testing a hypothesis. Scientists make predictions because predictions provide a way to test the validity of the hypothesis. If an experiment yields results inconsistent with the prediction, the hypothesis must be modified or rejected. If experimental results support the prediction, the hypothesis, in turn, is supported. The more experimentally supported predictions a hypothesis generates, the greater the validity of that hypothesis. What is the difference then between a fact and a hypothesis? Edward Teller (known as "the father of the hydrogen bomb") may have said it best when he penned: "A fact is a simple statement that everyone believes. It is innocent unless found guilty. A hypothesis is a novel suggestion that no one wants to believe. It is guilty until proven effective."

Experimentation With the realization that speculation and prediction must be tested in controlled experiments, modern science was born. Scientists, however, must never lose sight of the fundamental fact that experiments are done to test hypotheses, not to confirm them.

Conclusion What does the data mean? What conclusions (answers) may be drawn from the data? Things are not always clear cut. "Muddy data" can lead to "cloudy conclusions." The validity of any scientific conclusion rests squarely on the accuracy of the data and the interpretation of that data.

Communication After performing a series of experiments that support the hypothesis, there comes a point at which the scientist must communicate and throw his or her work into the bright light of scientific scrutiny. This communication usually comes in the form of a written paper describing the experiment and its results. The paper is then submitted for publication in a recognized academic journal. However, before it is published, the paper is subjected to a process known as **peer review** in which the work is subjected to the scrutiny of others who are experts in the field. The process of peer review is the cornerstone of the self-correcting nature of the scientific endeavor. With the announcement of an important discovery published in a reputable scientific journal the peer review process continues as other scientists attempt to reproduce the work and gather the same results as the original. This multi-layered peer review approach provides the checks on accuracy and honesty that are such an important part of the repeatability characteristic of science.

Modify or Trash If contradictions arise, a hypothesis should be modified to account for challenges to its validity. If, however, modification to the old hypothesis cannot bring it within the bounds of more recent findings and results, it must be trashed. (In reality, even experienced scientists are often reluctant to do this.) In science one must be comfortable with failure for failure, not success, is the norm. We learn more from being wrong than we do from being right because a wrong hypothesis opens up even more experimental pathways as we attempt to understand why our predictions were incorrect. Or as James Joyce so eloquently put it, "Mistakes are the portals of discovery."

One cannot underestimate the power of insight, intuition, and imagination in the turning of the science cycle. It is because these qualities play such a large role in scientific progress that some scientists are so much better at science than others, just as some composers, artists, and athletes are better at their craft than are their contemporaries.

Terminology of the Truth

Not only scientists but the general public must appreciate the idea that some knowledge (truth) is less likely to be (or has a lower *probability* of being) wrong than other knowledge. Unfortunately, many of the terms employed by investigative science to assign certain levels of truth at certain stages of the science cycle are often misused and misunderstood.

As curiosity leads to questions to be answered and problems to be solved, the science cycle begins to turn, and the search for the truth is on. A **hypothesis** is then offered as a tentative explanation or answer to the problem question. A hypothesis may be based on past observation or experiences of the scientist, the findings of other scientists, or simply a hunch. As such, a hypothesis is certainly more than mere opinion or a simple guess but lacking little evidential support, it must be considered as the lowest level or probability of truth.

If over time a hypothesis accumulates supporting evidence through observation and experimental results, it can be said to have achieved the level of a **theory**. Whereas a hypothesis has little factual support, a theory is supported by ever-increasing evidence. Therefore, a theory could be considered as having a median level of truth probability.

"Theory" can also be used to describe a collection of related concepts, supported by observation and experimental evidence that provides a framework for organizing a body of knowledge in some field of study. The quantum theory in physics and the cell theory and the theory of evolution in biology are examples. Thus scientists distinguish between an **experimental theory** that is specific to a certain experiment and has a median level of truth probability from an **integrational theory** that is general to a large body of evidence in some area of study and has a high level of truth probability. The general public often confuses the former meaning with the latter which results, not surprisingly, in confusion and misunderstanding. Critics outside the fields of science are adept at utilizing whichever definition fits their agenda. For example, to suggest that evolution is "just a theory" is a misleading and even deceptive ploy to reduce evolution to experimental status when in reality the fact that evolution has occurred is an accepted scientific fact supported by overwhelming evidence.

If evidence continues to mount over a considerable period of time and it seems likely that a theory will not be proven incorrect or falsified, it moves to the category of the highest probability level of truth known as a **scientific law**. Exactly how and when a theory advances to become a law is somewhat ambiguous and arbitrary, and as a result, there are few true scientific laws. Although a scientific law has the highest degree of truth probability, it cannot and should never be regarded as the absolute truth. By its very nature science should preclude any and all absolutes for what we regard as the truth is, at best, a momentary illusion seen through a veil of ignorance.

Designing the Good Experiment

We defined science earlier as the search for the truth by exacting individuals (scientists) using precise and reliable methods. What are these so-called scientific methods? How do scientists design and conduct experiments to get at the truth? Experiments are a series of components usually carried out in sequence:

Component 1: *The Problem*. Curiosity leads to questions to be answered and problems to be solved. The first step must be to specify exactly what problem is to be solved. This is done through the *problem question*. The *problem question* should:

- Be in the form of a question.
- Be stated in clear and concise language.
- Present only one problem to be solved at a time.

It is no easy task to formulate a strong problem question, and even professional scientists and researchers struggle with the process. It is, however, a crucial first step for as C.F. Kettering reminds us: "A problem well-stated is a problem half solved."

Component 2: *Form a Hypothesis*. A hypothesis may be formally defined as the possible explanation of some phenomenon. Realistically, a hypothesis is nothing more than an educated guess (but certainly more than a mere opinion) that the researcher makes as to the answer to the problem question. The hypothesis gives a prediction of the outcome of an experiment, and the analysis determines if the prediction is accurate. Often the experiment is set up using a **null hypothesis** which is the opposite of the actual hypothesis. Better stated, the null hypothesis predicts the results that would be obtained if an actual hypothesis is not true. To use a simple example, say that the hypothesis is that variable A directly affects variable B. In this case, the null hypothesis would be that B is not directly affected (is independent) by A. Analysis of the data then determines if the null hypothesis is or is not supported by the experimental results; if it is, then the actual (original) hypothesis can be rejected. If the null hypothesis is not supported, then the actual hypothesis can be accepted, at least from a statistical standpoint.

Don't let the fear of being wrong prevent you from freely expressing your ideas and hypotheses. In science, wrong doesn't matter, because right or wrong, a hypothesis can still get us to the truth. Any natural truth might be visualized as hidden in a thick concrete bunker of ignorance. Further visualize that the truth bunker has two doors. If over time and with scrutiny, a hypothesis proves correct, the researcher has entered the front door of the truth bunker. However, if a hypothesis proves false, the researcher has stilled entered the bunker, only through the back door. Either way, a truth has been learned for as John Dewey encourages us: "Failure is instructive. The person who really thinks learns quite as much from his failure as from his success."

Component 3: *Methods and Materials.* This component involves planning and then conducting the actual experiment and would include a step-by-step and highly detailed plan of exactly what would be done in the experiment. Within this plan would be found:

- Details on the specific kinds and amounts of materials and supplies that would be required to conduct the experiment.

- Details on manipulating variables and establishing control and test groups. Many factors or **variables** influence processes and outcomes. For example, to study the effect of a synthetic hormone on the growth rate of rats, rats would be put into identical cages, given identical amounts of water and food, and placed so they receive identical amounts of sunlight. The cages, water, food, and light are called the **independent (or control) variables**. The synthetic hormone would be called the **dependent (or test) variable** and might be dispensed to the test rats in their food or water. The rats receiving the hormone would be designated as the **test group** whereas the rats not receiving the hormone would be designated as the **control group**. If the test group grows larger faster than the control group, then it might be concluded that since the hormone was the only variable that differed between the two groups, the hormone did indeed increase the growth rate of the test rats. Without the control group, there is no way to guarantee that the hormone (dependent or test variable) produced an increased growth rate as there would be no group with which to compare results. Some types of experiments, such as field studies, do not lend themselves to controls. For example, suppose we hypothesize that a predator fish population determines the population of a prey species in a large lake and conduct an experiment to test this hypothesis. This may involve measuring the populations of predator and prey over a period of time and then trying to find a relationship between the two. In such an experiment, natural changes are merely monitored; establishing a control group is not practical or possible.

- Details on sample size and sampling methods. The rule of thumb on sample size is: *The larger the sample, the more accurate the results.* Would you be suspicious of a new medicine that was advertised to have miraculous powers if you discovered that the drug had only been tested on 20 people? 200 people? Certainly, as well you should. There is a practical limit to sample size that must be considered, however. Although it may be practical and possible to sample one million bacteria for a specific genetic trait, it would not be practical or possible to sample one million salamanders to determine the variations of a common gene.

- Details on what data will be collected and how the data will be organized. In our rat growth experiment example, one would carefully weigh the rats in both the test group and the control group over a given period of time organizing the weights obtained into a data table. Scientists often distinguish between **qualitative data** (or soft data) and **quantitative data** (or hard data). We will consider hard data to be numerical data, such as the exact weights of the rats in the hormone experiment. Soft data will be considered to be observational or anecdotal data regarding both the control and test rats feeding in our rat growth experiment.

Component 4: *Analyze the Data.* Once you have the observed results (known at this point as "raw" data), what do you do with it? How can we make heads or tails out of what the numbers may be trying to tell us? Just looking at a jumble of numbers will probably reveal nothing of significance so scientists must analyze the data. One method of analysis is to graph the data. Graphs turn numbers into graphics which is useful because pictures are more easily understood than lists of numbers. Another analysis scientists perform is to run statistical tests on their raw numerical data. Statistical tests are mathematical processes that help determine if the data collected is significant and if so, how significant.

Component 5: *Draw a Conclusion.* Analysis of the data will hopefully reveal whether our original hypothesis was correct (supported by the data) or incorrect (not supported by the data). That, in turn, should answer the original problem question. In this component, scientists attempt to put the results in perspective, establish the significance of the results, and explain how the experiment fits into existing knowledge.

Statistically speaking, there are two major errors that can be made in the interpretation of experimental results. A *Type I error* is to reject a hypothesis that is correct, and a *Type II error* is to accept a hypothesis that is wrong. An important part of experimental design is to plan an experiment that minimizes the likelihood of both kinds of error. This is one reason the results of single studies should not be considered as conclusive evidence for some theory. The hard fact is that erroneous conclusions are always a possibility in scientific research and error is an integral part of the process.

It is also important to understand that negative results, where the data does not support the hypothesis, are valuable. Knowing what is not correct can be as useful as knowing what is correct. Whether the results are positive or negative should not determine the value of the research. However, the reality is that landing large grants, getting work published, and even entire careers often hinge on positive results.

Statistical Tests—Making Sense of the Data

The general public in general often has the mistaken idea that "with statistics, you can prove anything." This is incorrect. With statistics one cannot *prove* anything, but statistics, if used properly, can indeed demonstrate certain degrees of relationship among factors or variables. However, as John Fennick cautions us, "No discipline has been so misused as statistics."

Statistics can be manipulated, however, and connections made that are artificial if not outright ridiculous at best. Did you know that living near the equator is dangerous to your health? Studies show that on average, people live about a half year longer for each 100 miles of resident distance from the equator. People who live 1,000 miles north of your present location have a life expectancy five years greater than yours. That is equal to what we have supposedly gained from the last thirty years of medical progress.

Can the microscopic parasite *Toxoplasma gondii* which infects the brains of rats, cats, and possibly 60 million symptom-free Americans, skew our gender rations, increase the odds of having a traffic accident, and even "affect the cultures of nations?" You may scoff at the artificial contrivance and connection made in each of these cases but the evidence supporting these contentions is as good (and in some cases, better) than that supporting many statistical "truths" that we sustain with billions of dollars of grant money each year. For example, it is better than the evidence that low-fat diets and exercise have reduced deadly heart attacks. Studies proclaiming doom or glory bombard us one after the other with the last often contradicting the previous. What choices are we to make? What lifestyle changes should we consider to improve and safeguard our health? At present, there is no clear-cut answer or easy choice.

What then is the best and proper use of statistics in biology? Statistical procedures are used to determine if the hypothesis is supported or rejected by the experimental data. The appropriate statistical test properly applied can verify whether our data shows any significant difference between our control group and our test group or if such difference could be attributed to mere chance.

What kinds of statistical tests do biologist conduct and in what experimental situations are each type of test appropriately used?

The **Chi-square Test** (or goodness of fit test) is used to determine whether or not two or more samples (of seeds, bacteria, or humans) are significantly different enough from each other in some characteristic, trait, or behavior that we can generalize that the populations from which our samples were taken are also significantly different. Chi-square tests are most often applied to data generated in genetics experiments or situations where there may be two possible outcomes. For example, what feather color(s) are dominant in a wild population of certain birds? The Chi-square test could be used to answer this question.

The **t-test** assesses whether the means (simple average) of two groups are significantly different from each other in some characteristic or trait. For example, in our rat growth experiment, is the growth rate of the rats fed synthetic hormone greater than that of rats fed no hormone? The t-Test could be used to answer this question.

ANOVA (or Analysis of Variance) is a test applied when you wish to compare samples among more than two and up to any manageable number of groups. For example, a researcher wishes to determine the best concentration of synthetic hormone to be dispensed to rats to generate the fastest rate of growth. The rats would be divided into three equal groups—low hormone concentration, medium hormone concentration, and high hormone concentration. The total weight gain over a select period of time of the rats from each group is determined and recorded. The ANOVA test could verify which concentration of the hormone, if any, yielded a significantly faster growth rate. ANOVA essentially tests whether there is greater variance among than within groups than would be expected by chance, but it does not tell you which particular samples are different from each other.

In Summary

- Biology is the study of all aspects of living things. Zoology and it companion discipline botany were among the first investigations of the biological world undertaken by humankind.
- Biology as a science arose in the time of the ancient Romans, Greeks, and Muslims.
- With the rise of the Renaissance, biology as an investigative science was established.
- Biology is one of the broadest fields in all of science. The science of biology is composed of many branches and subdisciplines.
- To fully understand the art and craft of biology one must understand the nature of science.
- What is science? The uniquely human enterprise known as science has two components: investigation (the search for the truth) and accumulation (acquiring facts and knowledge).
- Science is (1) grounded in material causes, (2) testable, (3) repeatable, and (4) self-correcting.
- Scientists apply logical thought to solve problems and answer questions using two methods of reasoning: deductive reasoning and inductive reasoning.
- The working of science is most accurately viewed as a circular enterprise—the science cycle.
- Different scientific terms—hypothesis, theory, and scientific law—are used to describe the probability of truth achieved through scientific investigation.

- A well-designed experiment has a series of components carried out in sequence: problem question, hypothesis, methods and materials, data analyzation, and conclusion.
- Various types of statistical tests properly applied can verify whether data collected show any significant difference between control and test groups.

Review and Reflect

1. *Inspired* Your younger sister has been looking through this book and has been so inspired by it as to state, "I am going to become a biologist." Knowing something of the depth and breadth of biology you ask, "OK, but what kind of biologist do you want to be?" Why would you ask that?

2. *Define*

    A. You have been commissioned to write a dictionary of scientific terms. When it comes time to write a definition for the word science, what would you say? Write your personal definition of science.

    B. Some terms associated with the scientific process are often confusing to students and the general public and may be used in misleading ways. Develop a personal definition for the terms: hypothesis, prediction, theory, and law? Word your definitions in such a way as to show the scientific relationships between these terms and the truth probability of each term.

3. *Think Like a Scientist* Empiricism, the notion that reason alone is not sufficient and that ideas must be tested, appeared in Western culture at the time of the Renaissance. Scientists apply empiricism to solve problems and answer questions in two ways: inductive reasoning and deductive reasoning. Explain the difference between inductive and deductive reasoning and give an action example of both.

4. *Characteristics of Science*

    A. As part of the requirements for passing this course, your instructor has required you to attend an evening lecture by a visiting scientist. The lecture is entitled; Materialism vs. Magic and Mysticism in Science. In general terms, explain what this lecture will be about.

    B. Explain this statement—"Modern science was born with the realization of the need for testability."

    C. Explain this statement—"Science should be self-correcting."

5. *Ride the Cycle* Why is it more realistic to think of science as a cyclic rather than a linear enterprise? Explain

6. *Figures Don't Lie* During a lecture on the concepts covered in this chapter, the person sitting next to you whispers, "I don't get this statistics stuff. What good is it anyway?" You quietly and politely promise to explain after the lecture. What will you say?

Create and Connect

1. *Slimy Aardvarks* From a meteorite crash site, NASA scientists have collected a small quantity of a sticky green substance thought to have come from within the meteorite itself. For reasons they will not reveal; NASA has turned to you to design an experiment to test the problem question: *What are the effects of exoslime on aardvarks?*

Guidelines

A. Following the tenets of a well-constructed experiment, your design should include the following components in order:

  ➤ The *Problem Question.* State exactly what problem you will be attempting to solve.

  ➤ The *Hypothesis.* Although this is a fictitious experiment, word your hypothesis as realistically as possible.

  ➤ *Methods and Materials.* Explain exactly what you will do in your experiment including the materials necessary to accomplish the task. Be specific, take nothing for granted, and do not expect people to read your mind as they read your work.

  ➤ *Collecting and Analyzing Data.* Explain what type(s) of data will be collected, and what statistical tests might be performed on that data. It is not necessary to concoct fictitious data or imaginary observations.

B. Your instructor may provide additional details or further instructions.

2. ***Blowin' in the Wind?—Analyzing Data and Drawing Conclusions*** Herpetologists collected data concerning migration and wind direction for a certain species of frog, as shown in **Table 1.1**. Construct a bar graph of this data and use the graph to answer the questions that follow.

A. Specifically, what did you graph?

B. What general conclusion can you draw from this data?

C. This data has been shown to be quite significant and not due to chance or coincidence. Would this conclusion hold true for other species of frogs? For the same species of frogs in different locales? Explain.

Table 1.1 Data Collected by Herpetologists Concerning Migration and Wind Direction for a Certain Species of Frog	
Wind Direction	**Frogs Migrating in That Direction**
N	3
NNE	0
NE	0
ENE	0
E	4
ESE	6
SE	0
SSE	25
S	32
SSW	7
SW	21
WSW	1
W	6
WNW	0
NW	3
NNW	4

3. **_Refute This_** Accurate observation is the cornerstone of science. However, less than accurate interpretation of those observations has led to numerous incorrect theories about life and the planet. For example, people saw frogs coming out of the mud in the spring and interpreted that as meaning the mud was turning into frogs (Spontaneous Generation theory). As people looked from horizon to horizon Earth appeared flat and to them the logical interpretation was that Earth was flat (Flat Earth theory). It took scientists years even centuries to refute what seem to us to be not only wrong, but laughable theories. Consider the Dark Sucker Theory detailed here and explain how you would attempt to refute it. Refuting incorrect interpretations of observations is a difficult task so may the force be with you as you take up this challenge.

The Dark Sucker Theory

For years, it has been believed that electric bulbs emit light, but recent information has proved otherwise. Electric bulbs don't emit light; they suck dark. Thus, we call these bulbs Dark Suckers. The Dark Sucker Theory and the existence of dark suckers prove that dark has mass and is heavier than light.

The basis of the Dark Sucker Theory is that electric bulbs suck dark. For example, take the Dark Sucker in the room you are in. There is much less dark right next to it than there is elsewhere. The larger the Dark Sucker, the greater its capacity to suck dark. Dark Suckers in the parking lot have a much greater capacity to suck dark than the ones in this room.

A candle is a primitive Dark Sucker. A new candle has a white wick. You can see that after the first use, the wick turns black, representing all the dark that has been sucked into it. If you put a pencil next to the wick of an operating candle, it will turn black. This is because it got in the way of the dark flowing into the candle. One of the disadvantages of these primitive Dark Suckers is their limited range. As with all things, Dark Suckers don't last forever. Once they are full of dark, they can no longer suck. This is proven by the dark spot on a full Dark Sucker.

There are also portable Dark Suckers. In these, the bulbs can't handle all the dark by themselves and must be aided by a Dark Storage Unit. When the Dark Storage Unit is full, it must be either emptied or replaced before the portable Dark Sucker can operate again.

Dark has mass. When dark goes into a Dark Sucker, friction from the mass generates heat. Thus, it is not wise to touch an operating Dark Sucker. Candles present a special problem as the mass must travel into a solid wick instead of through clear glass. This generates a great amount of heat and therefore it's not wise to touch an operating candle. Also, dark is heavier than light. If you were to swim just below the surface of the lake, you would see a lot of light. If you were to slowly swim deeper and deeper, you would notice it getting darker and darker. When you get really deep, you would be in total darkness. This is because the heavier dark sinks to the bottom of the lake and the lighter light floats at the top. That is why it is called light.

Finally, we must prove that dark is faster than light. If you were to stand in a lit room in front of a closed, dark closet, and slowly opened the closet door, you would see the light slowly enter the closet. But since dark is so fast, you would not be able to see the dark leave the closet.

CHAPTER 2

PROPERTIES AND ORGANIZATION OF LIFE

The most wonderful mystery of life may well be the means by which it created so much diversity from so little physical matter.

—E. O. Wilson

Introduction

A rmed with only a few basic types of living building blocks known as **cells**, nature has managed to fashion millions to perhaps tens of millions of different types of creatures—plants, animals, and microbes—that occupy every nook and cranny of this place. As a result, we are blessed to live on a planet bubbling and swarming with a riotous multitude of life forms. The mere contemplation of the beauty, mystery, and wonder of it all cannot help but invoke awe and reverence.

Defining Life

Throughout time, humankind has raised and pondered questions of life and death. As our scientific wit has grown sharper, our life questions have become more focused and defined: Biologically speaking, exactly what is "life"? When, where, and how did this life force appear on this planet? When does an individual organism become living? How does life at the individual cell level translate into life at the organismal level? Conversely, what is "death" biologically speaking? When an organism dies what happens to its life force? All legitimate questions that impact each and every one of us in one way or another on any number of different levels—biologically, medically, socially, philosophically, and religiously.

Unfortunately (or fortunately depending on your faith in humankind to properly handle such monumentally important knowledge), we have no definitive answers for any of these questions at the present time. It is a challenge for scientists to define **life** in specific terms because life is a process, not a substance or

object. Any definition must be general enough to both encompass all known life and any unknown life that may be different from life on Earth. We may not be able to yet define life scientifically, but we all recognize it when we see it or at least we do most of the time. Study **Figure 2.1** of Norman, the bike-riding dog. We know that Norman is clearly alive, but Norman's bicycle is not, nor is the asphalt on which Norman is riding his bike, but we cannot scientifically delineating in exact detail why Norman possesses life and his bike does not.

Figure 2.1

Even consulting a biological dictionary provides no specific answer. In fact, one of the definitions for *life* is: "The absence of death." Life is sometimes defined as the sum of all life properties, but some inanimate objects display many life properties. For example, an automobile moves as do living things, it takes in energy and gives off wastes as do living things, and it has a definite organization. An icicle grows as do living things, but if an icicle falls and breaks into pieces has it asexually reproduced as do some living things? Clearly, life is more than the sum of its parts.

Physicists view living beings as thermodynamic systems with an organized molecular structure that can reproduce themselves and evolve as survival dictates. Thermodynamically, life has been described as an open system, which makes use of gradients in its surroundings to create imperfect copies of itself. Hence, life is a self-sustained chemical system capable of undergoing Darwinian evolution. A major strength of this definition is that it distinguishes life by the evolutionary process rather than its chemical composition, but it still does not provide the biological definition of life we seek.

Will biologists ever be able to actually create a living cell(s) from the requisite inorganic components? Is it possible to create an artificial life form? As unbelievably science-fictionish or even biblically profound as that prospect sounds, we may be inching ever closer to doing exactly that. The question at this point seems not to be "Will we ever create life from scratch?" but simply how much creating the life would cost. There is the belief among some that the number of steps that might be real potential roadblocks to such an undertaking has declined almost to zero. In fact, some researchers optimistically suggest that for less than a third of the approximately $500 million dollars spent to sequence (map) the

Figure 2.4 This toad represents life organized to the organ system level.

human genome, it is conceivable that organic chemicals could be transformed into a single-celled organism that would grow, divide, and evolve and that such an amazing and profound feat could be accomplished in 3 to 5 years time.

The essence of life is a statistical improbability on a colossal scale.

—Richard Dawkins

Properties of Life

Since there is no unequivocal definition of life, most current definitions in biology are descriptive. Although most questions dealing with the biological nature of life still cannot yet be answered, we do know that all living organisms, regardless of their complexity (or the lack thereof), share a set of basic characteristics or properties. These commonly-held life properties should be regarded as the characteristics or signs of life rather than the definition of life.

- ***Life arises through biogenesis*** That is, life comes from life (**Figure 2.2**). This was not always known to be the case. For centuries people believed in **abiogenesis** (also known as **spontaneous generation**), a theory that proposed living things could originate from the transmutation of nonliving or inorganic things. For example, mud could turn into frogs, dirty rags could become rats and mice, and decaying meat could generate fly maggots to name a few. Later chapters will detail how scientists such as Spallanzani, Redi, and Pasteur disproved the idea of abiogenesis in a number of landmark experiments. We now know that any organism originates from an organism (asexual) or organisms (sexual) of the same species, or in the Latin: *Omne vivum ex vivo*, "all life is from life." Living things do not spontaneously spring into life. We must clarify that by saying that living things do not spontaneously spring into life *any more*. As you will come to understand in later chapters, there was a time on early Earth when inanimate molecules did organize and eventually form living cells. Thus, life is thought to have originated through abiogenesis, but it continues generationally through **biogenesis.**

Figure 2.2 Life Begets Life. Any living organism comes from the cell(s) of another living organism.

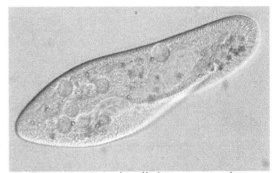

Figure 2.3 A single-celled organism such as this Paramecium displays all the properties of life and conducts the same processes of life as large multi-cellular organisms.

- ***Life has a cellular organization*** All matter is composed of **atoms**. Those atoms, in turn, bond to make **molecules**. It is the types of atoms and the molecules they produce that determines if the matter is a rock, a piece of glass, or a living organism such as you dear reader. A living thing is composed of molecular assemblages called cells. The "body" of some organisms consists of only one cell (**Figure 2.3**) whereas others are multi-cellular (**Figure 2.4**). In a multi-cellular creature, similar cells are organized into **tissues**. Tissues are organized into **organs** (body structures with a distinct function), and organs work together to form **organ systems**.

- *Life responds to the internal and external environment*
  All organisms respond to **stimuli** (**Figure 2.5**) be they plants, animals, or singled-celled organisms. Multi-cellular organisms can manage more complex responses than can single-celled organisms. The response may be positive (movement toward the stimuli) as shown in Figure 2.5 or negative (movement away from the stimuli). For example, if a few grains of salt are placed to one edge of a drop of water containing different types of protozoa, the protozoa will move away from the salt grains. Responses may not be visible to us if they occur internally or very slowly.

Figure 2.5 A plant responding to an external stimuli—the direction of light.

- *Life utilizes energy* All organisms require energy (food) and produce waste products. Food energy is used to maintain internal order and grow. Food provides nutrients that are used as building blocks or for energy. Energy is the capacity to do work, and it takes a great deal of work to maintain the organization of individual cells. Using nutrient molecules to build new organelles or cells or to produce chemical energy requires many sequences of chemical reactions. Collectively, all the chemical reactions that occur in a cell are termed **metabolism.**

- *Life maintains homeostasis* All organisms maintain stable internal conditions that are different than the surrounding environment, a process called **homeostasis**. To survive, it is imperative that an organism maintain a steady state of biological balance in which all physiological factors of that organism remain within the tolerance range of the organism. To maintain homeostasis, organisms have intricate feedback and control mechanisms. Calcium homeostasis in humans is an example (**Figure 2.6**). The body regulates calcium homeostasis through two feedback loops: a positive feedback loop is activated when blood calcium levels are too low whereas a negative feedback loop is activated when blood calcium levels are too high. Some organisms use behavior to help regulate their internal environment. For example, a lizard may bask in the sun in the early morning (**Figure 2.7**)

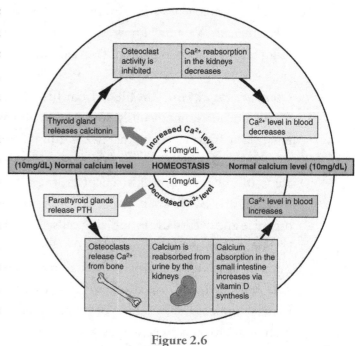

Figure 2.6

to help elevate its temperature which had fallen during a cold night but then move from the sun into shade when its internal temperature gets too high.

- *Life transmits genetic information* All living things have the potential to reproduce or make another organism. Even bacteria replicate themselves as Francois Jacob succinctly stated when he said: "The dream of every bacterium is to become two bacteria." When living organisms reproduce sexually, their genes or genetic information are passed on to their offspring. Random combinations of sperm and egg, each of which contains a unique suite of genes, ensures that the offspring has slightly new and different traits. In all organisms, the genes are made of **DNA (deoxyribose nucleic acid)** molecules. DNA provides the blueprint or set of instructions for the organization, maintenance, and metabolism for every single organism only the planet. Genes of DNA is probably the greatest single unifying characteristics of living things.

Figure 2.7 When basking, lizards will orient their body to receive the sun's rays at an optimal angle for heating.

- *Life grows and develops* The **embryo** that develops from the fusion of sperm and egg develops into a human being, a red rose, or a toad because of the specific **genome** it inherited from its parents. As organisms develop from an embryo they go through definite stages—beginning, growth, maturity, decline, and death (**Figure 2.8**).

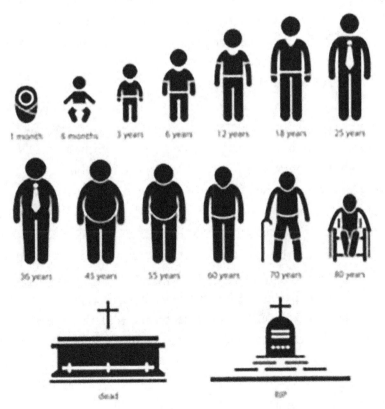

Figure 2.8

Box 2.1
Designing Life—Synthetic Biology

Synthetic biology is an exciting technology that can be defined as the artificial design and engineering of biological systems and living organisms for purposes of improving applications for industry or biological research. Synthetic biology integrates biology with engineering and draws from a number of different disciplines—biotechnology, evolutionary biology, genetic engineering, molecular biology, molecular engineering, systems biology, electrical engineering, organic chemical engineering, biophysics, and computer engineering.

Instead of the simple manipulation of single genes, researchers in this emerging science seek to engineer many genes to work in harmony, like the components of an electronic circuit board. The goal is to build living machines from off-the-shelf chemical components, employing many of the same strategies that electrical engineers use to make computer chips. Drawing upon the powerful techniques for automated synthesis of DNA molecules, bioengineers envision creating artificial genes and entire genomes ("genetic circuits") and inserting them into microorganisms such as bacteria and yeast thus bending the metabolic pathways of the microbes to our will and purpose.

The potential applications of this technology are astonishing:

- **Cell transformation** Currently, entire organisms are not being created from scratch, but instead, living cells are being transformed with inserts of new DNA. There are several ways of constructing synthetic DNA components and even entire synthetic genomes, but once the desired genetic code is obtained, it is integrated into a living cell that is expected to manifest the desired new capabilities or phenotypes. Cell transformation can lead to the creation of bioengineered microorganisms that can produce pharmaceuticals, break down pollutants, detect toxic chemicals, repair defective genes, destroy cancer cells, and generate fuels such as hydrogen and perhaps cellulose.

- **Information storage** Scientists can encode vast amounts of digital information onto a single strand of synthetic DNA. In 2012, George M. Church encoded one of his books about synthetic biology in DNA. The 5.3Mb of data from the book is more than 1000 times greater than the previous largest amount of information to be stored in synthesized DNA. A similar project had encoded the complete sonnets of William Shakespeare in DNA.

- **Biosensors** A **biosensor** refers to an engineered organism, usually a bacterium, which is capable of reporting some ambient phenomenon such as the presence of heavy metals or toxins. One such sensor created in Oak Ridge National Laboratory, and named "critter on a chip," consisted of a bioluminescent bacterial coating on a photosensitive computer chip to detect certain petroleum pollutants. When the bacteria sense the pollutant, they begin to luminesce.

- **Materials production** By integrating synthetic biology approaches with materials sciences it would be possible to envision cells as microscopic molecular foundries to produce materials with properties that can be genetically encoded. Recent advances towards this include re-engineering curli fibers, a component of extracellular material of biofilms, as a platform for programmable nanomaterial. These nanofibers have been genetically constructed for specific functions, including adhesion to substrates; nanoparticle templating; and protein immobilization.

- **Industrial enzymes** Researchers and companies utilizing synthetic biology aim to synthesize enzymes with high activity, to produce products with optimal yields and effectiveness. These synthesized enzymes aim to improve products such as detergents and lactose-free dairy products, as well as make them more cost effective.
- **Space exploration** Synthetic biology has piqued NASA's interest as it could help to produce resources for astronauts from a restricted number of compounds sent from Earth. On Mars, in particular, synthetic biology could also lead to production processes based on local resources, making it a powerful tool in the development of manned outposts with minimal dependence on Earth.

Although advances in biotechnology present the promise of profound practical and beneficial applications to man and beast, there are those who have voiced caution and concerns of possible perils from the inception of these endeavors. One such concern is the re-creation of known pathogens (such as the Ebola virus) or the development of new synthetic pathogens as biological warfare agents. Another worry is unpredictable risks to public health and the environment posed by synthetic microorganisms.

Hierarchy of Life

At first glance, one might not perceive much, if any, order to the chaos of life that inhabits this planet. However, through the work of many biologists in different disciplines over long periods of time, we now realize that the biological world is indeed organized and that this organization is hierarchal. That is, each level builds on the level below it. From the atoms and molecules that build cells to the planet-sized ecological systems that sustain whole species, there is an ordered regularity to this pandemonium we call life (**Figure 2.9**).

Atomic and Molecular Levels

Ninety-two naturally-occurring **elements** (types of **atoms**) form the physical structure of this planet and its atmosphere. However, the main bulk of any living thing is comprised primarily of carbon atoms, some hydrogen, oxygen, and nitrogen atoms, and trace amounts of a few other elements.

Atoms **bond** (join) together into clusters known as **molecules.** The molecules that constitute the cells of living things are large and complex assemblages called **macromolecules**. There are four basic types of biological macromolecules: carbohydrates, lipids, proteins, and nucleic acids.

> *Almost all aspects of life are engineered at the molecular level, and without understanding molecules we can only have a very sketchy understanding of life itself.*
> —Francis Crick

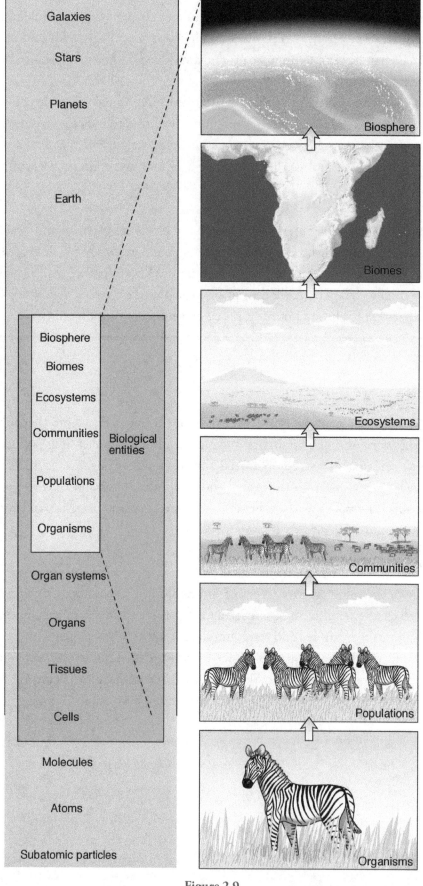

Figure 2.9

Cellular Level

Any multicellular organism is vested with life at two levels—the cellular level and the organismal level. Each cell in a creature is alive, but somehow the collective lives of each of those cells are transposed to a higher plane, and as a consequence, the entire organism lives. This amazing extrapolation of the life force results in any multicellular organism being far more than just the sum of its parts.

Organismal Level

Cells are grouped into tissues, such as muscle cells, which are grouped into organs, such as the heart, blood vessels, and blood, and organs are coordinated into organ systems to form a complex multicellular organism. A total of eleven different organ systems are found in **metazoans** (complex multicellular organisms): integumentary, muscular, skeletal, respiratory, digestive, circulatory, excretory, nervous, endocrine, immune, and reproductive. Only with the single-celled protists and bacteria does the totality of life and death exist solely at the cellular level.

Ecological Level

Individual organisms are intertwined into several hierarchical levels—*population, community, ecosystem,* and *biome*—within the larger physical world of this planet:

- A **population** is a group of organisms of the same species living in the same area at the same time.
- A **biological community** consists of all the populations of different species living together in the same place at the same time.
- An **ecosystem** would be a complex of communities and their interactions with the physical environment—air, water, and soil—in a particular area.
- A **biome** is the largest geographical biotic unit and consists of various ecosystems and their attendant communities with similar requirements of environmental conditions.

Contemplating a grasshopper gives us an example of the entire hierarchy in action. Atoms mainly of hydrogen, oxygen, carbon, and nitrogen are arranged into biological macromolecules of carbohydrates, proteins, fats, and nucleic acids which, in turn, are organized into living animal cells. These animal cells of various types are arranged into tissues, organs, and organ systems ultimately resulting in the complex multicellular organism we know as a grasshopper. A *population* of grasshoppers lives in a ravine *community* that is rich with many other species of animals and plants. This particular ravine is located within the mixed-grass prairie *ecosystem*. Finally, the three prairie ecosystems—short, mixed-grass, and tall—on this continent meld into the gigantic grassland *biome* of North America.

One should understand from Figure 2.9 that each level of biological organization builds upon the level below it and is more complex than the previous level. Moving up the hierarchy, each level acquires new emergent properties or new, unique characteristics that are determined by the interactions between the individual parts. For example, a change in the concentration of carbon dioxide molecules can affect cells, organs, organisms, and entire ecosystems.

We shall walk together on this path of life, for all things are part of the universe and are connected with each other to form one whole unity.

—Maria Montessori

In Summary

- Life cannot be defined or delineated in specific biological terms.
- Living things must display all the properties of life. Possessing only some of the properties of life does not make an object alive.
- Life is more than the sum of its parts.
- The properties of life are:
  - Life arises through biogenesis.
  - Life has a cellular organization.
  - Life responds to the internal and external environment.
  - Life utilizes energy.
  - Life maintains homeostasis.
  - Life transmits genetic information.
  - Life grows and develops.
- The biological world is organized, and this organization is hierarchal.
- The hierarchy of life from simplest level to most complex level is:
  - Atoms and molecules
  - Cells
  - Organism
  - Ecological
- There is a hierarchy to the ecological level from smallest unit to largest unit:
  - Population
  - Community
  - Ecosystem
  - Biome

Review and Reflect

1. *Signs of Life* Trees move in the wind and boulders tumble downhill. Icicles and crystals grow. Are trees and boulders alive because they can move? Are icicles and crystals alive because they can grow? How many of the characteristics or signs of life must something demonstrate to be considered truly alive? Discuss.

2. *Prove It* You take for granted that you are alive, but can you prove it? Use the characteristics of life discussed earlier in this chapter to make the case that you are, if fact, alive.

3. *What's in a Quote?* We opened this chapter with a quote from E. O. Wilson. React to Wilson's quote and explain in your own words what you believe Wilson is trying to say.

4. ***Synthetic Life*** Synthetic biology is a new field of biotechnology that holds great promise. Explain what synthetic biology is and detail the possibility of this emerging biological field.

5. ***Sum It Up*** In this chapter is was stated that "Life is more than the sum of its parts." Explain what that statement means.

6. ***Beldar's Confusion*** Beldar of Remlac was sent to Earth by the High Command to find life forms. While on Earth, Beldar and his away team captured an automobile and took it back to Remlac. Beldar thought he had found a good example of life on Earth. Now, however, he's on trial for not carrying out his duties. The prosecuting lawyer says that the car is not alive. Beldar's defense attorney claims the car is alive and is a good example of life on Earth. The following is what Beldar said at his trial:

    "I first found these life forms rolling along the roads in great numbers. They were giving off thick clouds of poisonous wastes as they moved. They appeared to have a great deal of energy. Some of them moved faster than 80 km per hour. When one of these life forms stopped or slowed down, all the others behind responded. They slowed down and gave off a reddish light from the back. Some made honking noises. Because of all the energy their activity required, some of them stopped to feed on a liquid substance. After feeding, these Earthlings became active again. "Finally, I spotted one of these creatures by a house. I lowered my spacecraft to talk to it, but the creature didn't respond. So I took it prisoner and brought it back to Mars."

Take the part of Beldar's defense attorney and make a good case for the cars being alive then become the prosecutor and show that the car is nonliving. Make a two-column table and head one column *Defense* and the other *Prosecution*. Under the *Defense* column make the case as to why the car is alive and under the *Prosecution* column make the case as to why the car in not alive.

Create and Connect

1. ***Aliens Among Us*** The starship LecLero 2 has returned from a biological survey mission of the planets orbiting a nearby star. In cryogenic suspension aboard the ship are 20 purplish orbs. The orbs are soft and jelly-like and range in size from 1-4 inches in diameter. As the head of the United Space Federation's Xenomorph Division, it is you and your team's job to determine if these alien globs are actually alive. Write a report to your superiors in which you detail any and all tests you would perform to determine if the globs were living or inanimate.

2. ***Teach the Alien*** Having secretly identified you as one of the superior biological minds in all humankind, an alien has come to you and asked you to explain the structure, nature, and hierarchy of life. What would you say?

3. ***Pull the Plug?—A Position Paper*** There have been several well-publicized situations in the last several years that have created a firestorm of controversy about whether terminally ill individuals should be kept alive solely because we have the technically sophisticated life support equipment to do so. Such incidents impact society on a number of levels— legally, religiously, scientifically, and even legislatively. What are your feelings and beliefs on this issue? Write a position paper in which you take a stand on these sorts of situations and defend the stand you take.

Guidelines

A. Write a position paper, NOT an opinion paper. Defend your position with as many facts, figures, quotes, and pertinent information as possible.

B. When completed your paper should detail: (1) Are there any situations in which terminally ill patients should be kept alive indefinitely by life support machines? (2) Are there any situations in which terminally ill patients should not be kept alive indefinitely by life support equipment? (3). Who should decide whether someone should be kept on life support for an extended period of time? (4) What is **euthanasia** and is it ever justified?

C. Your work will be evaluated not on the "correctness" of your position but the quality of the defense of your position.

CHAPTER 3

BASIC CHEMISTRY

You mix a bunch of ingredients, and once in a while, great chemistry happens.
—Bill Watterson

Introduction

Life is chemistry. In fact, the definition of a living thing might be given as: *Life is a self-sustaining chemical system capable of Darwinian evolution.* Every instant in the life of a cell is biology, chemistry, and physics all working together. If we reduce anything, including you dear reader, to its elemental parts, it is nothing but atoms and molecules. And those atoms and molecules are the focus of the science of **chemistry**, a branch of the physical sciences that studies the composition, structure, properties and change of matter.

Rise of Chemistry

Early civilizations, such as the Egyptians, Babylonians, and Indians established practical chemical knowledge through their arts of metallurgy, pottery, and dyes, but didn't develop a science or scientific theory to explain the workings of those things.

A basic chemical hypothesis first emerged in classical Greece with the theory of Empedocles' four elements in which he theorized that fire, air, earth, and water were the fundamental elements from which everything is formed as a combination. The idea of the atom as a basic particle dates back to 440 BC in Greece, arising in works by philosophers such as Democritus and Epicurus. In 50 BC, the Roman philosopher Lucretius (**Figure 3.1**) expanded upon the theory in his epic philosophical poem *De rerum natura* (*On the Nature of Things*). Unlike modern concepts of science, Greek atomism was purely philosophical in nature, with little concern for empirical observations and no possibility of chemical experiments.

Figure 3.1 Titus Lucretius Carus (99 BC-55 BC)

In the Hellenistic world around the Mediterranean (323-31 BC), the art of *alchemy* first appeared, mingling magic and occultism into the study of natural substances with the ultimate goal of transmuting elements into gold and discovering the elixir of eternal life, alchemy was discovered and practiced widely throughout the Arab world after the Muslim conquests, and from there, diffused into medieval and Renaissance Europe.

Chemistry as a science began with the scientific revolution that began in Europe towards the end of the Renaissance period and continued through the late 18th century. Using newly-developed empirical methods (or scientific methods), Sir Francis Bacon, Robert Boyle, and John Mayow began to reshape the old alchemical traditions into a scientific discipline. Boyle (**Figure 3.2**) in particular is regarded as one of the founding father of chemistry due to his most important work, the classic chemistry text *The Sceptical Chymist* where the differentiation is made between the claims of alchemy and the empirical scientific discoveries of the new chemistry. He formulated Boyle's law, rejected the classical "four elements" and proposed a mechanistic alternative of atoms and chemical reactions that could be subject to rigorous experiment.

Figure 3.2 Robert Boyle (1627-1691)

The 18th century saw the growth and expansion of chemistry as a science. The Scottish chemist Joseph Black (the first experimental chemist) and the Dutchman J. B. van Helmont discovered carbon dioxide in 1754; Henry Cavendish discovered hydrogen and elucidated its properties and Joseph Priestley and, independently, Carl Wilhelm Scheele isolated pure oxygen, and the English scientist John Dalton proposed the modern theory of atoms; that all substances are composed of indivisible 'atoms' of matter and that different atoms have varying atomic weights.

The French chemist Antoine Lavoisier (**Figure 3.3**) elucidated the principle of conservation of mass, recognized and named hydrogen and oxygen, wrote the first extensive list of elements, and developed a new system of chemical nomenclature still used today. All of which accords him the title of one of the founding fathers of chemistry.

Figure 3.3 Antoine Lavoisier (1743-1794)

The development of the electrochemical theory of chemical combinations occurred in the early 19th century as the result of the work of two scientists in particular, J. J. Berzelius and Humphry Dav. Davy also discovered nine new elements including the alkali metals by extracting them from their oxides with electric current.

Figure 3.4 Dmitri Mendeleev (1834-1907) predicted the existence of 7 new elements and placed all 60 elements known at the time in their correct places on the periodic table.

British chemist William Prout first proposed ordering all the elements by their atomic weight as all atoms had a weight that was an exact multiple of the atomic weight of hydrogen. J. A. R. Newlands devised an early table of elements, which was then developed into the modern periodic table of elements in the 1860s by Dmitri Mendeleev (**Figure 3.4**) and independently by several other scientists. The inert gases, later called the noble gases were discovered by William Ramsay in collaboration with Lord Rayleigh at the end of the century, thereby filling in the basic structure of the periodic table.

Organic chemistry was developed by Justus von Liebig and others, following Friedrich Wöhler's synthesis of urea that proved that living organisms were, in theory, reducible to chemistry. Other crucial 19th century advances were an understanding of valence bonding (Edward Frankland in 1852) and the application of thermodynamics to chemistry (J. W. Gibbs and Svante Arrhenius in the 1870s).

At the turn of the twentieth century, the theoretical underpinnings of chemistry were finally understood due to a series of remarkable discoveries that succeeded in probing and discovering the very nature of the internal structure of atoms. In 1897, J. J. Thomson discovered the electron and soon after the French scientist Becquerel as well as the couple Pierre and Marie Curie investigated the phenomenon of radioactivity. In a series of pioneering scattering experiments Ernest Rutherford discovered the internal structure of the atom and the existence of the proton, classified and explained the different types of radioactivity and successfully transmuted the first element by bombarding nitrogen with alpha particles.

His work on atomic structure was improved on by his students, the Danish physicist Niels Bohr and Henry Moseley. Around the same time, the electronic theory of chemical bonds and molecular orbitals was developed by the American scientists Linus Pauling and Gilbert N. Lewis.

Principles of Modern Chemistry

Chemistry is sometimes called the "central science" because it bridges other natural sciences, including physics, geology, and biology. Chemistry studies the properties of individual atoms, how atoms form chemical bonds to create chemical compounds, the interactions of substances through intermolecular forces that give matter its general properties, and the interactions between substances through chemical reactions to form different substances.

In essence, chemistry is the study of **matter**, what it is composed of and how it interacts. Matter is defined as anything that has mass and takes up space. Matter exists in only four distinct forms: solid, liquid, gas, or plasma.

Nature of Atoms and Atomic Structure

Figure 3.5 Using a scanning-tunneling microscope, IBM researchers can not only see atoms but manipulate individual atoms as well. Atom manipulation has many practical applications, but having a little fun with the process never hurts.

All matter is composed of ultra-small particles called **atoms**. An atom may be defined as the tiniest particle of an element that cannot be further divided. Modern technology allows us to not only observe atoms (**Figure 3.5**) but to manipulate individual atoms as well.

The atom is the basic unit of chemistry. The modern view of an atom suggests it consists of a dense core called the **nucleus** surrounded by a space called the **electron cloud**. The nucleus is made up of positively charged **protons** and uncharged **neutrons** (together called **nucleons**), while the electron cloud consists of negatively charged electrons that orbit the nucleus (**Figure 3.6**). In a neutral atom, the negatively charged electrons balance out the positive charge of the protons. The nucleus is dense; the mass of a nucleon is 1,836 times that of an electron, yet the radius of an atom is about 10,000 times that of its nucleus. The majority of an atom is empty space. If an atom could be expanded to the size of a football field, the nucleus would be a pea in the center of the field, and the electrons would be tiny specks whirling about in the upper stands (**Figure 3.7**). Another way of visualizing it is to imagine that if the nucleus were the size of a golf ball, the closest electron orbital would be a mile away.

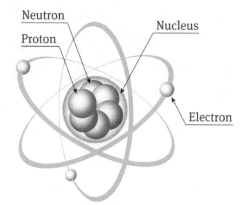

Figure 3.6 The Rutherford model of atomic structure.

Figure 3.7 If pea were placed in the center of the red N representing the nucleus of an atom, the electrons would be tiny blips much smaller than the nucleus whirling around the upper stands at 4,921,260 mph. Everything in-between is empty space

Scientists do not know the exact size of electrons. They are smaller than we can currently measure and may not have a size at all!

Within the nucleus, the cluster of proton and neutrons (nucleons) is held together by the **strong nuclear force**. Each proton carries a positive electrical charge (+), and each neutron has no charge (⊠). The nuclear attraction between two protons is not quite strong enough to keep two protons bound together against the electrical repulsion, but with the addition of a neutron or two, there is more overall attractive force to overcome the repulsion. For nuclei with more and more protons, the repulsive forces grow more than the attractive forces do, as the repulsive forces have much greater range than the attractive forces. To form stable or nearly stable nuclei more than one neutron is needed per proton to provide the attractive force needed.

Typically, an atom has one electron for each proton and is, thus, electrically neutral. Atoms may be defined by the number of protons they possess, a quantity called the **atomic number**. Atoms with the same atomic number (same number of protons) have the same chemical properties and are said to belong to the same element. An **element**, therefore, is any substance that cannot be broken down into any other substance by ordinary chemical means.

The **atomic mass** of an atom is equal to the sum of the masses of its protons and neutrons in the nucleus. Protons and neutrons are assigned one atomic mass unit (AMU) each. Electrons are so small their AMU is considered zero in most calculations (**Figure 3.8**). It is important to note that the terms *mass* and *weight* are often used interchangeably, but they have different meanings. For example, your mass and weight on Earth and the Moon would be the same because mass = amount of something. However, although your mass on the Moon would be the same as on Earth, your weight on the Moon would be only 1/6 your Earth weight because the force of gravity is much less on the smaller Moon.

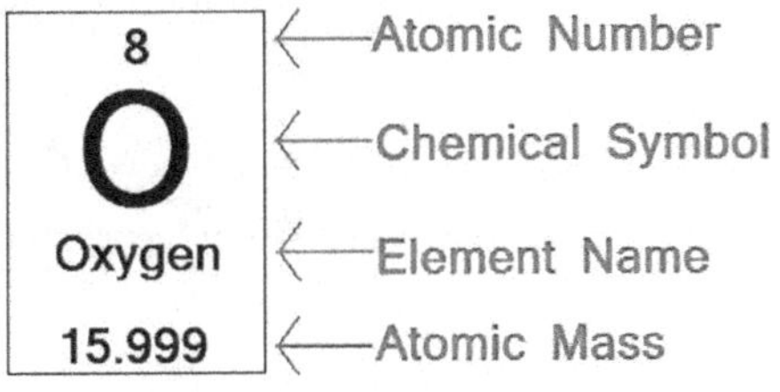

The positive charges in the nucleus of an atom are neutralized by the negatively charged electrons that are located in regions called **orbitals** that lie at varying distances away from the nucleus. Atoms with the same number of protons and electrons have no net charge and are therefore called **neutral atoms**.

Nucleons are held together by the strong atomic force whereas the negatively charged electrons are held in their orbits by their attraction (negatives attract) to the positively charged protons. Sometimes other forces overcome this balance of forces, and an atom loses one or more electrons. In other cases, atoms gain additional electrons. Atoms in which the number of protons does not equal the number of electrons are said to be **ions**, and they behave as charged particles. For example, an atom of sodium (Na) that has lost one electron becomes a sodium ion (Na^+), known as a **cation**, with a charge of +1. A chlorine atom (Cl) that has gained one electron becomes a chloride ion (Cl^-), known as an **anion**, with a charge of ⊠1.

Although all atoms of an element have the same number of protons, they may not have the same number of neutrons. Atoms of the same elements that possess different numbers of neutrons are called **isotopes** of that element. Most elements in nature exist as mixtures of different isotopes. Hydrogen (H), for example, has three isotopes, all containing one proton (**Figure 3.9**). The most common form of hydrogen is 1H protium whereas the rarest is 3H tritium. Tritium is unstable causing its nucleus to break up into elements with lower atomic numbers. This breakup is called **radioactive decay** and isotopes that decay in this fashion are **radioactive isotopes**.

8	← Atomic Number
O	← Chemical Symbol
Oxygen	← Element Name
15.999	← Atomic Mass

Figure 3.8 The element oxygen has an atomic number (number of protons) of 8 and an atomic weight (protons + neutrons) of 16. The number of neutrons in an atom can be determined by subtracting the atomic number from the atomic weight. Thus, oxygen has 8 neutrons.

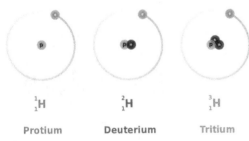

^1_1H ^2_1H ^3_1H

Protium   Deuterium   Tritium

Figure 3.9 The three naturally-occurring isotopes of hydrogen. The fact that each isotope has one proton makes them all variants of hydrogen: the identity of the isotope is given by the number of neutrons. From left to right, the isotopes are protium (1H) with zero neutrons, deuterium (2H) with one neutron, and tritium (3H) with two neutrons."

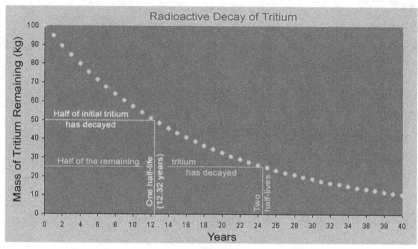

Figure 3.10

Some radioactive isotopes are more unstable than others, and therefore they decay more rapidly. For any given isotope, however, the rate of decay is constant. The decay time is expressed as **half-life**, the time it takes for one-half of the atoms in a sample to decay. For example, the half-life of 3H (tritium) is 12.32 years (**Figure 3.10**) whereas the half-life of radioactive C^{14} is 5,730 years.

Radioactivity has many useful applications in modern biology and medicine. Radioactive isotopes are one way to "tag" a specific molecule and then follow its progress, either in chemical reactions within living cells or its movement through the ecosystem. The dark side of radiation is that high energy subatomic particles emitted by radioactive substances breaking down have the potential to severely damage or kill living cells and produce genetic mutations. Consequently, exposure to radiation is carefully controlled and regulated. For example, when women of child-bearing age go the dentist and X-rays are taken, a lead apron is placed over the patient to protect the woman's eggs and possibly any embryos that might be present from radiation damage (**Figure 3.11**).

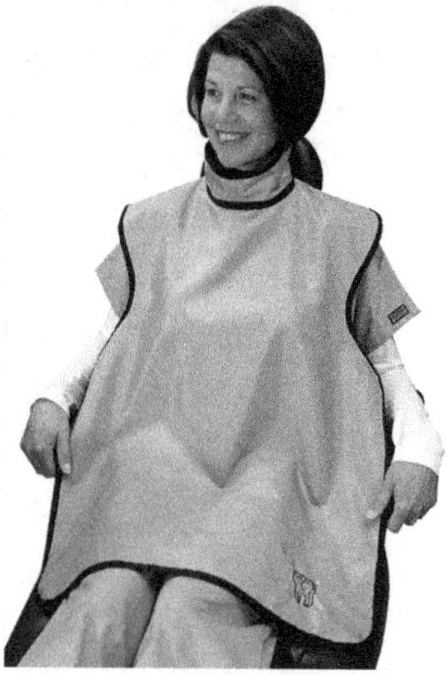

Figure 3.11

Electrons and Energy

The key to the chemical behavior of an atom lies in the number and position of its electrons in their orbitals. The problem is modern physics cannot pinpoint the position of any individual electron at any given time. In fact, an electron could be anywhere, from close to the nucleus to infinitely far away from it. A particular electron, however, is more likely to be in some areas (orbitals) than in others. These orbitals represent probability distributions for electrons, that is, regions more likely to contain an electron. Some orbitals near the nucleus are spherical (*s* orbitals) whereas those further out are dumbbell-shaped (*p* orbitals), and those further out yet (d orbitals) may have different shapes (**Figure 3.12**).

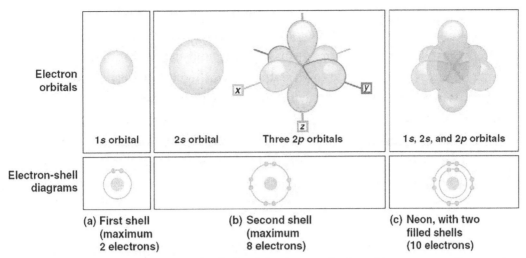

Figure 3.12 These diagrams represent the shape of the electron cloud would assume at each energy level.

Because the negatively charged electrons are attracted to the positively charged nucleus, it takes energy to keep them in their own orbital. The more distant the orbital (energy level), the more energy it takes. Electrons around a nucleus have discrete **energy levels**, and electrons can move between these energy levels. Because the amount of energy an electron possesses is related to its distance from the nucleus, electrons that are the same distance from the nucleus have the same energy, even if they occupy different orbitals. The energy levels are denoted with letters K, L, M, N, and so on (**Figure 3.13**).

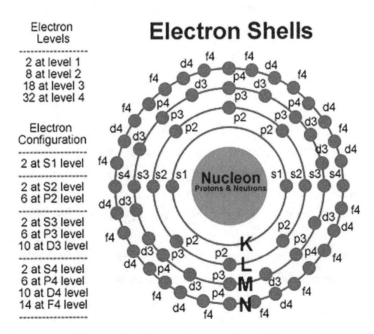

Figure 3.13 Each energy level is populated by a certain number of electrons: K(2), L(8), M(18), N(32).

Electrons have energy of position. The lowest energy level an electron can occupy is called the **ground state**. The higher orbitals represent higher excitation states. The higher the excitation state, the more energy the electron contains. When an electron absorbs energy, it jumps to a higher orbital. An electron in an excited state can release energy and 'fall' to a lower state. When it does, the electron releases a photon of electromagnetic energy. The energy contained in that photon corresponds to the difference between the two states the electron moves between. When the electron returns to the ground state, it can no longer release energy but can absorb quanta of energy and move up to excitation states (higher orbitals) (**Figure 3.14**). An example can be seen in the process of photosynthesis when atoms absorb the energy of sunlight, electrons are boosted to an excitation state, and they move further from the nucleus. Later, as the electrons fall back to the ground state (original energy level), a quantum packet of energy is released and transformed into chemical energy. The chemical energy released by excited electrons falling back to their ground state during photosynthesis powers all life on Earth; thus, our very existence is dependent on the changing energy states of electrons.

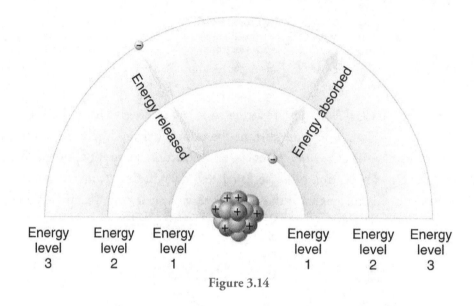

Figure 3.14

The Periodic Table

As mentioned previously, Dmitri Mendeleev and others, developed what came to be known as the **periodic table** in the mid to late 1800s. The table was developed to group the elements (types of atoms) according to certain chemical and physical characteristics. In doing so, Mendeleev discovered that the elements exhibit a pattern of chemical properties that repeats itself in groups of eight. This periodically repeating pattern lent the table its name: the periodic table of elements (**Figure 3.15**). In the periodic table, the horizontal rows are called *periods,* and the vertical columns are called *groups*. The atomic number of every atom in a period increases by one if you read from left to right. All the atoms in a group share similar chemical characteristics, namely in the type of chemical bonds they form. Note the groups are color-coded in Figure 3.15.

Group

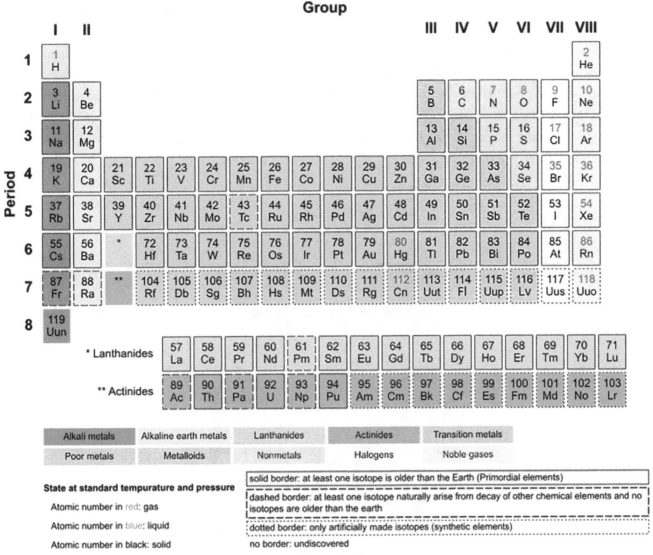

Figure 3.15

The eight-element periodicity that Mendeleev found is based on the interactions of the electrons in the outermost energy level of the different elements. Known as **valence electrons**, their interactions are the basis for the elements' differing chemical properties. The outermost energy level can contain no more than eight valence electrons; the chemical behavior of an element reflects how many of the eight positions are filled. Elements possessing all eight electrons in their outer energy level (two for helium) are inert, or nonreactive. On the periodic table the inert elements are in Group VIII, and because they are gases by nature, Group VIII is known as the *noble gases*.

In sharp contrast, elements with seven valence electrons (one less than the maximum number of eight) in their outer energy level vigorously seek to *gain* one more electron. Fluorine (F), chlorine (Cl), and bromine (Br) are highly reactive elements that need one more electron to reach the magic number of eight. Elements with only one electron in their outermost energy level, such as Lithium (Li), sodium (Na), and potassium (K), are also very reactive as they vigorously seek to *lose* one electron from their outer level.

What happens if two nonreactive atoms encounter each other? Nothing really, because each atom has its maximum number of eight valence electrons and does not seek to gain or loose any of its outer energy level electrons. What happens when two reactive atoms encounter each other? If a reactive atom that wishes to lose one electron, such as sodium Na), encounters a reactive atom that wishes to gain one electron, such as chlorine (Cl), the outermost electron of NA transfers to Cl. Once that happens, both atoms have their outermost energy level filled with eight electrons. Both atoms also become electrically charged in the process. Because Na now has one more proton (+) than electrons (−), it becomes a positively charged ion (Na^{+}) cation. Because Cl now has one more electron than protons, it becomes a negatively charged ion $Cl^{−}$) anion (**Figure 3.16**).

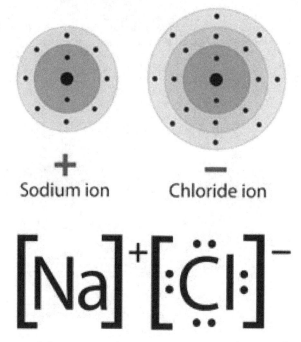

Figure 3.16 Losing or gaining an outer electron causes atoms to become electrically charged.

Of the 90 naturally occurring elements on Earth, only 12 (C, H, O, N, P, S, Na, K, Ca, Mg, Fe, and Cl) are found in living things in more than trace amounts (0.01% or higher). Of these elements, the first four—C, H, O, N—constitute 96.3% of the weight of your body. Because most of the compounds found in the body of a living thing contain carbon, Earthly life is said to be carbon-based. Compounds of carbon are called **organic compounds** whereas those compounds containing little or no carbon are called **inorganic compounds**. Glucose (blood sugar, sap sugar) $C_6H_{12}O_6$ is an organic compound, but water H_2O is an inorganic compound because water does not contain carbon atoms.

Molecules and Bonding

Few free atoms exist in nature. The natural order of things is for atoms to bind (bond) together to form molecules. A **molecule** exists when two or more elements bond together. When a molecule contains atoms of more than one element, it is called a **compound**. The molecule we call water (H_2O) is a compound because it contains 2 atoms of hydrogen (H) and 1 atom of oxygen (O). The type, number, and arrangement of atoms in a molecule may be expressed as a formula. A *molecular formula* reveals the type and number of atoms in a molecule whereas a *structural formula* details exactly how the atoms in a molecule are bonded together. The molecular formula for water is the familiar H_2O denoting two atoms of hydrogen and one atom of oxygen. The structural formula for water is H-O-H denoting that the two atoms of hydrogen are bonded to the oxygen and not to each other.

Molecules are held together by **chemical bonds**. There are two categories of chemical bonds: ionic bonding and covalent bonding.

Ionic bonding As the name indicates, **ionic bonds** are those that form between ions in which opposite electrical charges attract. Let's revisit sodium (Na) and chlorine (Cl). Both are reactive atoms but in opposite ways. Sodium wishes to lose one electron from its outer energy level whereas chlorine wishes to gain one electron. When Na loses an electron, it becomes electrically charged Na^+ (1 more proton than electrons), and when Cl gains an electron it becomes electrically charged Cl^- (1 more electron than protons). In nature, opposite electrical charges attract so the sodium and chlorine ionically bond together to form a molecule of sodium chloride Na^+Cl^- (**Figure 3.17**).

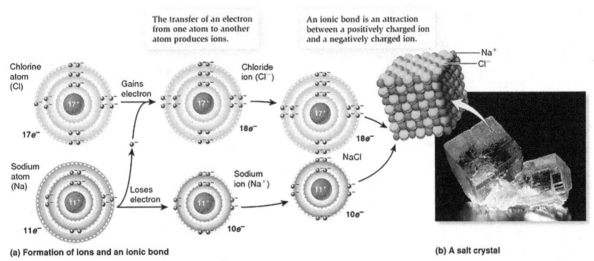

(a) Formation of ions and an ionic bond   **(b) A salt crystal**

Figure 3.17 Formation of Sodium Chloride (table salt). (a) During the formation of sodium chloride, an electron is transferred from the sodium atom to the chlorine atom. In the end, each atom has eight electrons in their outer shell, but each also carries an electrical charge as shown. (b) Sodium chloride molecules assume a three-dimensional lattice in which each sodium ion is surrounded by six chlorine ions, and each chlorine ion is surrounded by six sodium ions. The result is a crystal of salt.

If a salt such as NaCl is placed in water, the electrical attraction of the water molecules disrupts the forces holding the ions in their crystal matrix, causing the salt to dissolve into a roughly equal mixture of free Na^+ and Cl^- ions. Because living systems always include water, ions are more important than ionic crystals.

Covalent bonding In ionic bonding, there is a give and take of electrons, but in covalent bonding, there is a sharing of electrons. Unlike ionic bonds, **covalent bonds** are formed between two individual atoms, giving rise to true, discrete molecules. Consider gaseous hydrogen (H_2) as an example. The outer shell is complete

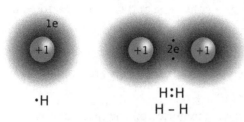

Figure 3.18 Because hydrogen atoms share a pair of electrons, they are said to form a single covalent bond

when it contains two electrons, but atomic hydrogen only has one electron in its outer shell. However, if two atomic hydrogen atoms overlap their electron shells, the electrons are shared. Because they share the electron pair, each atom has a completed outer shell and thus become molecular hydrogen (**Figure 3.18**).

Hydrogen atoms form **single covalent bonds** but sometimes atoms share more than one pair of electrons to complete their octet. **Double covalent bonds** satisfy the *octet rule* by allowing two atoms to share two pairs (four) of electrons whereas in **triple covalent bonds** two atoms share three pairs (six) of electrons (**Figure 3.19**). Although a single hydrogen bond is more easily broken that a single covalent bond, multiple hydrogen bonds are collectively quite strong. Hydrogen bonds between cellular molecules help maintain their proper structure and function. For example, hydrogen bonds hold the two strands of DNA firmly together. However, when DNA replicates, hydrogen bonds easily break, allowing DNA to unzip. Similarly, the shape of protein molecules is often maintained by hydrogen bonding between different parts of the same molecule. Maintaining the proper shape of a protein is often critical for the protein to function properly.

A vast number of biological compounds are composed of more than two atoms. An atom that requires two, three, or four additional electrons to fill its outer energy level completely may acquire them by sharing its electrons with two or more atoms. The carbon atom (C) seems to have a great affinity for sharing electrons. Carbon contains four unpaired electrons in its outer energy level. To satisfy the octet rule, a carbon atom must form four covalent bonds. As a result, carbon atoms are found in many different kinds of biologically important molecules.

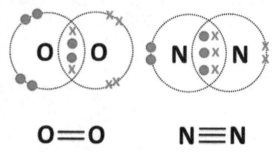

Figure 3.19 By sharing electrons where the shells overlap, each atom fills its outer shell with 8 electrons. (a) oxygen shares 2 pair (4) forming a double covalent bond. (b) Nitrogen shares 3 pair (6) forming a triple covalent bond.

Atoms differ in their affinity for electrons, a property called **electronegativity**. In the periodic table, electronegativity increases left to right across a column and decreases down the column. Thus, the elements in the upper-right corner have the highest electronegativity. For bonds between atoms of equal electronegativity, the electrons are shared equally, and such bonds are termed *nonpolar*. For bonds between atoms of differing electronegativity, the electrons are not shared equally, and such bonds are termed *polar* (**Figure 3.20**).

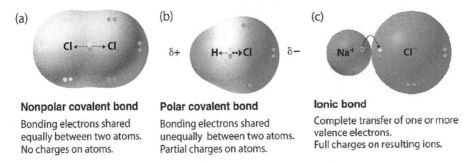

Nonpolar covalent bond
Bonding electrons shared equally between two atoms. No charges on atoms.

Polar covalent bond
Bonding electrons shared unequally between two atoms. Partial charges on atoms.

Ionic bond
Complete transfer of one or more valence electrons. Full charges on resulting ions.

Figure 3.20 Bonding. In a purely covalent bond (a), the bonding electrons are shared equally between the atoms. In a purely ionic bond (c), an electron has been transferred completely from one atom to the other. A polar covalent bond (b) is intermediate between the two extremes: the bonding electrons are shared unequally between the two atoms and the electron distribution is asymmetrical with the electron density being greater around the more electronegative atom. Electron-rich (negatively charged) regions are shown in blue; electron-poor (positively charged) regions are shown in red."

The shape of a molecule may also influence whether it is polar or nonpolar. In methane (CH_4), for example, although the carbon is larger and has more protons than a hydrogen atom, the symmetrical nature of a methane molecule cancels out any polarities. Water, on the other hand, has polarity because its shape prevents the polar bonds from canceling each other out (**Figure 3.21**). The polarity of molecules affects how they interact with other molecules.

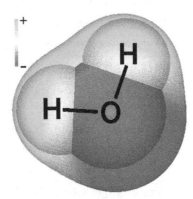

Figure 3.21 The approximate shape and charge distribution of a water molecule. The more electronegative oxygen end of the molecule is designated slightly negative, and the hydrogens are designated slightly positive."

Chemical Reactions

The essence of chemistry is the formation and breaking of chemical bonds; a process termed *chemical reactions*. All chemical reactions involve the shifting of atoms from one molecule or ionic compound to another without any change in the number or identity of the atoms. In a chemical reaction, the original molecules before the reaction starts are referred to as **reactants**, and the molecules resulting from the chemical reaction as **products**. For example:

$$6H_2O + 6CO_2 \leftrightarrow C_6H_{12}O_6$$

(reactants)   (products)

This is a simplified form of the photosynthesis reaction in which water and carbon dioxide combine to form glucose. Many reactions in nature are reversible with the reactants sometimes becoming the product and visa versa. The double arrow in the diagram above indicates that that reaction is reversible. Glucose is produced during photosynthesis and then broken down (reversed) during respiration.

Chemical reactions are influenced by three important factors:

1. **Concentration of reactants and products** Reactions proceed more quickly when more reactants are available, allowing for more contact between reactant molecules. An accumulation of products typically slows the reactions. In the case of reversible reactions, product accumulation may speed the reaction in the reverse direction.

2. **Temperature** Heating increases the reaction rate because the reactants become more energized and contact each other more often. Cooling has the opposite effect.

3. **Catalysts** A **catalyst** is a substance that increases the rate of a reaction sometimes dramatically without becoming part of the reaction. In living systems, proteins called enzymes catalyze almost every chemical reaction.

Water Supports Life

Water is such an important biological molecule that it bears closer scrutiny. Not only is three-fourths of the planet covered by liquid water, life itself is thought to have originated in water. Life remains inextricably tied to water. About two-thirds of any organism's body is composed of water, and all organisms require a water-rich environment either internally or externally to survive.

Water has a number of unique properties:

- **High heat capacity** The many hydrogen bonds that link water molecules together help water absorb heat without a great change in temperature. Water retains heat, and its temperature falls more slowly than that of other liquids. This property of water is not only important for aquatic organisms, but for all life.

- **High heat of vaporization** As water changes from a liquid to a gas it requires energy in the form of heat to break many hydrogen bonds. Water's high heat of vaporization gives animals in a hot environment an efficient way to release body heat. When an animal sweats, or gets splashed with water or mud, body heat is used to vaporize water, thus cooling the animal (**Figure 3.22**).

- **Acts as a solvent** Due to its polarity, water facilitates chemical reactions, both outside and within living systems. As a solvent, it dissolves a great number of substances, especially those that are polar. A **solution** contains dissolved substances called **solutes**. When ionic salts—sodium chloride (NaCl), for

Figure 3.22

example—are placed in water, the negative ends of the polar water molecules are attracted to the sodium ions, and the positive ends of the water molecule are attracted to the chloride atoms. This attraction causes the sodium and chlorine ions to separate, or disassociate, in water (**Figure 3.23**).

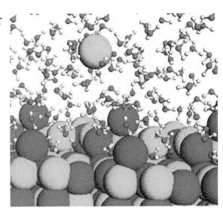

- **Cohesive-Adhesive** The ability of water molecules to cling to each other because of polarity is known as **cohesion**. Because of cohesion, water exists as a liquid under the conditions of temperature and pressure present at Earth's surface. The strong cohesion of water molecules allows water to flow freely without the water molecules separating from each other. **Adhesion** refers to the ability of water molecules to cling to other polar surfaces. Therefore, water can be said to be "sticky" in that they cling to each other and to surfaces. Biologically, cohesion and adhesion of water allow the flow of liquids through the tubular vessels of the cardiovascular system of animals and the xylem transport system of plants.

Figure 3.23 Water molecules disassociating sodium (green) and chlorine (blue) ions from a sodium chloride crystal matrix. As the water molecules break the weak ionic bonds holding the sodium and chlorine ions together, the ions disassociate.

- **Density differences** Water is one of the few substances in which the density of the molecules becomes less as temperature drops. Water expands as it reaches 0° and freezes, which is why frost heaves break up roadways during northern winters and why a can of soda placed in the freezer will burst. Frozen water is less dense than liquid water, and therefore ice floats on liquid water (**Figure 3.24**). If ice did not float on water, it would sink to the bottom, and ponds, lakes, and perhaps shallower areas of the ocean would freeze solid, making aquatic life impossible.

- **Water ionizes** Even though water molecules are held together by double covalent bonds, they can ionize. When water does ionize, it releases equal numbers of *hydrogen ions* (H^+) and *hydroxide ions* (OH^-). Substances that disassociate in water, releasing hydrogen ions (H^+) are called **acids**. The acidity of a substance depends on how fully it disassociates in water. Hydrochloric acid (HCl) is a strong acid that almost completely disassociates. On the other hand, lemon juice, vinegar, tomatoes, and coffee are mild acids because they only partially disassociate in water. **Bases** are substances

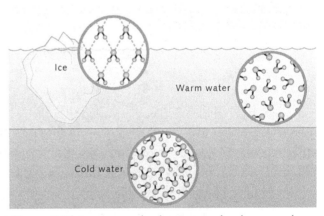

Figure 3.24 As water cools, the water molecules move closer together (become denser), but at the freezing point they align in a crystal matrix that is far less dense than cold water.

that either take up hydrogen ions or release hydroxide ions (OH^-). Sodium hydroxide (NaOH) is a strong base that disassociates almost completely in water whereas human blood, egg whites, and sea water are weak bases.

pH Scale

The pH scale is a convenient way of expressing the acidity (H^+) or basicity (alkalinity) (OH^-) of a solution (**Figure 3.25**). The scale is logarithmic so each unit is 10 times greater than any unit above neutral pH 7 (basic) or below it (acidic). Therefore, tomatoes are a 1000 times more acidic than urine.

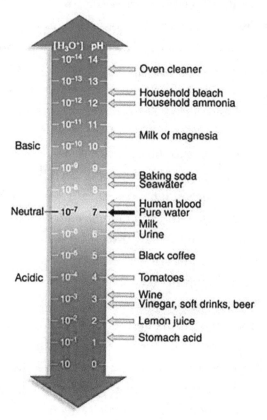

Figure 3.25 The pH scale ranges from 0 to 14. Pure water is 7 (neutral) because it releases equal amounts of hydrogen and hydroxide ions. Any substance that releases more hydrogen ions (H+) than hydroxide ions (OH⁻) is considered acidic whereas any substance that releases more hydroxide ion is considered basic (alkaline).

The pH inside almost all living cells, and the extracellular fluid surrounding cells in multicellular organisms, is close to neural pH 7. Most of the enzymes in living systems are extremely sensitive to pH levels, and even a small change in pH can alter their shape thereby disrupting their function. To prevent swings in pH, living things produce buffers. A **buffer** is a substance that resists changes to pH. Buffers act by releasing hydrogen ions when a base is added and absorbing hydrogen ions when acid is added.

Within organisms, most buffers consist of pairs of substance, one acid, and the other base. For example, the key buffer in human blood is an acid-base pair consisting of carbonic acid (acid) and bicarbonate (base). First, carbon dioxide (CO_2) and water (H_2O) bond to form carbonic acid (H_2CO_3) that in a reversible reaction disassociates to yield bicarbonate ion (HCO_3^-). If something acidic adds H^+ to the blood, the HCO_3^- acts as a base and removes the excess H^+ by forming H_2CO_3. Similarly, if a basic substance removes H^+ from the blood, H_2CO_3 disassociates releasing more H^+ into the blood. The forward and reverse reactions stabilize the blood's pH.

In Summary

- Chemistry is a branch of the physical sciences that studies the composition, structure, properties and change of matter.

- The first chemist practiced the art of alchemy, mingling magic and occultism into the study of natural substances with the ultimate goal of transmuting elements into gold and discovering the elixir of eternal life.

- Matter is defined as anything that has mass and takes up space. Matter exists in only four distinct forms: solid, liquid, gas, or plasma.

- All matter is composed of ultra-small particles called atoms.

- An atom consists of a dense core called the atomic nucleus surrounded by a space called the electron cloud. The nucleus is made up of positively charged protons and uncharged neutrons (together called nucleons), while the electron cloud consists of negatively charged electrons that orbit the nucleus.

- Within the nucleus, the cluster of proton and neutrons (nucleons) is held together by the strong nuclear force. Each proton carries a positive electrical charge (+), and each neutron has no charge (⊠).

- Atoms may be defined by the number of protons they possess, a quantity called the atomic number.

- Atoms with the same atomic number (same number of protons) have the same chemical properties and are said to belong to the same element.

- The atomic mass of an atom is equal to the sum of the masses of its protons and neutrons in the nucleus.

- The positive charges in the nucleus of an atom are neutralized by the negatively charged electrons that are located in regions called orbitals that lie at varying distances away from the nucleus.

- Atoms in which the number of protons does not equal the number of electrons are said to be ions, and they behave as charged particles.

- Atoms of the same elements that possess different numbers of neutrons are called isotopes of that element.

- Electrons around a nucleus have discrete energy levels, and electrons can move between these energy levels.

- The lowest energy level an electron can occupy is called the ground state. The higher orbitals represent higher excitation states.

- The periodic table was developed to group the elements (types of atoms) according to certain chemical and physical characteristics.

- The eight-element periodicity that Mendeleev found is based on the interactions of the electrons in the outermost energy level of the different elements. Known as valence electrons, their interactions are the basis for the elements' differing chemical properties.

- Compounds of carbon are called organic compounds whereas those compounds containing little or no carbon are called inorganic compounds.

- A molecule exists when two or more elements bond together. When a molecule contains atoms of more than one element, it is called a compound.

- The type, number, and arrangement of atoms in a molecule may be expressed as a formula. A molecular formula reveals the type and number of atoms in a molecule whereas a structural formula details exactly how the atoms in a molecule are bonded together.

- Molecules are held together by chemical bonds. There are two categories of chemical bonds: ionic bonding and covalent bonding.

- Ionic bonds are those that form between ions in which opposite electrical charges attract.

- In covalent bonding, there is a sharing of electrons.

- Covalent bonds may be single, double, or triple bonds.

- Atoms differ in their affinity for electrons, a property called electronegativity.

- For bonds between atoms of equal electronegativity, the electrons are shared equally, and such bonds are termed nonpolar. For bonds between atoms of differing electronegativity, the electrons are not shared equally, and such bonds are termed polar covalent bonds.

- The essence of chemistry is the formation and breaking of chemical bonds, a process termed chemical reactions.

- Water has unique properties: high heat capacity, high heat of vaporization, acts as a solvent, cohesive-adhesive, density differences, and water ionizes.

- Substances that disassociate in water, releasing hydrogen ions (H^+) are called acids whereas bases are substances that either take up hydrogen ions or release hydroxide ions.

- The pH scale is a convenient way of expressing the acidity (H^+) or basicity (alkalinity) (OH^-) of a solution.

Review and Reflect

1. **Does It Matter?** Your body is matter, but your body is atoms also. What is the relationship between matter and atoms?

2. **Empty Space** Your study buddy turns to you and asks, "If atoms are mainly empty space, why can't I just stick my hand through this textbook?" How would you respond?

3. **My Planet, My Atom** How might the structure of an atom be compared to the structure of the solar system?

4. **It's Elemental** You have been given two vials each containing a white powder. The white powder is an element or two different elements that have been finely ground. If you were challenged to determine if the powder in each vial was the same element or two different elements, what would you have to know about the atomic structure of each vial of powder?

5. **Run the Numbers** The atoms of an element have a nucleon with 8 protons and 8 neutrons. What are the atomic number and atomic mass of that element? What is the specific element in question?

6. **Let's Bond** In nature, atoms bind (bond) together. With that in mind, compare and contrast covalent and ionic bonding and give examples of each.

7. **Glow Little Glowworm** Relate the bioluminescent light given off by fireflies to changes in electron energy levels.

8. **Run the Table** Consider Figure 3.15. If two elements are in the same row, what do they have in common? If two elements are in the same column, what do they have in common?

9. ***Back and Forth*** Many of the chemical reactions that occur in living systems are reversible reactions. Why is this important?

10. ***Find the Water*** NASA is searching Mars for signs of water instead of looking for biological organisms (living or fossilized). Explain this strategy.

Create and Connect

1. ***Stain Busters*** Proteases are enzymes that break down protein. Bacterial enzymes often are used in detergents to help remove stains such as egg, blood, grass, and sweat from clothes. In this chapter, you learned that enzyme activity (chemical reaction) can be influenced by concentration (amount)of the enzyme and the temperature of the environment in which the chemical reaction is occurring. This raises the question—which has the greater influence on the effectiveness of the enzymes in detergents: temperature or concentration? Design an experiment to test this problem question.

   Guidelines
   A. Following the tenets of a well-constructed experiment, your design should include the following components in order:
   - The *Problem Question*. State exactly what problem you will be attempting to solve.
   - Your *Hypothesis*. Although this is a fictitious experiment, word your hypothesis as realistically as possible.
   - *Methods and Materials*. Explain exactly what you will do in your experiment including the materials necessary to accomplish the task. Be specific, take nothing for granted, and do not expect people to read your mind as they read your work.
   - *Collecting and Analyzing Data*. Explain what type(s) of data will be collected, and what statistical tests might be performed on that data. It is not necessary to concoct fictitious data or imaginary observations.

   B. Your instructor may provide additional details or further instructions.

2. ***Wonders of Water*** You are the science editor of a large metropolitan newspaper. Your editor has given you the following assignment: Write an article entitled *The Wonders of Water*. How would you proceed and what would you say? Format your work as if it were an actual newspaper article. Your article should link the chemical properties of water to living systems.

3. ***Whacky Water*** Knowing you have at least somewhat of an understanding of chemistry, the local swimming pool manager has come to you with a problem. Swimmers have been complaining that the water in the pool is burning their eyes. The manager doesn't know if the chlorine level is too high or if the pH of the water is out of whack. What measurements might you need to take and how would you attempt to solve the problem?

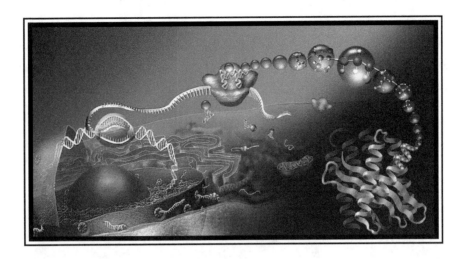

CHEMISTRY OF ORGANIC MOLECULES

Knowing without seeing is at the heart of chemistry.

—Bill Watterson

Introduction

Life is organic chemistry, the chemistry of molecules that contain large amounts of carbon atoms. There are only four categories of organic molecules in any living organism: carbohydrates, lipids, proteins, and nucleic acids. Collectively known as **biomolecules**, these organic molecules perform a diverse number of functions in a cell despite being few in number. The diversity of life is made possible by the diversity of biomolecules.

Carbon is Life

At the heart of every biomolecule are carbon atoms, so much so that life on this planet is said to be carbon-based. Every single creature from the tiniest bacterium to the tallest redwood tree to a lumbering rhinoceros is carbon-based, including you. If you totally incinerate the body of any creature, you will be left with a pile of pure carbon. Therefore, before we investigate biomolecules we need to take a closer look at the element that forms the backbone of those molecules.

A carbon atom is quite small, with a total of six electrons: two electrons in the first shell and four electrons in the outer shell. To acquire the four electrons to complete the outer shell, a carbon atom forms covalent bonds with as many as four other elements (**Figure 4.1**). It is this unique ability to form stable covalent bonds with up to four different elements that makes carbon the perfect element to serve as the base on which all biomolecules are built.

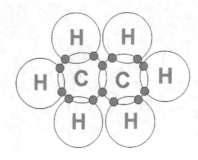

Figure 4.1 In this molecule of ethane (C2H4), each carbon atom forms a single covalent bond with three hydrogen atoms.

Because the C-C bond is very strong and allows for covalent bonds in various locations within a biomolecule, it allows the formation of straight chains, branches, rings, balls, tubes, and coils. In addition to forming single covalent bonds, carbon can form double bonds and even triple covalent bonds with itself and other atoms. Branches may also form at any carbon atom, allowing for the formation of long, complex carbon chains (**Figure 4.2**).

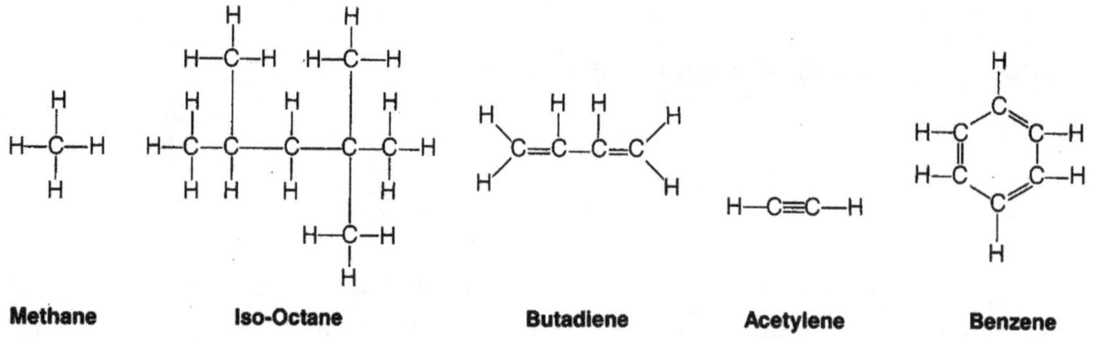

Methane   **Iso-Octane**   **Butadiene**   **Acetylene**   **Benzene**

Figure 4.2 The bonding flexibility of the carbon atom allows for the formation of biomolecules of many shapes.

Carbon and hydrogen atoms both have very similar electronegativity so for this reason molecules composed of only carbon and hydrogen (known as **hydrocarbons**) are nonpolar molecules (carrying no electrical charge). However, most biological molecules produced by cells also contain other atoms and therefore different electronegativities than carbon resulting in a polar molecule (carrying an electrical charge). These molecules may be thought of as a C-H core to which specific molecular groups—functional groups—are attached (**Figure 4.3**).

Functional Group	Structure	Properties
Hydroxyl	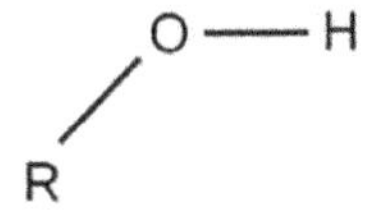 O — H / R	Polar
Methyl	 R — CH₃	Nonpolar
Carbonyl	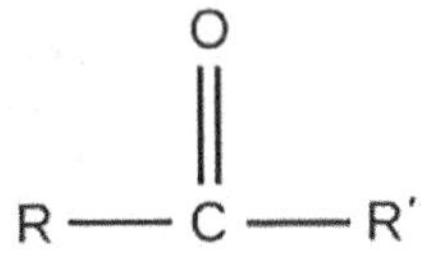 O ‖ R — C — R′	Polar
Carboxyl	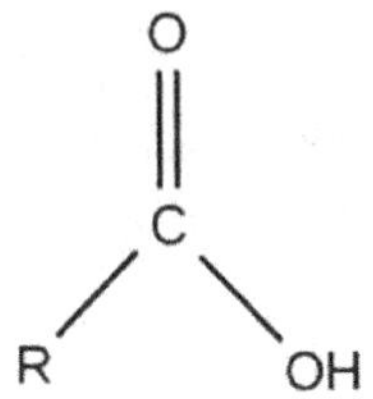 O ‖ C / \ R OH	Charged, ionizes to release H^+. Since carboxyl groups can release H^+ ions into solution, they are considered acidic.
Amino	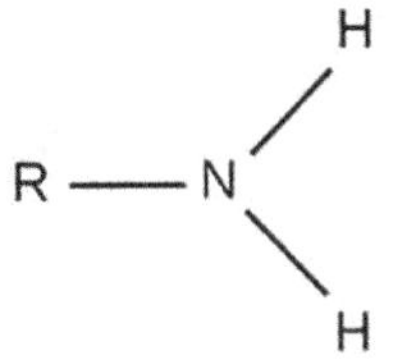 H / R — N \ H	Charged, accepts H^+ to form NH_3^+. Since amino groups can remove H^+ from solution, they are considered basic.
Phosphate	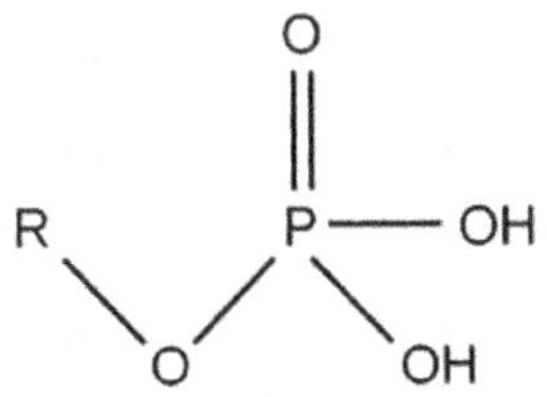 O ‖ R — P — OH / \ O OH	Charged, ionizes to release H^+. Since phosphate groups can release H^+ ions into solution, they are considered acidic.
Sulfhydryl	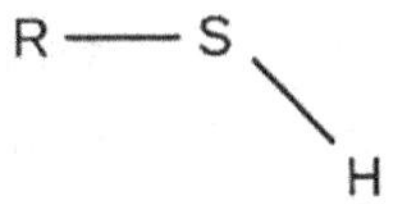 R — S \ H	Polar

Figure 4.3 Primary Functional Groups. These groups tend to act as units during chemical reactions and give specific chemical properties to the molecules that possess them. These functional groups are widely distributed in biomolecules.

The configuration of the functional groups determines the properties of the biomolecule. For example, the addition of an –OH (hydroxyl group) to a carbon skeleton turns that molecule into an alcohol. The addition of functional groups can alter the water solubility of biomolecules making them **hydrophobic** (not soluble in water) or **hydrophilic** (soluble in water), an important function given that cells are 70-90% water.

Some organic molecules, known as **isomers**, have identical molecular formulas but different arrangements of atoms. If there are differences in the actual structure of their carbon skeleton, we call constitutional isomers or structural isomers. Another form of isomer, called **stereoisomers**, have the same carbon skeleton but differ in how the groups attached to this skeleton are arranged in space (**Figure 4.4**). Isomers are another example of how the chemistry of carbon leads to variation in the structure of organic molecules.

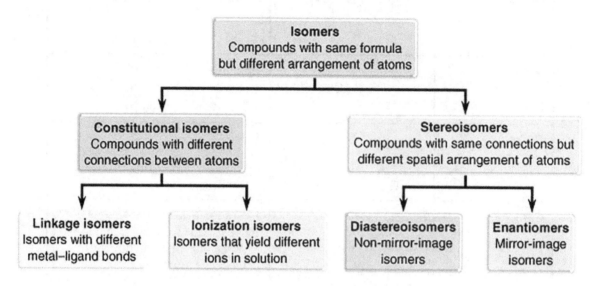

Figure 4.4 A general scheme for classifying isomers.

Biomolecules of Cells

Biomolecules—carbohydrates, lipids, proteins, and nucleic acids—are macromolecules, meaning they contain smaller subunits that are joined. Biomacromolecules are referred to as **polymers** since they are constructed of linking together a large number of the same type of subunit or **monomers**. Lipids are not polymers because they contain two different types of subunits (glycerol and fatty acids).

To synthesize a macromolecule, the cell uses a **dehydration synthesis** reaction in which the equivalent of a water molecule—an –OH (hydroxyl group) and an –H (hydrogen atom) is removed as subunits are joined. To break down macromolecules, dehydration synthesis is reversed as an –OH from a water molecule attaches to one subunit and the –H from that same water molecule attaches to the other subunit, a process called a **hydrolysis** reaction (**Figure 4.5**). These reactions do not occur spontaneously and require special enzymes to act as a catalyst to allow the reaction to occur or to speed up the reaction.

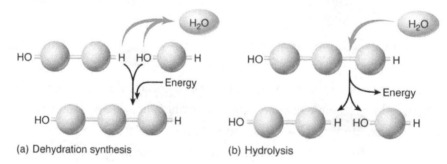

Figure 4.5 By adding or removing a single molecule of water, cells can construct or destruct large biomolecules.

Carbohydrates

Carbohydrates are compounds of carbon, hydrogen, and oxygen bonded as H—C—OH and occurring in the ration of 1 C: 2 H: 1 O. **Carbohydrates** perform numerous roles in living organisms. Polysaccharides serve for the storage of energy (starch in plants and glycogen in animals), and as structural components (cellulose in plants and chitin in arthropods). The monosaccharide ribose is an important component of coenzymes (ATP, FAD, and NAD), and is the core of the genetic molecule known as RNA (ribonucleic acid). The related monosaccharide deoxyribose is the core of the genetic molecule known as DNA (deoxyribose nucleic acid). Different saccharides play key roles in the immune system, fertilization, preventing pathogenesis, blood clotting, and development.

Monosaccharides Through the wondrous process of photosynthesis, plants and some photosynthetic bacteria and algae assemble the carbon, hydrogen, and oxygen molecules of water (H-0-H)) and carbon dioxide (0-C-0) into simple sugars such as glucose, galactose, and fructose known as **monosaccharides** (one sugar). Simple sugars may contain as few as three carbon atoms to five (pentose sugars) as many as seven carbons, but those that play the central role of energy storage have six carbons (hexose sugars). Of the hexose sugars, **glucose** is the most biologically significant (**Figure 4.6**). Despite the fact that glucose has several isomers, such as fructose and galactose, we usually think of $C_6H_{12}O_6$ as glucose. Monosaccharide glucose is critical to biological functions and is the molecule that is broken down and converted into stored chemical energy (ATP) during cellular respiration in nearly all types of organisms. Plants, animals, and most bacteria depend on glucose as their immediate source of biological energy. Because nearly all life forms on the planet depend on it to provide the chemical energy that drives their bodies, glucose is the single most important biological molecule on the planet. Plants produce the glucose initially through photosynthesis whereas animals satisfy their carbohydrate requirements by eating the plants or by eating the animals that ate the plants and digested out their glucose.

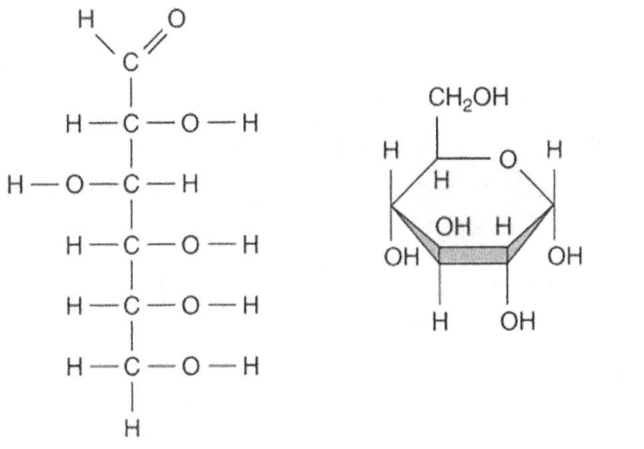

(a) Chain structure of glucose   **(b)** Ring structure of glucose

Figure 4.6 The Monosaccharide Glucose. Monosaccharides have a large number of hydroxyl groups, and the presence of this polar functional group makes them soluble in water.

Ribose and deoxyribose are pentose sugars of great biological importance because make up the core structure of the nucleic acids RNA and DNA respectively. Both of these will be discussed in more detail later in this chapter

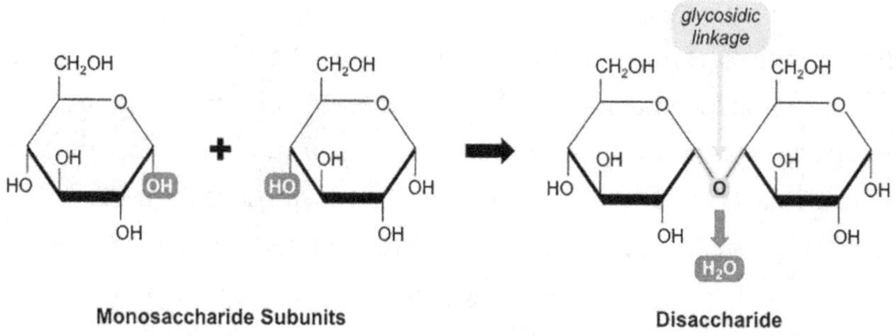

Figure 4.7 The oxygen atom that remains between the two monosaccharide subunits after a molecule of water is spun off forms a connection known as a glycosidic linkage.

Disaccharides A **disaccharide** consists of two monosaccharides that have been joined by a dehydration synthesis reaction (**Figure 4.7**). When the enzymes in our digestive system break the glycosidic linkage, the result is two monosaccharide glucose molecules. The disaccharide sucrose, formed from the monosaccharides glucose and fructose, is the familiar table sugar used in and on many foods. Sucrose is also the form in which sugar is transported throughout plants in sap. Another familiar disaccharide is lactose. Formed from glucose and galactose, lactose is found in milk and cheese. Individuals who are lactose intolerant lack the enzyme lactase responsible for the breakdown of lactose.

Polysaccharides Glucose may also be chemically modified in such a way to make it suitable for long term storage and as a structural component for plants. To do this plants bond many glucose molecules together into large complex sugars known as **polysaccharides**. The most important of these polysaccharides are starch and cellulose. **Starch** molecules are more chemically stable (not soluble in water) than glucose and thus provide a long-term storage solution for excess glucose. With a different twist to their configuration starch molecules become **cellulose**, a polysaccharide that is so chemically tough (indigestible by most animals' enzyme systems) and stable (not soluble in water) that plants use it structurally to form their very bodies. The wood that forms the stem of a 300-foot tall redwood tree is nothing more than untold cellulose molecules bonded together.

Cellulose is not only the most abundant carbohydrate on Earth but it is also the most abundant organic molecule on the planet—over 100 billion tons of cellulose are produced by plants every year. Cellulose provides many benefits to a large number of organisms as will be detailed in coming chapters of this book. In animals, excess glucose is stored short-term in the liver and muscle cells as the polysaccharide **glycogen** (**Figure 4.8**).

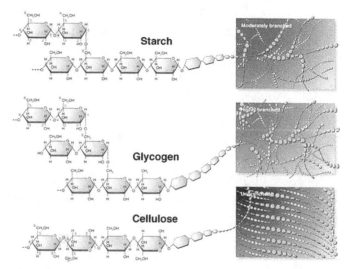

Figure 4.8 Biologically Important Polysaccharides. The branched nature of starch and glycogen allow for their easy reduction through hydrolysis reactions. The linear nature of cellulose makes it water resistant and very difficult to reduce chemically.

It has been said that the second most common polysaccharide in the world after cellulose is chitin. **Chitin** is a polysaccharide found in the cell walls of fungi and in the exoskeleton of arthropods (**Figure 4.9**). Like cellulose, chitin is impervious to water and nearly impossible for animals to digest because of its unique chemical structure and bonding.

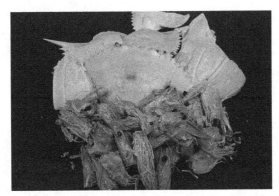

Figure 4.9 The chitin found in these shrimp and crab exoskeletons has many commercial and medical uses.

Lipids

The **lipids** are a large and diverse group of naturally occurring organic compounds that are related by one general chemical characteristic: they are insoluble in water. Insolubility is due to the fact that when they are placed in water, many lipid molecules spontaneously cluster together and expose what polar (hydrophilic) groups they have to the surrounding water while confining the nonpolar (hydrophobic) parts of the molecule together within the cluster. You have probably noticed this effect when cooking or salad oil is added to water and the oil beads up into cohesive drops on the water's surface.

Lipids comprise a group that includes waxes, steroids, fat-soluble vitamins (vitamins A, D, E, and K), monoglycerides, diglycerides, triglycerides (fats), and phospholipids. In living systems, lipids function as structural components of the plasma membrane that surrounds cells and the intracellular membranes of cell organelles. Triglycerides are major form of long-term energy storage in animals (fats) and plants (oils).

Triglycerides Triglycerides are commonly known as **fats**. Fats are the primary lipid used by animals for both long-term energy storage and insulation. Triglycerides in plants, however, are commonly referred to as **oils**.

Triglycerides are composed of two types of subunit molecules: fatty acids and glycerol. Fatty acids consist of a long hydrocarbon chain with an even number of carbons and a –COOH (carboxyl) group at one end. The fatty acid chains may be either saturated or unsaturated. *Saturated fatty acids* have no double bonds between the carbon atoms and contain as many hydrogen atoms as they can hold. *Unsaturated fatty acids* have double bonds in the carbon chain, which reduces the number of bonded hydrogen atoms. In addition, double bonds in unsaturated fatty acids may have chemical groups arranged on the same side (termed *cis configuration*) or opposite sides (termed *trans configuration*) of the double bond. A *trans fat* is a triglyceride that has at least one bond in the trans configuration (**Figure 4.10**).

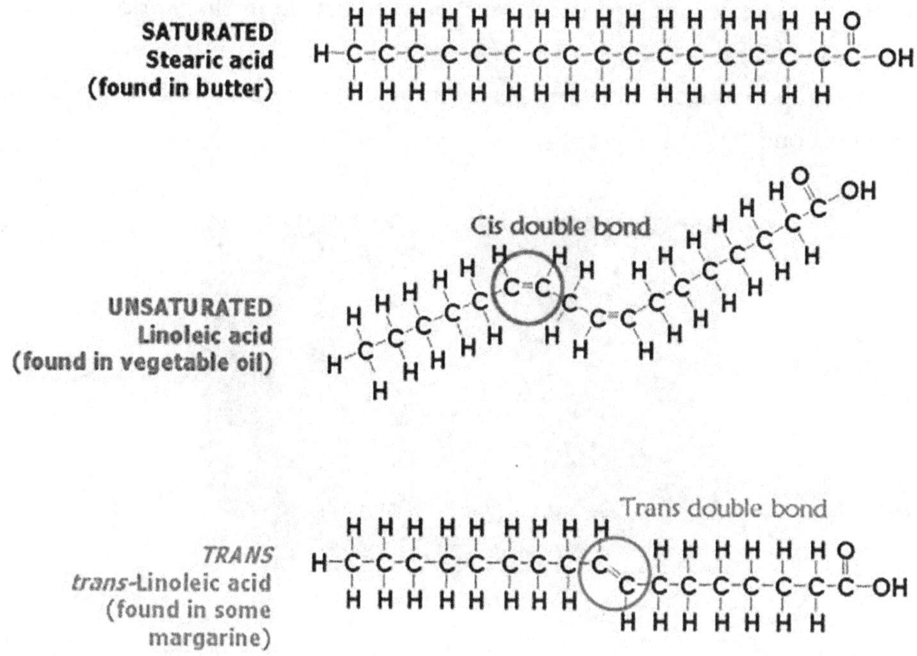

SATURATED
Stearic acid
(found in butter)
UNSATURATED
Linoleic acid
(found in vegetable oil)
Cis double bond
TRANS
trans-Linoleic acid
(found in some margarine)
Trans double bond

Figure 4.10 The cis or trans configuration of an unsaturated fatty acid affects its biological activity."

Glycerol is a 3-carbon compound with three polar –OH groups, making glycerol soluble in water. When a triglyceride fat or oil forms, the –COOH functional groups of three (tri) react with the –OH groups of glycerol during a dehydration synthesis reaction (**Figure 4.11**).

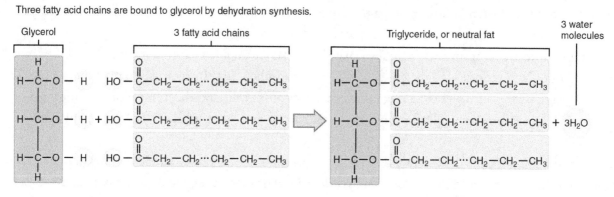

Three fatty acid chains are bound to glycerol by dehydration synthesis.
Glycerol
3 fatty acid chains
Triglyceride, or neutral fat
3 water molecules
+ 3H2O

Figure 4.11 Notice that triglycerides have many nonpolar C—H bonds; therefore, they do not mix with water. Even though cooking or salad oils and water are both liquids, they do not mix, even after shaking because the nonpolar oil and the polar water are chemically incompatible."

In general, fats, which most often come from animals, are solid at room temperatures whereas oils, which come from plants, are liquid at room temperature.

Box 4.1
Fats—the Good, the Bad, and the Not Quite so Bad

For years, *fat* was considered a dirty word. We were urged to banish it from our diets whenever possible. We switched to low-fat foods. But the shift didn't make us healthier, probably because we cut back on healthy fats as well as harmful ones.

Your body needs some fat from food. It's a major source of energy. It helps you absorb some vitamins and minerals. Fat is needed to build cell membranes, the vital exterior of each cell, and the sheaths surrounding nerves. It is essential for blood clotting, muscle movement, and inflammation. For long-term health, some fats are better than others. Good fats include monounsaturated and polyunsaturated fats. Bad ones include industrial-made trans fats. Saturated fats fall somewhere in the middle.

As we are learning in this chapter, all fats have a similar chemical structure: a chain of carbon atoms bonded to hydrogen atoms. What makes one fat different from another is the length and shape of the carbon chain and the number of hydrogen atoms connected to the carbon atoms. Seemingly slight differences in structure translate into crucial differences in form and function. The worst type of dietary fat is the kind known as trans fat. It is a byproduct of a process called hydrogenation that is used to turn healthy oils into solids and to prevent them from becoming rancid. When vegetable oil is heated in the presence of of hydrogen and a heavy-metal catalyst such as palladium, hydrogen atoms are added to the carbon chain. This turns oils into solids. It also makes healthy vegetable oils more like not-so-healthy saturated fats. On food label ingredient lists, this manufactured substance is typically listed as "partially hydrogenated oil."

Early in the 20th century, trans fats were found mainly in solid margarines and vegetable shortening. As food makers learned new ways to use partially hydrogenated vegetable oils, they began appearing in everything from commercial cookies and pastries to fast-food French fries.

Eating foods rich in trans fats increases the amount of harmful LDL cholesterol in the bloodstream and reduces the amount of beneficial HDL cholesterol. Trans fats create inflammation, which is linked to heart disease, stroke, diabetes, and other chronic conditions. They contribute to insulin resistance, which increases the risk of developing type 2 diabetes. Research from the Harvard School of Public Health and elsewhere indicates that trans fats can harm health in even small amounts: for every 2% of calories from trans fat consumed daily, the risk of heart disease rises by 23%. Trans fats have no known health benefits and that there is no safe level of consumption. Today, these mainly man-made fats are rapidly fading from the food supply.

Saturated fats are common in the American diet. They are solid at room temperature — think cooled bacon grease. Common sources of saturated fat include red meat, whole milk and other whole-milk dairy foods, cheese, coconut oil, and many commercially prepared baked goods and other foods. The word "saturated" here refers to the number of hydrogen atoms surrounding each carbon atom. The chain of carbon atoms holds as many hydrogen atoms as possible — it's saturated with hydrogens.

A diet rich in saturated fats can drive up total cholesterol, and tip the balance toward more harmful LDL cholesterol, which prompts blockages to form in arteries in the heart and elsewhere in the body. For that reason, most nutrition experts recommend limiting saturated fat to under 10% of calories a day.

Good fats come mainly from vegetables, nuts, seeds, and fish. They differ from saturated fats by having fewer hydrogen atoms bonded to their carbon chains. Healthy fats are liquid at room temperature, not solid. There are two broad categories of beneficial fats: monounsaturated and polyunsaturated fats.

Monounsaturated fats When you dip your bread in olive oil at an Italian restaurant, you're getting mostly monounsaturated fat. Monounsaturated fats have a single carbon-to-carbon double bond. The result is that it has two fewer hydrogen atoms than a saturated fat and a bend at the double bond. This structure keeps monounsaturated fats liquid at room temperature. Good sources of monounsaturated fats are olive oil, peanut oil, canola oil, avocados, and most nuts, as well as high-oleic safflower and sunflower oils.

Polyunsaturated fats When you pour liquid cooking oil into a pan, there's a good chance you're using polyunsaturated fat. Corn oil, sunflower oil, and safflower oil are common examples. Polyunsaturated fats are *essential* fats. That means they're required for normal body functions, but your body can't make them. So you must get them from food. Polyunsaturated fats are used to build cell membranes and the covering of nerves. They are needed for blood clotting, muscle movement, and inflammation.

A polyunsaturated fat has two or more double bonds in its carbon chain. There are two main types of polyunsaturated fats: omega-3 fatty acids and omega-6 fatty acids. The numbers refer to the distance between the beginning of the carbon chain and the first double bond. Both types offer health benefits.

Eating polyunsaturated fats in place of saturated fats or highly refined carbohydrates reduces harmful LDL cholesterol and improves the cholesterol profile. It also lowers triglycerides. Good sources of omega-3 fatty acids include fatty fish such as salmon, mackerel, and sardines, flaxseeds, walnuts, canola oil, and unhydrogenated soybean oil.

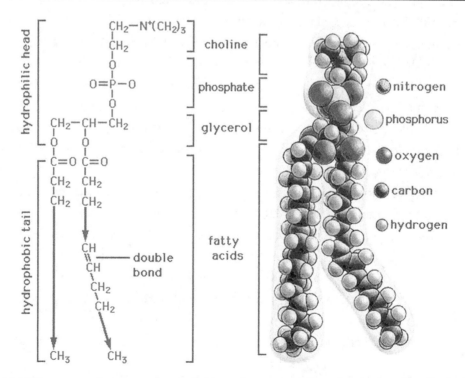

Figure 4.12 When exposed to water, phospholipids tend to arrange themselves so that the polar hydrophilic heads orient toward water and the nonpolar hydrophobic fatty acid tails are oriented away from water.

Phospholipids are triglycerides in which the third fatty acid attached to glycerol has been replaced by a polar phosphate group (**Figure 4.12**). In living organisms, which are made mostly of water, phospholipids tend to become a bilayer (double layer), because the polar heads interact with other polar molecules such as water. Thus, phospholipids arrange themselves into a sandwich-like configuration with the outfacing polar heads being the "bread" and the infacing nonpolar tails being the "filling" of the sandwich. The plasma membrane that surrounds the exterior of cells and the organelles within cells is a phospholipid bilayer (**Figure 4.13**). The presence of kinks in the tails of the phospholipids causes the plasma membrane to be fluid across a range of temperatures found in nature.

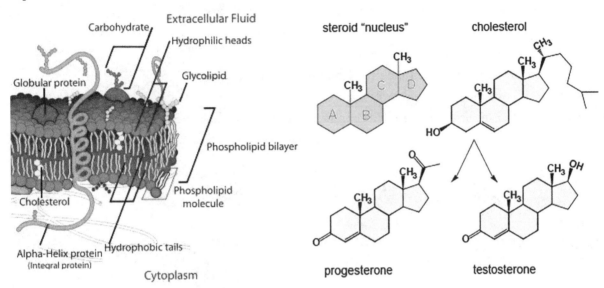

Figure 4.13 The lipid bilayer in the plasma membrane functions to keep extracellular fluids out and cytoplasmic fluids in.

Figure 4.14 Each type of steroid differs primarily by the types of functional groups attached to the carbon ring skeleton.

Steroids are lipids with structures completely different from those of triglycerides and phospholipids. Steroids have a skeleton of four fused carbon rings (**Figure 4.14**). Cholesterol is an essential component of the cell's plasma membrane, where it provides structural stability (Figure 4.13), and it is the precursor of several other steroids, such as the sex hormones testosterone produced in the testes and estrogen produced in the ovaries.

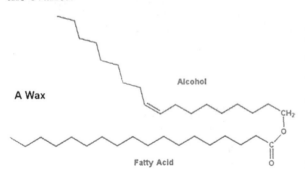

Figure 4.15 The simplified structure of a wax.

Waxes are long-chain fatty acids connected to carbon chains containing alcohol functional groups (**Figure 4.15**). Waxes are solid over a range of temperatures and thus have a high melting point. Being hydrophobic, waxes are waterproof and resistant to degradation.

In plants, waxes form a translucent covering called the cuticle over stems, leaves, and fruit (**Figure 4.16**). In animals, wax is produced in a number of ways. Animal skin produces a waxy substance called *sebum* that waterproofs the skin and hair of mammals. In humans, wax is also produced by the inner ear canal that functions to trap dust and bacteria, preventing these contaminants from reaching the eardrum. Honeybees produce beeswax in glands on the underside of their body. Beeswax is used to make the six-sided honeycombs in which honey is stored, and eggs are laid.

Proteins

Proteins are large biomolecules consisting of long chains of amino acid subunits. The fact that proteins are of primary importance to the structure and function of cells is reflected in the fact that as much as 50% of the dry weight of most cells consists of proteins. Several hundred proteins have been identified making them the most diverse group of biological macromolecules, both chemically and functionally. This multitude of proteins can be categorized by their function (**Figure 4.17**).

Figure 4.16 The shiny wax cuticle of this leaf is waterproof which helps reduce evaporative loss of water from the internal tissues of the leaf

Type of protein	Example	Function
Enzymes	Amylase	Digestion
Transport	Hemoglobin Myoglobin Albumin Lipoprotein	Transports O2 in blood Transports O2 in muscle Transports fatty acids Transports lipids
Storage	Ovalbumin Milk Ferritin	Egg-white protein Milk Iron storage in spleen
Contractile	Myosin, actin	Muscle movement
Protection	Antibodies Fibrinogen, thrombin	Fight infection Blood clotting
Hormones	Insulin Growth hormone	Carbohydrate metabolism Growth and regeneration
Structural	Glycoproteins Collagen Elastin	Cell walls, skin Tendons, bones, cartilage Ligaments
Toxins	Clostridium botulinum Ricin Snake venom	Botulism food poisoning Castor bean toxin Snake venom

Figure 4.17

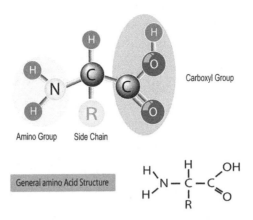

Figure 4.18 The R in the structural formula represents a functional side group.

Proteins are long chain polymers made up of 20 different amino acid subunits. **Amino acids**, as their name suggests, contain an amino group (–NH₂) and an acidic carboxyl (—COOH) (**Figure 4.18**). Amino acids differ according to their particular *R* group. The *R* groups range in complexity from a single hydrogen atom to complicated ring compounds. Some *R* groups are polar and associate with water whereas others are nonpolar and do not (**Figure 4.19**).

Functional Group	Structure	Properties
Hydroxyl	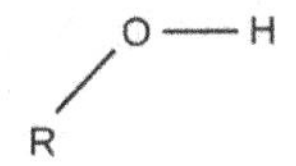 O — H / R	Polar
Methyl	 R —— CH₃	Nonpolar
Carbonyl	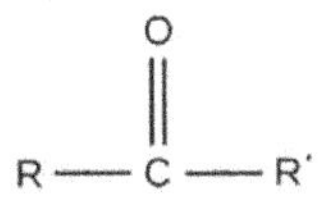 O ‖ R — C — R'	Polar
Carboxyl	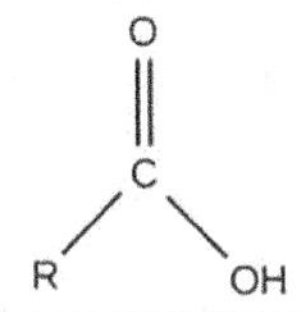 O ‖ C / R OH	Charged, ionizes to release H⁺. Since carboxyl groups can release H⁺ ions into solution, they are considered acidic.
Amino	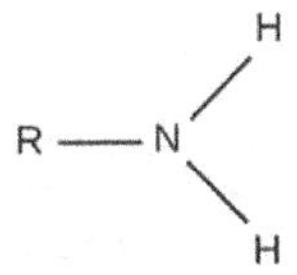 H / R —— N \ H	Charged, accepts H⁺ to form NH₃⁺. Since amino groups can remove H⁺ from solution, they are considered basic.
Phosphate	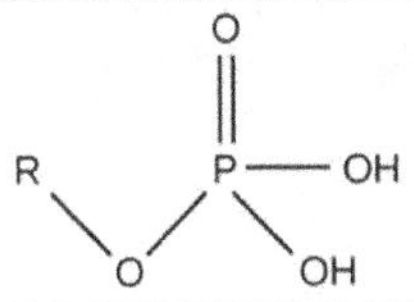 O ‖ R \ P — OH / \ O OH	Charged, ionizes to release H⁺. Since phosphate groups can release H⁺ ions into solution, they are considered acidic.
Sulfhydryl	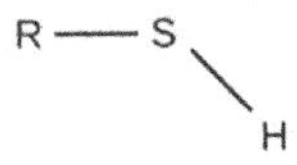 R — S \ H	Polar

Figure 4.19 Categories of Functional Units. The R in these diagrams represents the amino acid base unit to which the functional group is attached.

Amino acids are linked by dehydration synthesis reactions that connect the carboxyl group of one amino acid to the amino group of another amino acid. The resulting covalent bond between the newly joined amino acids is called a *peptide bond* (**Figure 4.20**). The peptide bond may be broken apart in a hydrolysis reaction. A *peptide* is two or more amino acids bonded together, and a *polypeptide* is a chain of many amino acids joined by peptide bonds. A protein is a polypeptide that has been coiled and folded into a particular shape that has some specific function. Think of a protein as a message to be read and the amino acid types and sequence as the individual letters and paragraphs of that message.

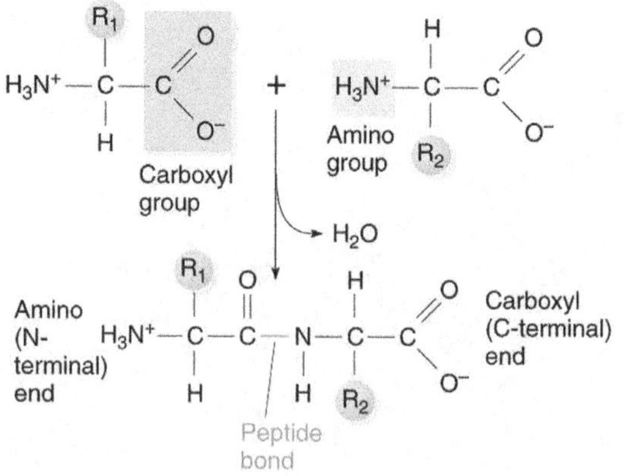

Figure 4.20

Proteins are large, complex macromolecules that four levels of structural organization: primary, secondary, tertiary, and quaternary (**Figure 4.21**).

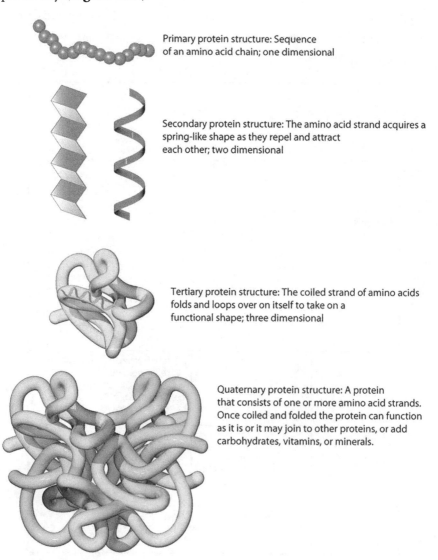

Primary protein structure: Sequence of an amino acid chain; one dimensional

Secondary protein structure: The amino acid strand acquires a spring-like shape as they repel and attract each other; two dimensional

Tertiary protein structure: The coiled strand of amino acids folds and loops over on itself to take on a functional shape; three dimensional

Quaternary protein structure: A protein that consists of one or more amino acid strands. Once coiled and folded the protein can function as it is or it may join to other proteins, or add carbohydrates, vitamins, or minerals.

Figure 4.21 Levels of Protein Organization. All proteins have a primary structure that can be built up into increasingly complex structures.

Primary structure of a protein is the linear sequence of amino acids. Just as millions of different words can be constructed from the 26 letters in the English alphabet, so can hundreds of thousands of proteins be constructed from just 20 amino acids. Just as varying the number and sequence of a few letters can change a word, changing the sequence of 20 amino acids can produce a huge array of different proteins.

Secondary structure of a protein occurs when the polypeptide coils or folds in a particular way. A coil is known as a α helix and a pleated sheet is called a β pleated sheet. Each polypeptide can have multiple α helices and β pleated sheets. The spiral shape of α helices is formed by hydrogen bonding between every fourth amino acid within the polypeptide chain whereas β pleated sheets are formed when the polypeptide turns back on itself, allowing hydrogen bonding to occur between extended lengths of the polypeptide.

Tertiary structure is the folding that results in the final three-dimensional shape of a globular protein, which tend to ball up into rounded shapes. Globular proteins fold into and maintain their final shape because of the orientation of hydrophilic polar amino acids to the surface and the hydrophobic nonpolar amino acids to the center along with hydrogen bonds, ionic bonds, and covalent bonds between the *R* groups. Strong disulfide linkages (—S—S—) in particular help maintain the tertiary shape. *Quaternary structure* occurs when two or more folded polypeptides interact to perform a biological function.

Nucleic Acids

> *DNA neither cares nor knows. DNA just is. And we dance to its music.*
>
> —Richard Dawkins

There are two types of *nucleic acids*: **DNA (deoxyribonucleic acid)** and **RNA (ribonucleic acid)**. DNA is the genetic code in chemical form for the formation of proteins whereas RNA serves as messengers and organizers in the construction of those proteins. Both DNA and RNA are polymers built of **nucleotides**.

The importance of DNA to the second-by-second life of a cell and by extension to the entire organism can be seen in the amount of it in each cell. If you put all the DNA molecules in just your body end to end, the DNA would reach from Earth to the Sun and back over 600 times! That DNA is ubiquitous across all forms and types of organisms is revealed by the fact that every human shares 99% of their DNA with every other human but we also share 96% of our DNA with a chimpanzee, 95% of our DNA with a mouse, 61% of our DNA with a fruit fly, and 60% of our DNA with a banana. The complexity of the DNA molecule is demonstrated by the fact that if you could type 60 words a minute, eight hours a day, it would take you approximately 50 years to type the human genome.

Not all nucleotides are made into DNA or RNA polymers. Some are directly involved in metabolic functions in cells, often as a component of *coenzymes* (nonprotein organic molecules that help regulate enzymatic reactions. One example is **ATP (adenosine triphosphate)**, a nucleotide that stores large amounts of chemical energy. Every nucleotide is comprised of three types of molecules: a pentose sugar, a phosphate

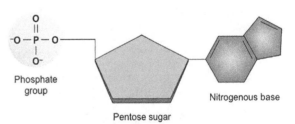

Figure 4.22 Nucleotide Structure. The deoxyribose pentose sugar lacks an oxygen atom found in ribose. Five different kinds of nitrogenous bases may be attached to the pentose sugar.

(phosphoric acid), and a nitrogen-containing base. In DNA, the pentose sugar is deoxyribose, and in RNA the pentose sugar is ribose (**Figure 4.22**).

Both DNA and RNA contain combinations of four nucleotides, but these vary somewhat between the two nucleic acids.

Table 4.1

Comparison of DNA and RNA

	DNA	**RNA**
Sugar	Deoxyribose	Ribose
Bases	Adenine, guanine, thymine, cytosine	Adenine, guanine, cytosine, uracil
Strands	Double-stranded with base pairing	Usually single-stranded
Double Helical Structure	Yes	No

Nucleotides (also known as *nucleobases*) that have a base with a single ring are called **pyrimidines**, and nucleotides with a double ring are called **purines**. In DNA, the pyrimidine bases are cytosine (C) and thymine (T); in RNA the pyrimidine bases are cytosine and uracil (U). Both DNA and RNA contain the purine bases adenine (A) and guanine (G) (**Figure 4.23**).

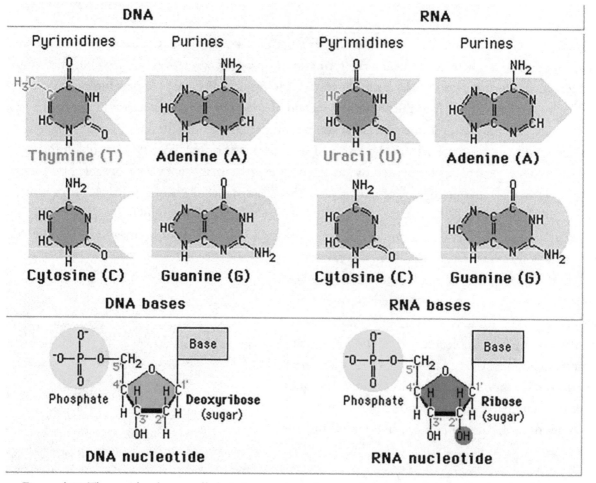

Figure 4.23 These molecules are called nitrogenous bases because their presence raises the pH of a solution.

Nucleotides are joined into a DNA or an RNA polymer by a series of dehydration synthesis reactions. The resulting polymer is a linear molecule called a strand, in which the backbone is made up of an alternating series of sugar-phosphate-sugar-phosphate molecules. The bases project from one side of the stand (**Figure 4.24**). In DNA, the bases can be in any order within a strand but between strands, thymine (T) is always paired with adenine (A) and guanine (G) is always paired with cytosine (C). This is called *complimentary base pairing* and results in the number of purine bases (A+G) always equaling the number of pyrimidine bases (T+C).

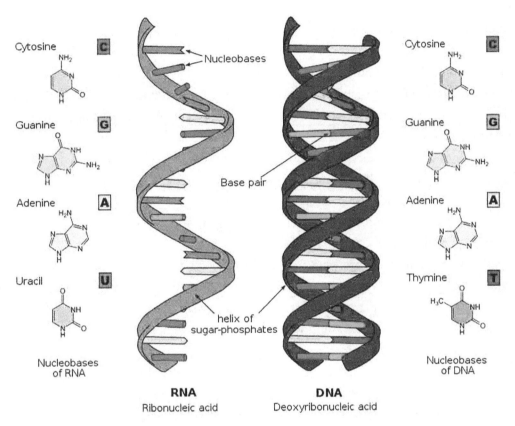

Figure 4.24 DNA vs. RNA structure.

As Table 4.1 indicates, RNA is a single-stranded polymer of nucleotides that is not a double helix in nature. The structure and function of both DNA and RNA will be discussed in more detail in coming chapters. It may sound unflattering, but in simplest organic chemistry parlance, a cell, and by extension, the entire body of an organism including you, is a watery bag of protein regulated by nucleic acid molecules and powered by the breakdown of carbohydrates and fats into CO_2 and H_2O.

In Summary

- There are only four categories of organic molecules in any living organism: carbohydrates, lipids, proteins, and nucleic acids.
- At the heart of every biomolecule are carbon atoms, so much so that life on this planet is said to be carbon-based.

- To acquire the four electrons to complete the outer shell, a carbon atom forms covalent bonds with as many as four other elements.

- Functional groups are attached to a hydrocarbon core.

- Some organic molecules, known as isomers, have identical molecular formulas but different arrangements of atoms. If there are differences in the actual structure of their carbon skeleton, we call constitutional isomers or structural isomers. Another form of isomer, called stereoisomers, have the same carbon skeleton but differ in how the groups attached to this skeleton are arranged in space.

- Biomolecules are macromolecules, meaning they contain smaller joined subunits. Biomacromolecules are referred to as polymers, since they are constructed of linking together a large number of the same type of subunit or monomers.

- To synthesize a macromolecule, the cell uses a dehydration synthesis reaction in which the equivalent of a water molecule—an –OH (hydroxyl group) and an –H (hydrogen atom) is removed as subunits are joined.

- To break down macromolecules, dehydration synthesis is reversed as an –OH from a water molecule attaches to one subunit and the –H from that same water molecule attaches to the other subunit, a process called a hydrolysis reaction.

- Carbohydrates are compounds of carbon, hydrogen, and oxygen bonded as H—C—OH and occurring in the ratio of 1 C: 2 H: 1 O.

- Simple sugars such as glucose, galactose, and fructose are known as monosaccharides (one sugar).

- Simple sugars may contain as few as three carbon atoms to five (pentose sugars) as many as seven carbons, but those that play the central role of energy storage have six carbons (hexose sugars).

- A disaccharide consists of two monosaccharides that have been joined by a dehydration synthesis reaction.

- Polysaccharides are many glucose molecules bonded together into large complex sugars.

- The lipids are a large and diverse group of naturally occurring organic compounds that are related by one general chemical characteristic: they are insoluble in water.

- Lipids comprise a group that includes waxes, steroids, fat-soluble vitamins (vitamins A, D, E, and K), monoglycerides, diglycerides, triglycerides (fats), and phospholipids.

- Triglycerides in animals are referred to as fats whereas in plants they are referred to as oils.

- Triglycerides are composed of two types of subunit molecules: fatty acids and glycerol.

- Phospholipids are triglycerides in which the third fatty acid attached to glycerol has been replaced by a polar phosphate group.

- Proteins are long chain polymers made up of 20 different amino acid subunits. Amino acids, as their name suggests, contain an amino group ($-NH_2$) and an acidic

- carboxyl (—COOH).

- The resulting covalent bond between the newly joined amino acids is called a peptide bond.

- Proteins are large, complex macromolecules that four levels of structural organization: primary, secondary, tertiary, and quaternary.

- There are two types of nucleic acids: DNA (deoxyribonucleic acid) and RNA (ribonucleic acid). Both DNA and RNA are polymers built of nucleotides.

- Nucleotides (also known as nucleobases) that have a base with a single ring are called pyrimidines, and nucleotides with a double ring are called purines. In DNA, the pyrimidine bases are cytosine (C) and thymine (T); in RNA the pyrimidine bases are cytosine and uracil (U). Both DNA and RNA contain the purine bases adenine (A) and guanine (G).
- Nucleotides are joined into a DNA or an RNA polymer by a series of dehydration synthesis reactions. The resulting polymer is a linear molecule called a strand, in which the backbone is made up of an alternating series of sugar-phosphate-sugar-phosphate molecules. The bases project from one side of the stand.
- Nucleotides are joined into a DNA or an RNA polymer by a series of dehydration synthesis reactions. The resulting polymer is a linear molecule called a strand, in which the backbone is made up of an alternating series of sugar-phosphate-sugar-phosphate molecules. The bases project from one side of the stand.

Review and Reflect

1. *Something Fishy* If you eat a piece of fish, what reactions must occur for the amino acid monomers in the protein of the fish to be converted to new proteins in your body?

2. *Sinking Fast* Oh, No! Your pleasure boat, the *Minnow*, is going down. You only have time to grab one food item to sustain you on what might be a long wait for rescue in a tiny rubber raft. Would you grab a 5-pound bag of sugar or a gallon container of lard? Defend your choice.

3. *The Horta* In one of Star Trek's most famous episodes, Dr. McCoy has a difficult time analyzing or understanding the bizarre (to humans) silicon-based Horta. Evaluate the possibility of silicon-based life given the chemical structure and potential for chemical bonding of a silicon atom. Remember that life on Earth is based on carbon.

4. *A Hate/Love Affair With Water* You have been given a solution of small biomolecules that are hydrophilic on one end and hydrophobic on the other. Predict through a labeled diagram the shape and orientation these molecules will take when the solution is slowly dripped into water.

5. *Water In-Water Out* In general terms, diagram a dehydration synthesis reaction and a hydrolysis reaction.

6. *Sugar on the Brain* "I'm confused," says your study buddy. "Mono-, di-, polysaccharides—I don't understand what a saccharide is let alone the different kinds of saccharides." How would you respond?

7. *What's in a Quote?* React to the quote by Richard Dawkins that opens the section on *Nucleic Acids*. Explain in your own words what you think Dawkins is trying to say.

8. *Phasers on Stun* Imagine you wish to invent a weapon that will disassociate animals. What biomacromolecules—carbohydrates, lipids, proteins, or nucleic acids—would you attack and how would you attack them? Explain

9. *Nucleic Acid Compare and Contrast* Compare and contrast DNA with RNA both in their structures and in their functions.

Create and Connect

1. ***Gluten Free*** Gluten is a mixture of proteins abundant in wheat, barely, and rye. Although these grains are staples in the Western diet, almost 1 percent of Americans have celiac disease—a disorder of the intestines caused by an abnormal immune response after eating gluten. This immune response damages the finger-like villi of the small intestine leading to intestinal upset and malnourishment.

    Write a short report on celiac disease in which you detail the cause of the disorder, the effects of the disorder, and any treatment for the disorder.

    Guidelines

    A. Format your report in the following manner:

    Title page (including your name and lab section)

    Body of the Report (include pictures, charts, tables, etc. here as appropriate). The body of the report should be a minimum of 2 pages long—double-spaced, 1 inch margins all around with 12 pt font.

    Literature Cited A minimum of 2 references required. Only <u>one</u> reference may be from an online site. The *Literature Cited* page should be a separate page from the body of the report, and it should be the last page of the report. Do NOT use your textbook as a reference.

    B. The instructor may provide additional details and further instructions.

2. ***Teaching to the Test*** Believe it or not, a test is a teaching tool, and it has been said that the best way to learn something is to teach it. To that end, prepare a 10 question test over the concepts and facts you learned in this chapter. The questions should be in essay form, and you should provide the answers for each question.

3. ***Designer Molecules*** Advancements in molecular engineering have gotten to the point where scientists can create practically any molecule they can dream up. If you could artificially create any biomacromolecule, what would you come up with? Diagram and label your molecule and explain its function. Your diagram should be general in form so don't try and draw every atom. Your instructor may give more guidance along those lines.

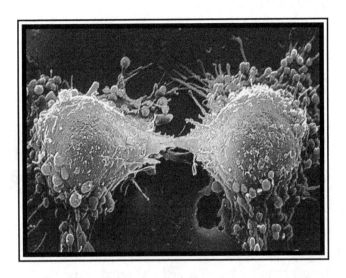

CELL STRUCTURE AND FUNCTION

A cell is regarded as the true biological atom.

—George Henry Lewes

Introduction

If you've ever tried to stuff a week's worth of clothing, toiletries, and other necessities into a piece of carry-on luggage, you would be amazed at what cells can pack into a space smaller than the period at the end of this sentence. Your body is composed of trillions of cells working in harmony. They possess an internal skeleton that gives them shape and controls their movement. They harbor microscopic assembly lines that churn out a wide array of proteins. Every cell even has it own power stations that produce energy. And nestled among such structures, usually walled off from the rest of the cell, is the all-important chromatin that stores the instruction manual (genes) for everything the cell is and does.

Cells are the building blocks of life because any living thing is composed of at least one cell. Only when all the chemistry and biomolecules we have discussed are brought together within the confines of a cell is life possible. Some life-forms exist as single cells whereas others are complex, interconnected systems of many, many cells. The human body, for example, has 37.2 trillion cells and this is not counting the multitude of microorganisms that live in us and on us.

Discovering the Cellular Nature of Life

In an age when we can peer outward telescopically to nearly the edge of the known universe and microscopically inward to the atomic level, it is difficult for us to imagine a time when humankind did not even suspect the existence of a world smaller than what the unaided eye can discern. The actuality of this unseen world was first revealed nearly 350 years ago when Robert Hooke (1635-1703) used one of the first primitive compound microscopes to observe the fine structure of cork (**Figure 5.1**). In his major written work, *Micrographia,* Hooke described the honeycomb of cork cell walls as seen through his microscope by the term *cellulae* (L., small rooms) or cells as we term them today. Microscopists that followed Hooke came to realize that the cork cells that fired his imagination were actually dead plant cells with only resistant box-like cell walls remaining.

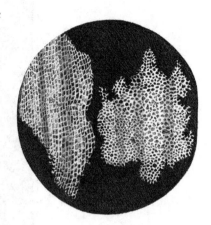

Figure 5.1 Hooke's drawing of cork cells.

Hobbyist lens-grinder Anton von Leeuwenhoek, a contemporary of Hooke, would be the first to glimpse many types of living cells such as single-celled algae, bacteria, protozoans in pond water, sperm cells, and red blood cells. In a series of letters to the Royal Society of London from 1673 to 1723, Leeuwenhoek detailed his microscopic observations, often referring to the cells he had observed as "little animalcules" or "wee beasties" (**Figure 5.2**). The realization quickly set in that cells were not the hollow boxes first seen by Hooke in cork, but rather, cells were filled will all manner of tiny exquisite structures; they were filled with life. It would take almost 200 years of technical improvements in the design of microscopes and lenses before Theodor Schwann, Matthias Schleiden, and Rudolf Virchow would state and refine one of the central pillars of biology now known as the *cell theory*.

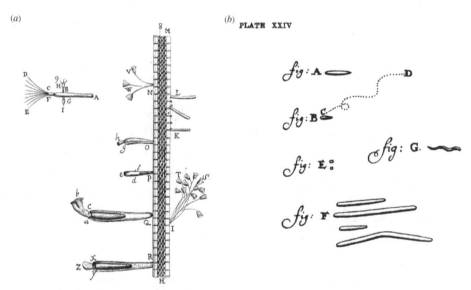

Figure 5.2 Leeuwenhoek's "wee beasties." (a) Rotifers, hydra and vorticellids associated with a duckweed root, from a Delft canal; (b) Bacteria from Leeuwenhoek's mouth; the dotted line portrays movement.

Figure 5.3
Jakob Schleiden (1804-1881)

Figure 5.4
Theodor Schwann (1810-1882)

Figure 5.4
Rudolph Virchow (1821-1902)

In 1824, Rene Dutrochet discovered that "The cell is the fundamental element in the structure of living bodies, forming both animals and plants through juxtaposition." In 1838 German botanist, Matthias Schleiden, (**Figure 5.3**) proposed that all plant tissue was composed of cells. In 1834 Schleiden's countryman, Theodor Schwann (**Figure 5.4**), postulated that all animal tissue was composed of cells. In 1858, another German, Rudolf Virchow (**Figure 5.5**), advanced the proposition that all cells came from preexisting cells. Thus were laid the foundations for what has become known as **cell theory**—a proposition that all living organisms are composed of cells and cell products and that all cells come from preexisting cells. Cell theory initially stated that:

1. Cells are the basic unit of life.
2. All living things are composed of at least one cell.
3. Cells form by free-cell formation similar to the formation of crystals.

We know today that the first two tenets were correct but that the third was clearly wrong. The correct interpretation of cell formation as a result of cell division was enunciated by cytologist Rudolph Virchow in a powerful dictum, *"Omni cellula e cellula"*...."All cells arise only from preexisting cells." Modern cell theory then states that:

1. Cells are the basic unit of life.
2. All living things are composed of at least one cell.
3. All cells come from preexisting cells.

Box 5.1
Microscopes—Revealing Unseen Worlds

The first living cells may have been seen by Leuwenhoek using what is today regarded as a powerful magnifying glass (Figure 20.6). Microscopy has advanced greatly in the intervening years since the first true microscope of Hooke (Figure 20.4) and the discovery that living cells have contents. Improvements in light microscopes and the invention and perfection of the various types of electron microscopes have revealed the astonishing intricacies of cell structure.

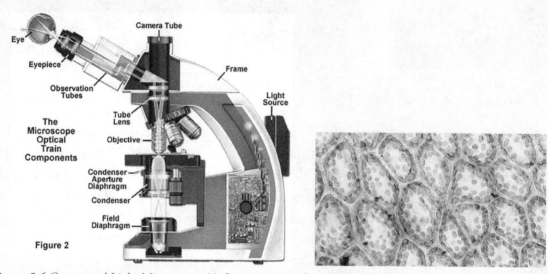

Figure 5.6 Compound Light Microscope. (a) Cutaway view of a compound light microscope showing the various parts and the path of the light. (b) Plant cells with chloroplasts as seen through a compound light microscope 400x.

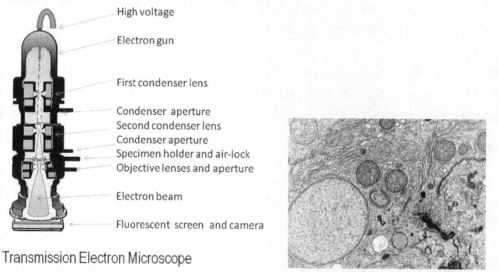

Figure 5.7 Transmission Electron Microscope. (a) Cutaway view of a TEM showing the various parts and the path of the electron beam. (b) A thymus gland cell taken through a TEM 10,000x.

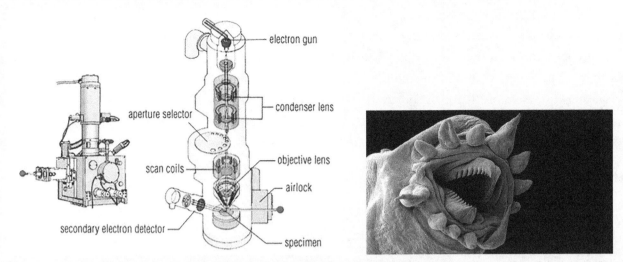

Figure 5.8 Scanning Electron Microscope. (a) Schematic diagram of a SEM showing the various parts of the scope. (b) SEMs can show great depth and clarity of image than TEMs as with this 500x image of a hydrothermal vent worm.

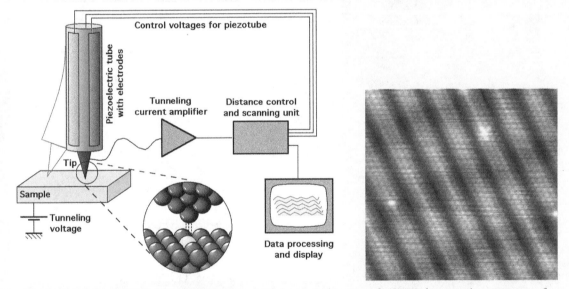

Figure 5.9 Scanning Tunneling Microscope. (a) Schematic diagram of a STM showing the operation of the probe tip. (b) Image of the surface of a piece of gold. The individual gold atoms are visible.

Modern microscopes are of four main types: compound light microscopes, transmission electron microscopes (TEM), scanning electron microscopes (SEM), and scanning tunneling microscopes (STM). A *compound light microscope* uses a set of glass lenses to focus light rays passing through a specimen to produce an image that is viewed directly by the observer's eye (**Figure 5.6**). A *transmission electron microscope* uses a set of electromagnetic lenses to focus electrons passing through a specimen to produce an image that is projected onto a fluorescent screen or photographic film (**Figure 5.7**). A *scanning electron microscope* uses a narrow beam of electrons to scan over the surface of a specimen that is coated with a thin layer of metal atoms (**Figure 5.8**). A *scanning tunneling microscope* obtains images of the atoms on the surface of a specimen. It works by detecting electrical forces with a probe that tapers down to a point only a single atom across (**Figure 5.9**).

The magnification power of a microscope is the ratio between the size of an image and the actual size of an object. Electron microscopes magnify to a far greater extent than do light microscopes. A light microscope can magnify up to 1500x whereas electron microscopes can magnify hundreds of thousands of times. The difference lies in the mode of illumination. Both light and electrons travel as waves, but the wavelength of electrons is much shorter than the wavelength of light. This allows for greater *resolution* (ability to distinguish between two points) by the electron microscope.

Size of Cells

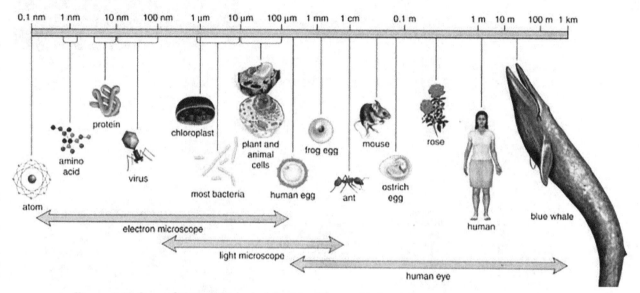

Figure 5.10 Sizes of Various Biological Entities. The unaided eye can see macroscopic organisms and a few large cells. Cells and some organelles are visible with the light microscope, but an electron microscope is needed to see most cell organelles, viruses and large biomolecules.

Although all cells are microscopic, they do come in a range of sizes (**Figure 5.10**). When it comes to the minute-by-minute life of every cell two mathematical quantities rule: their surface area and their volume. All cells must take in oxygen and food molecules through their plasma membranes to power the metabolism, growth, and repair of the internal cell structures. In the process of doing this, carbon dioxide and waste liquids must be given off by the cell. As the cell grows larger, its volume (inside) grows faster than its surface area (outside). Eventually, a point is reached where you have more inside (volume) demanding food and oxygen, and waste release than the outside (surface area) can keep up with. At this point, the cell must replicate into two smaller cells or die.

The relationship between surface area and volume is called *the surface-area-to-volume ratio*. Calculations show that a 1-cm cube has a surface-to-volume ratio of 6:1 whereas a 2-cm cube has a surface-to-volume ratio of 3:1, and a 4-cm cube a ratio 2:1 (**Figure 5.11**). That means that the surface area is growing arithmetically (2+2+2) whereas the volume is growing geometrically (2x2x2). Eventually, the cell reaches a size where the services demanded by the inside (volume) become too great for the surface area (outside) to meet.

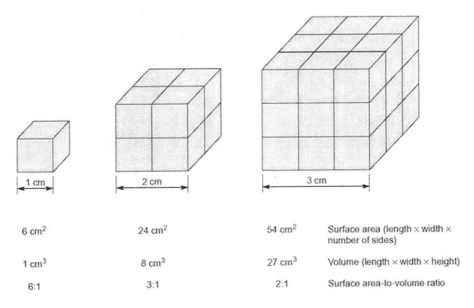

6 cm²	24 cm²	54 cm²	Surface area (length × width × number of sides)
1 cm³	8 cm³	27 cm³	Volume (length × width × height)
6:1	3:1	2:1	Surface area-to-volume ratio

Figure 5.11

Types of Cells

We now fully accept the tenet of the cell theory that any living thing is composed of cells and cell products (such as bone or shell), be it a one-celled protozoan or algae or the trillions of cells making up your body. If, as taxonomists suspect, there truly are tens of millions of different species of living things on this planet, the total number of cells that inhabit this planet could not even be imagined, let alone accurately determined. However, nature has smiled on biologists for all the untold multitude of cells on earth fall into only two general categories— *prokaryote* or *eukaryote*—based on inherent internal properties.

A **prokaryotic cell** (Gr. *pro*, before + *karyon*, kernel or nucleus) possesses DNA, but its DNA not isolated from the rest of the cell inside a membrane-bound nucleus. Instead, the DNA is a single loop floating free in the cytoplasm. In addition, the prokaryotic cell is less complex and much smaller than it eukaryotic counterpart. Prokaryotic cells range in size from 1 to 10μm and lack internal membrane-bound organelles, structures devoted primarily to the many metabolic tasks required to maintain the life of the cell. Most prokaryotic life forms are single cells, but some types are colonial (**Figure 5.12**).

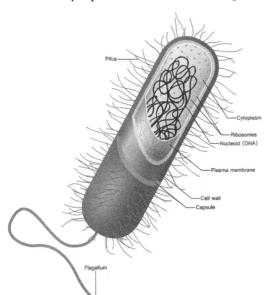

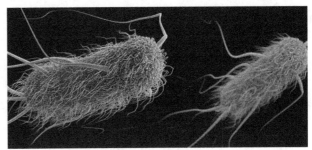

Figure 5.12 Prokaryotic Cell. (a) Diagram of a typical prokaryotic bacterial cell. (b) SEM of *Escherichia coli* bacteria from a stool sample.

The prokaryotes, represented by the domain *Bacteria* and the domain *Archaea*, are an extremely successful and diverse group that is thought to be the most ancient life form on earth. Bacteria occupy all terrestrial and aquatic habitats and even exist in great numbers on and inside the bodies of other living organisms, including humans. In our hubris, we declare humankind to be master of this place, but in biological actuality, prokaryotes rule this planet.

A **eukaryotic cell** (Gr. *eu*, true + *karyon*, kernel or nucleus) contains DNA that is complexed with DNA-binding proteins in complex linear chromosomes and located within a membrane-enclosed organelle known as the nucleus (**Figure 5.13**. Eukaryotic cells, ranging from 10 to 100µm, are larger than prokaryotic cells,

Figure 5.13 Micrograph of stained eukaryotic human cheek cells. The large purple orb in each cell is the nucleus.

and they contain numerous membrane-bound organelles not found in prokaryotic cells (**Figure 5.14**).

CHARACTERISTIC	PROCARYOTIC CELLS	EUCARYOTIC CELLS	MITOCHONDRIA AND CHLOROPLASTS
Size	1–10 μm	10–100 μm	1–10 μm
Nuclear envelope	Absent	Present	Absent
Chromosomes	Single, circular, with no nucleosomes	Multiple, linear, wound on nucleosomes	Single, circular, with no nucleosomes
Golgi apparatus	Absent	Present[e]	Absent
Endoplasmic reticulum, lysosomes, peroxisomes	Absent	Present[e]	Absent
Mitochondria	Absent	Present[e]	
Chlorophyll	Not in chloroplasts	In chloroplasts	
Ribosomes	Relatively small	Relatively large	Relatively small
Microtubules, intermediate filaments, microfilaments	Absent	Present	Absent
Flagella	Lack microtubules	Contain microtubules	

Figure 5.14 A comparison of prokaryote vs. eukaryote structure.

Cells let us walk, talk, think, make love and realize that the bath water is cold.

—Lorraine Lee Cudmore

Animal Cell Structure

As the origin and structure of prokaryotic cells will be detailed at length in Chapter 20, the remainder of this chapter will focus on eukaryotic cells—plant and animal. The internal architecture of the eukaryotic cell may be described according to two general features: *compartmentalization* and *macromolecular assemblages*. Most cells are divided into a number of compartments known as *organelles* (L., tiny organ). Organelles are usually surrounded by a membrane that serves to regulate the movement of molecules in and out of the organelle. Examples of organelles are the nucleus, mitochondria, Golgi apparatus, endoplasmic reticulum, and lysosomes (**Figure 5.15**).

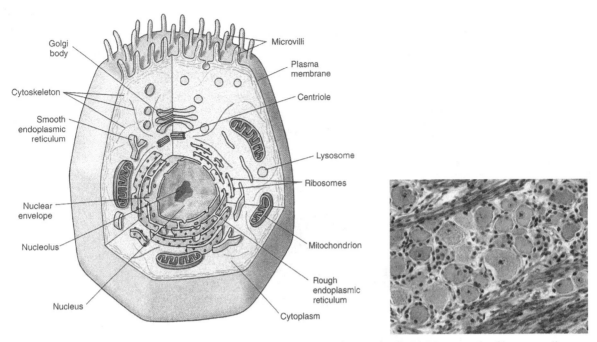

Figure 5.15 Typical Animal Cell. (a) Drawing of a generalized animal cell. (b) Micrograph of human cells.

Pervasive throughout the cell are macromolecular assemblages composed of specific macromolecules organized in a specific three-dimensional arrangement. Examples of macromolecular assemblages are the cytoskeleton, ribosomes, membranes, and the chromosomes.

In simplest terms, an animal cell is constructed of (1) a *system of membranes* that surround the cell and each of its organelles, (2) gelatinous *cytoplasm* that contains and supports the organelles as well as macromolecular assemblages and food storage products, and (3) the various *organelles* and *macromolecular assemblages* within the cytoplasmic matrix.

Membrane System

Visualize membranes as closed bags. That is, membranes have an inner and outer surface but no edges. Membranes serve to separate an interior space from its surroundings and to separate the entire cell from its external environment. In a eukaryotic animal cell, the membrane system is two-fold and consists of the **plasma membrane** which surrounds and contains the entire cell, and internal membranes which surround and contain the organelles.

According to current understanding, known as the *fluid mosaic model*, the plasma membrane is constructed of a bilayer of phospholipids. Phospholipids are lipids (fats) that have a phosphate group at the "head" end and a lipid "tail." Each layer of the plasma membrane consists of the hydrophilic phosphate groups aligned so they all point out resulting in all the hydrophobic lipid tails pointing in (**Figure 5.16**). Associated with the plasma membrane are a number of diverse proteins. Some of these proteins simple adhere to the surface of the membrane whereas others are embedded in the bilayer. Many, such as the transmembrane proteins, act as carriers or channels that passively or actively transport molecules through the bilayer. Membrane receptor proteins (glycoproteins) function as receptor sites first binding information molecules, such as hormones, and then transmitting signals from the bound molecules to the interior of the cell.

Glycoproteins serve as a connection between the interior of the cell and the external environment around it. Attached to the microfilaments of the cytoskeleton, structural proteins insure the stability of the cell, and cell adhesion proteins allow cells to interact and identify each other. Membrane proteins may also exhibit enzymatic activity, catalyzing various reactions related to the plasma membrane.

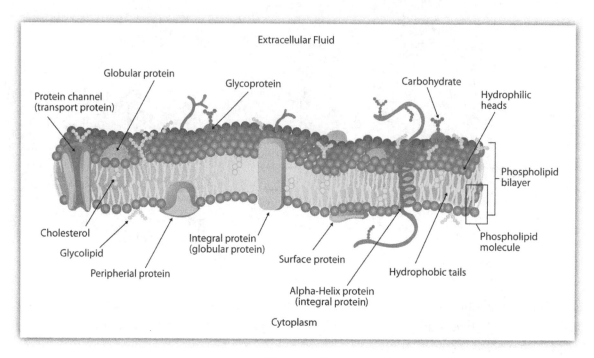

Figure 5.16

Eukaryotic cells also have membranes enclosing the organelles. Similar in structure and function to the plasma membrane, the membranes surrounding the organelles allow the cell to segregate its chemical functions into discrete internal compartments.

Cytoplasm

Cytoplasm is the part of the cell enclosed by the plasma membrane. Imagine a balloon filled with semi-solid gelatin. The balloon represents the plasma membrane, and the gelatin would represent the cytoplasm. Cytoplasm is a complex matrix consisting of different components: *cytosol, cytoskeleton,* and *inclusions.*

- *Cytosol* is a translucent gel consisting of a complex mixture of water, salts, and organic molecules that compose around 70% of the total volume of a cell. The other cell components are suspended within the cytosol. The viscosity of cytosol is constantly changing (**Figure5.17**).

- The *cytoskeleton* is a dynamic cellular scaffolding composed of protein filaments and tubules found within the cytosol. The cytoskeleton plays several important roles including maintaining the shape and stability of the cell, enabling intracellular transport, allowing some cellular

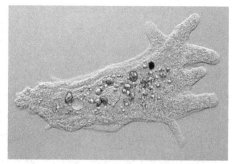

Figure 5.17 By changing the viscosity of its cytosol, this *Amoeba proteus* oozes along (amoeboid movement).

movement, and assisting in cell replication (**Figure 5.18**). The cytoskeleton has three main structural components: microfilaments, intermediate filaments, and microtubules.

○ *Microfilaments* are fine thread-like protein fibers composed primarily of the contractile protein actin. In association with myosin, microfilaments help generate the forces used in basic cell movements such as contraction, cell crawling, amoeboid movement, and cytokinesis.

○ *Intermediate filaments* are larger than microfilaments but smaller than microtubules. Intermediate fibers provide tensile strength for the cell and anchor organelles to parts of the cell.

○ *Microtubules* are tubular structures composed of a protein known as tubulin. Microtubules form the **mitotic spindle** which plays a vital role in the movement of chromosomes during cell replication. Microtubules are also involved in intracellular transport and are critical components of eukaryotic cilia and flagella. Microtubules radiate out from a micro-tubule organizing center, the **centrosome**, located near the nucleus.

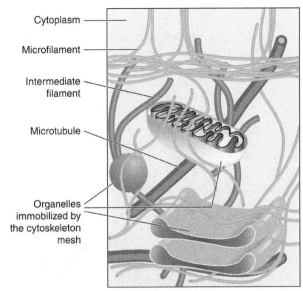

Figure 5.18

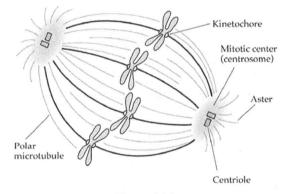

Figure 5.19

Within the centrosome is a pair of **centrioles**. Each centriole is composed of nine triplets of microtubules (**Figure 5.19**).

• *Inclusions* are tiny particles or droplets of insoluble materials within the cytosol that are not bound by membranes. Although a wide range of different types of inclusions exists, the most common inclusion may be lipid droplets that are found in both prokaryotic and eukaryotic cells.

Organelles

The membranous organelles embedded within the cytosol are diverse in structure and function:

Nucleus Bound by the nuclear envelope, the **nucleus** is the largest of the organelles. Within the nucleus resides the genetic material responsible for controlling and coordinating all the activities within the cell. This genetic material, known as **chromatin**, consists of long strands of DNA (Deoxyribose Nucleic Acid) complexed with proteins known as **histones**. Once a cell begins to replicate (prophase) the chromatin coils and thickens and is then referred to as **chromosomes**. (**Figure 5.20**). Pores in the nuclear envelope allow molecules to move from the nucleus out into the cytoplasm. Within the nucleus lies the **nucleolus**, a structure composed of proteins and nucleic acids not bound by membranes. The nucleolus functions to produce ribosomes (ribosomal RNA). Completed ribosomes detach from the nucleolus and move out of the nucleus through pores in the nuclear envelope.

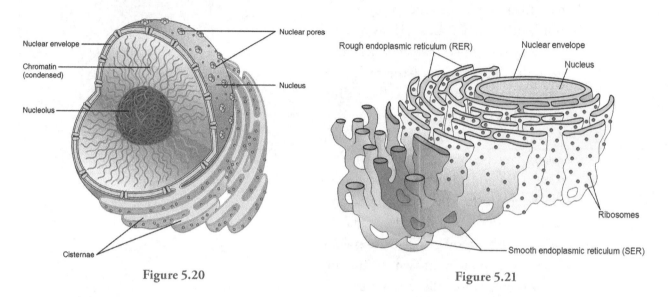

Figure 5.20   Figure 5.21

Endoplasmic Reticulum Structurally, the endoplasmic reticulum is an extensive network of highly folded membranous tubules and flattened sacs supported and stabilized by the cytoskeleton (**Figure 5.21**). The cavities inside the tubules are interconnected, and their membranes are continuous with the outer membrane of the nuclear envelope. The space between the membranes of the nuclear envelope "communicates" with the space between the membranes in the ER.

The complex of membranes that is the ER is either covered on their outer surface with ribosomes (*rough ER*) or lacking ribosomes (*smooth ER*) Rough endoplasmic reticulum (RER) is so named because its surface is studded with protein manufacturing **ribosomes**. However, the ribosomes bound to the RER at any one time are not an integral part of the RER membrane. Ribosomes attach to the RER membrane only during the process of protein formation and then release from the membrane once the protein has been formed. Smooth endoplasmic reticulum (SER) is so named because ribosomes do not attach to its membranes. ER performs a number of general functions in the cell:

1. Formation of transmembrane proteins and secreted proteins.
2. Transport system for proteins.
3. Synthesis and storage of steroids and lipids.
4. Metabolism of carbohydrates.
5. Regulation of calcium concentration.

Ribosomes, complexes of RNAs and proteins that are produced in the nucleolus, are cellular components that serve as the site of protein synthesis in a cell. Ribosomes are divided into two subunits, one larger, one smaller (**Figure 5.22**). During protein synthesis the smaller subunit binds to **messenger RNA** (mRNA) whereas the larger subunit binds to **transfer RNA** (tRNA) and the amino acids (**Figure 5.23**). Once a ribosome attaches to the ER, mRNA from the nucleus attaches to the ribosome. Transfer RNA molecules holding specific amino acids then attach to the mRNA in a precise manner dictated by the code on the mRNA. The amino acids carried by the tRNA bond together into a protein chain. The newly synthesized proteins are sequestered in small transport vessicles and then moved from the ER complex to the **Golgi complex**.

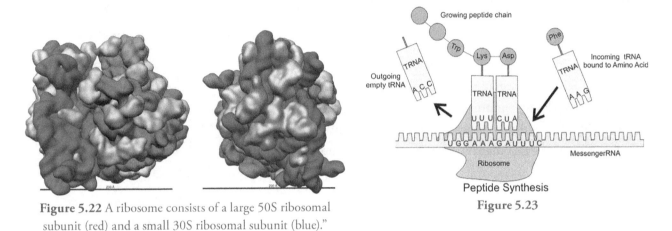

Figure 5.22 A ribosome consists of a large 50S ribosomal subunit (red) and a small 30S ribosomal subunit (blue)."

Figure 5.23

Mitochondria (sing., **mitochondrion**) are organelles present in nearly all eukaryotic cells. They vary in shape, size, and number. Some are elongated whereas others are more or less spherical. The number of mitochondria in a cell varies widely by type of organism and tissue. Unicellular organisms may contain a single mitochondrion, but human liver cells may contain 1,000 to 2,000 mitochondria per cell, 1/5 of the cell volume. Mitochondria are the only animal cell organelles that contain their own DNA and thus are the only animal cell organelle that is self-replicating. Mitochondrial DNA, which hints at the possibility of the endosymbiotic origin of mitochondria, is in the form of a tiny, circular genome, much like the circular genome of prokaryotes only smaller. Mitochondria may be thought of as cellular power plants because it is here that a cell produces chemical energy in the form of ATP (adenosine triphosphate) from the oxidation of glucose molecules.

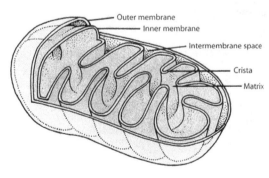

Figure 5.24

Structurally, a mitochondrion contains both outer and inner membranes (**Figure 5.24**). The outer membrane is similar in structure to the plasma membrane and contains transmembrane proteins known as porins. Materials pass in and out of the mitochondria through the porins. The inner membrane is highly convoluted into complex folds known as *cristae* (sing., crista). The cristae greatly increase the inner membrane's surface area. It is on the cristae that glucose is combined with oxygen to produce molecules of adenosine triphosphate (ATP), the primary energy source for the cell.

Golgi complex The Golgi complex typically consists of a stack of 3 to 20 slightly curved, flattened sacs that serve to store, modify, and package protein products, especially secretory products.

Small vesicles of the ER containing protein pinch off forming transport vesicles. The transport vesicles fuse with sacs on the *cis* face ("forming face") of a Golgi complex. After modification, the proteins are packaged into secretory vesicles on the *trans* face ("maturing face") of the complex (**Figure 5.25**). Secretory vesicles may contain contents from a glandular cell destined to be expelled to the outside of the cell. Some may contain proteins for incorporation into the plasma membrane whereas others may contain powerful enzymes and remain in the same cell that produces them.

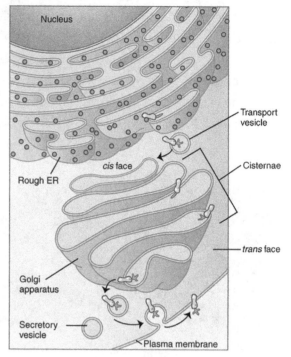

Figure 5.25

Lysosomes are vesicles involved in the breakdown of foreign protein or microbes, injured or diseased cells, and worn-out cell components. The interior of a lysosome is quite acidic (pH 4.8) compared to the slightly alkaline cytosol (pH 7.2) and contains enzymes for the digestion of starch, lipids, proteins, and nucleic acids. Lysosomal vesicles may also pump their enzymes into a larger membrane-body such as a food vacuole (*phagosome*) in the process forming a (**Figure 5.26**).

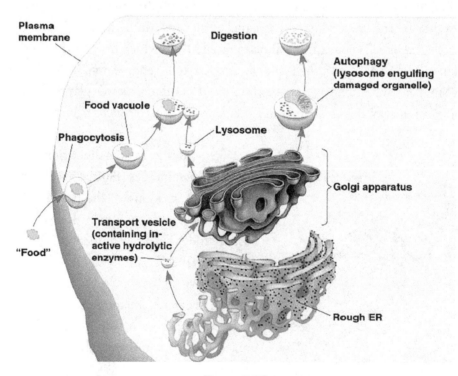

Figure 5.26

82

Vacuoles are membranous sacs present in all plant and fungal cells but only some bacteria, protozoa, and animal cells. Vacuoles are essentially enclosed compartments (sacs) that are filled with water containing inorganic and organic molecules including enzymes in solution, though in some cases they may contain solids (food) which have been engulfed. Vacuoles have no basic shape or size and are usually used to store excess food, water, or wastes.

Plant Cell Structure

Eukaryotic plant cells differ from eukaryotic animal cells in three ways: plant cells are bounded by a cell wall, animal cells are not; plant cells contain organelles called plastids, animal cells do not, and plant cells have a large central vacuole; animal cells do not (**Figure 5.27**).

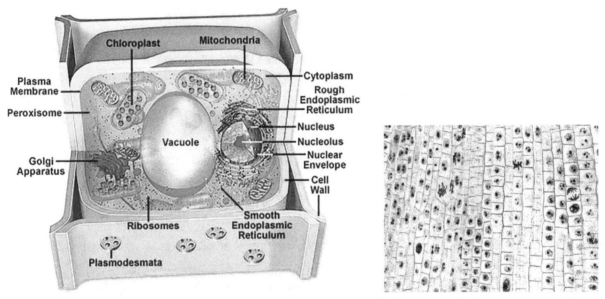

Figure 5.27 Typical Plant Cell. (a) Drawing of a generalized plant cell. (b) Micrograph of onion root tip cells.

Cell walls are a structural layer that surrounds the cells of plants, fungi, and prokaryotic cells. Cell walls are tough and sometimes rigid. The primary cell wall of land plants is composed of the fibrous polysaccharides cellulose, hemicellulose, and pectin (**Figure 5.28**). A plant cell wall consists of up to three layers. From the outside in, these layers are identified as the middle lamella, primary cell wall, and secondary cell wall. Although all plant cells have a middle lamella and primary cell wall, not all have a secondary cell wall.

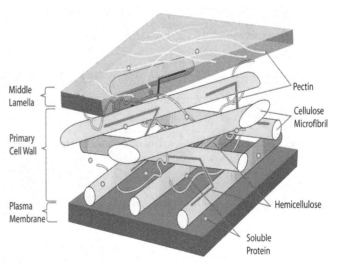

Figure 5.28

- *Middle lamella* is the outer cell wall layer that contains polysaccharides called pectins. Pectins aid in cell adhesion by helping the cell walls of adjacent cells to bind to one another.
- *Primary cell wall* is a layer formed between the middle lamella and the plasma membrane in growing plant cells. It is primarily composed of cellulose microfibrils contained within a gel-like matrix of hemicellulose fibers and pectin polysaccharides. The primary cell wall provides the strength and flexibility needed to allow for cell growth.
- *Secondary cell wall* is a layer formed between the primary cell wall and plasma membrane in some plant cells. Once the primary cell wall has stopped dividing and growing, it may thicken to form a secondary cell wall. This rigid layer strengthens and supports the cell. In addition to cellulose and hemicellulose, some secondary cell walls contain lignin. Lignin strengthens the cell wall and aids in water conductivity in plant vascular tissue cells.

Cell walls in plants have various functions:

Support The cell wall provides mechanical strength and support. It also controls the direction of cell growth. Cellulose fibers, structural proteins, and other polysaccharides help to maintain the shape and form of the cell. That, in turn, provides support for the whole plant. The next time you look at a giant sequoia growing hundreds of feet tall, remember that that plant is held up by the individual cell walls of every cell in its body.

Withstand turgor pressure Turgor pressure is the force exerted against the cell wall as the contents of the cell push the plasma membrane against the cell wall. This pressure helps a plant to remain rigid and erect, but can also cause a cell to rupture.

Regulate growth The cell wall sends signals for the cell to enter the cell cycle in order to divide and grow.

Regulate diffusion The cell wall is somewhat porous allowing some substances, including proteins, to pass into the cell while keeping other substances out.

Communication Cells communicate with one another via *plasmodesmata* (pores or channels between plant cell walls that allow molecules and communication signals to pass between individual plant cells).

Protection The cell wall provides a physical barrier to protect against plant viruses and other pathogens. It also helps to prevent water loss.

Storage Cell walls stores carbohydrates for use in plant growth, especially in seeds.

Plastids are plant organelles surrounded by a double membrane that can be categorized into three groups: *chloroplasts, chromoplasts, and leucoplasts.* Some algae cells have only one **chloroplast** whereas some plant cells have as many as a hundred. Chloroplasts can be quite large, being twice as wide and as much as five times the length of a mitochondrion (**Figure 5.29**). Chloroplasts use light to produce carbohydrates from water and carbon dioxide through the process of photosynthesis. The mitochondrion liberates the chemical energy in the carbohydrates through the process of respiration.

Chromoplasts produce and store pigments. Found in flowers, leaves, roots and ripe fruits, they contain carotenoids (lipid-soluble pigments ranging from yellow to red in color), that lend color to the plant tissues containing them. For example, it's the carotenoid pigment lycopene that makes a ripe tomato red.

There are hundreds of carotenoids, each with its own characteristic color. Carotenoids are divided into two categories, those that contain oxygen (*xanthophylls*) and those that lack it (*carotenes*). Xanthophylls are yellow, while carotenes are orange. The xanthophylls violaxanthin and neoxanthin are responsible for making the flowers of a tomato plant yellow. The carrot is an example of a root that gets its bright orange color from carotenes, particularly β-carotene, as well as lesser amounts of α-carotene and γ-carotene (with a dash of lutein and zeaxanthin). β-carotene is a precursor (inactive form) to vitamin A and is also found in sweet potatoes and pumpkins.

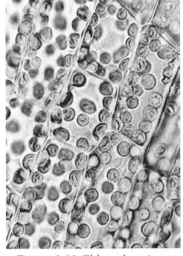

Figure 5.29 Chloroplasts in an Elodea leaf. Note the cell walls surrounding each cell and even though you cannot see it, the presence of the central vacuole in each cell can be discerned.

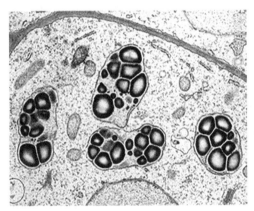

Figure 5.30 Amyloplasts in a Plant Cell. Each amyloplast contains several grains of starch.

Leucoplasts are organelles in plant cells that store starch, lipids or proteins, or have biosynthetic functions creating a variety of organic compounds. Leucoplasts used for storage fall into three categories. Leucoplasts that store starch are called *amyloplasts* (**Figure 5.30**). Leucoplasts that store lipids are called *elaioplasts.* Those that store proteins are called *proteinoplasts.* Elaioplasts and proteinoplasts are often found in seeds. However, in many plant cell types, leucoplasts are not used much for storage. In these cells, they create organic compounds rather than storing them. Some synthesize fatty acids, others amino acids and some generate more specialized compounds. Immature chloroplasts, as well as chloroplasts that have been deprived of light, do not have pigmentation and can also be considered leucoplasts. Once exposed to light, however, they produce the compounds necessary for photosynthesis and cease being leucoplasts.

Unlike other plastids, leucoplasts have no colored pigments. Leucoplasts are generally much smaller than chloroplasts and have a variable shape. In root cells and others, different leucoplasts are often connected in complex networks by structures known as *stromules.* Leucoplasts are often clustered around the nucleus of the cell during certain stages of development.

Central vacuole Animal cells contain relatively few vacuoles, but in plant cells vacuoles are essential. Plant vacuoles contain not only water, sugars, and salts but also water-soluble pigments and toxic

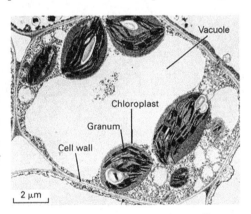

Figure 5.31 TEM of a plant cell showing the relationship of the central vacuole to the rest of the cell.

substances. The toxic substances help protect a plant from herbivores whereas the colored pigments produce many of the red, blue, or purple colors of flowers and some leaves.

Essentially, plant cells have a large central vacuole that may take up 90% of the volume of the cell. The vacuole is filled with a watery fluid called *cell sap* that provides additional support to the cell (**Figure 5.31**). The central vacuole maintains hydrostatic pressure or turgor pressure in plant cells that provides structural support for the entire plant. A plant wilts because its cells have lost turgor pressure. The central vacuole functions in storage of both nutrients and waste products. Metabolic waste products are pumped across the vacuole membrane and stored permanently in the central vacuole, where digestive enzymes break them down.

Table 5.1
Cellular Components—Structure and Function

Component	Description	Function
Plasma membrane	A double-layered outer boundary composed of protein, and phospholipids.	Separate cell from other cells or the environment; protect cell; regulate passage of molecules into and out of the cell
Cytoplasm	Viscous gel-like substance filling the interior of a cell; consists of fluid cytosol and organelles	Houses all of a cell's organelles and dissolves certain substances
Cytoskeleton	Interconnected filaments and tubules forming a flexible cellular scaffolding	Provides support; assists in cell movement; aids in transport of vesicles; assists in chromosome movement during cell replication.
Centrosome (Micro-tubule organizing center)	An area near the nucleus where microtubules are produced. Within the centromere is a pair of centrioles with each centriole being composed of a ring of nine groups of microtubules.	Form microtubules that, in turn, function as conveyor belts within the cell moving vesicles, organelles, and chromosomes via special attachment/motor proteins
Nucleus	Large spherical structure contained by the nuclear membrane; contains nucleolus and DNA.	DNA in the nucleus serves as template for the production of proteins that coordinate and control cell processes.
Endoplasmic reticulum	A network of membranous tubules continuously extending throughout the cytoplasm from the nuclear membrane to the plasma membrane.	Storage and internal transport; rough ER is a site for ribosome attachment; smooth ER makes lipids.
Ribosomes	Composed of two subunits-one large and one small. Each subunit is composed of RNA and protein.	Link amino acids together into protein chains in the order specified by messenger RNA transcribed from the DNA in the nucleus.

Mitochondria	Spherical or rod-shaped; bound by a double membrane system with the inner membrane highly folded.	Generate ATP, the source of chemical energy for the cell.
Golgi complex (Golgi apparatus)	Stacks of disklike membranes	Sorts, packages, and routes cell's synthesized products.
Lysosome (Animal cells only)	Small membranous sacs; containing enzymes.	Digest nonfunctional organelles and incoming food particles.
Cell walls (Plant cells only)	Box-like structure around plant cells composed mainly of cellulose fibers.	Support, withstand turgor pressure, regulate growth, regulate diffusion, communication, protection, and storage.
Plastids (Plant cells only)	Double membrane-bound organelle involved in the synthesis or storage of carbohydrates.	Chloroplasts synthesize carbohydrates, amyloplasts store starch, elaioplasts store fats, and proteinoplasts store protein.
Central vacuole (Plant cells only)	Membrane-bound organelle filling up most of the volume of plant cells.	Storing of salts, nutrients, pigments, minerals, proteins, and playing a vital role in the structural stability of plants.

In Summary

- Cells are the building blocks of life because any living thing is composed of at least one cell.
- Modern cell theory then states that:
    1. Cells are the basic unit of life.
    2. All living things are composed of at least one cell.
    3. All cells come from preexisting cells

- The untold multitude of cells on earth fall into only two general categories— *prokaryote* or *eukaryote*—based on inherent internal properties.
- Prokaryotic cells do not have their genetic material inside a distinct nucleus, and they lack most cell organelles.
- Eukaryotic cells have their genetic material inside a distinct nucleus and they possess many distinct organelles.
- A eukaryotic animal cell is constructed of (1) a *system of membranes* that surround the cell and each of its organelles, (2) gelatinous *cytoplasm* that contains and supports the organelles as well as macromolecular assemblages and food storage products, and (3) the various *organelles* and *macromolecular assemblages* within the cytoplasmic matrix.
- The plasma membrane is constructed of a bilayer of phospholipids. Each layer of the plasma membrane consists of the hydrophilic phosphate groups aligned so they all point out resulting in

all the hydrophobic lipid tails pointing in. Associated with the plasma membrane are a number of diverse proteins.

- Cytoplasm is a complex matrix consisting of different components: *cytosol, cytoskeleton,* and *inclusion*s.

- Inclusions are tiny particles or droplets of insoluble materials within the cytosol that are not bound by membranes.

- The nucleus is the largest of the organelles. Within the nucleus resides the genetic material responsible for controlling and coordinating all the activities within the cell. This genetic material, known as chromatin, consists of long strands of DNA (Deoxyribose Nucleic Acid) complexed with proteins known as histones.

- The endoplasmic reticulum is an extensive network of highly folded membranous tubules and flattened sacs supported and stabilized by the cytoskeleton.

- The complex of membranes that is the ER is either covered on their outer surface with ribosomes (rough ER) or lacking ribosomes (smooth ER).

- Ribosomes, complexes of RNAs and proteins that are produced in the nucleolus, are cellular components that serve as the site of protein synthesis in a cell. Ribosomes are divided into two subunits, one larger, one smaller.

- Mitochondria may be thought of as cellular power plants because it is here that a cell produces chemical energy in the form of ATP (adenosine triphosphate) from the oxidation of glucose molecules.

- The Golgi complex typically consists of a stack of 3 to 20 slightly curved, flattened sacs that serve to store, modify, and package protein products, especially secretory products.

- Lysosomes are vesicles involved in the breakdown of foreign protein or microbes, injured or diseased cells, and worn-out cell components.

- Vacuoles are essentially enclosed compartments (sacs) that are filled with water containing inorganic and organic molecules including enzymes in solution, though in some cases they may contain solids (food) which have been engulfed.

- Eukaryotic plant cells differ from eukaryotic animal cells in three ways: plant cells are bounded by a cell wall, animal cells are not; plant cells contain organelles called plastids, animal cells do not, and plant cells have a large central vacuole, animal cells do not.

- Cell walls are a structural layer that surrounds the cells of plants, fungi, and prokaryotic cells. Cell walls are tough and sometimes rigid.

- Plastids are plant organelles surrounded by a double membrane that can be categorized into three groups: *chloroplasts, chromoplasts, and leucoplasts.*

- Plant cells have a large central vacuole that may take up 90% of the volume of the cell. The vacuole is filled with a watery fluid called cell sap that provides additional support to the cell

Review and Reflect

1. ***What's in a Quote?*** React to the quote by George Henry Lewes that opens this chapter. In your own words, what do you think he is trying to say?

2. ***Mystery Cells*** Your colleague believes she has just discovered a new species of microorganism. This previously unknown microbe is single-celled, and the cells range in size from 9-11 μm in size. What further information must she gather to determine if these newly-discovered cells are prokaryotic or eukaryotic?

3. ***Cell Analogies*** "Analogies prove nothing that is true," wrote Sigmund Freud, "but they can make one feel more at home." Toward that end, we use analogies here to simplify the function of each of the organelles found within a cell. Answer the following:

    A. How is the nucleus of a cell analogous to *city hall*?
    B. How is a mitochondrion analogous to a *power plant*?
    C. How is the endoplasmic reticulum analogous to *highways and roads*?
    D. How are ribosomes analogous to *manufacturing plants*?
    E. How is the Golgi apparatus analogous to *packing and shipping facilities*?
    F. How are lysosomes analogous to *waste disposal and recycling plants*?

4. ***Cell Structure Rap*** Read each description and then identify the cell part being discussed in the lyric.

    A. I'm a real "powerhouse"
    that's plain to see.
    I break down food
    to release energy.

    B. I'm strong and stiff
    and pretty tough.
    I wrap plant cells
    but that's enough.

    C. I'm clear and runny
    and hard to see.
    I fill the cell
    so it can be.

    D. I'm the "brain" of the cell
    or so they say.
    I regulate activities
    from day to day.

    E. Found only in plant cells
    I'm green as can be.
    I make food for the plant
    using the sun's energy.

F. I'm a series of tunnels
found throughout the cell.
I transport proteins
and other things as well.

G. I'm full of holes
flexible and thin.
I control what gets out
and what comes in.

H. Proteins are made here
even though I'm quite small.
You can find me in the cytoplasm
or attached to the E.R.'s wall.

I. I've been called a "storage tank"
by those with little taste.
I'm a really a sac of water,
food or wastes.

5. *Gateway to the Cell* The plasma membrane is described as being "selectively permeable." What does this mean? Why is the permeability of its cell membrane critical to the life of any cell?

6. *Fear the Blob?* In the 1958 movie, *The Blob*, a small blob from outer space consumes living things (including humans) until it grows to gigantic size. Is it possible to take a single animal cell and nurture it until it grew as large as a car? Could you grow a super-size single animal cell in an orbiting laboratory? Explain

7. *An Unseen World* It has been said the invention of first the light microscope and then electron microscopes may well be the most important devices ever invented by humans. Write a short essay in which you defend this statement.

8. *Tough Stuff* Explain how a giant sequoia tree can grow to be over 300 feet tall and 25 feet in diameter without a skeleton to hold it up.

9. *Pretty Little Plastid* Only plant cells have plastids. What are plastids and what function(s) do they perform in plant cells?

Create and Connect

1. *Can You Patent a Living Thing?* A patent is a claim that is made for the ownership of an invention or a process. A government grants a license that gives the patent owner rights to control the making and selling of the patented object or process. Can you patent a living organism or products of living organisms? Is it right to patent a plant? What about cells or chemicals derived from cells taken from tissue removed during surgery?

   An organism that already exists in nature cannot be patented. But suppose genes from one species of plant are placed into another kind of plant. By law, this plant can be patented if it is useful to humans in some way. In April 1988, the United States Patent and Trademark Office granted Harvard University the first patent for a transgenic animal. A transgenic animal has one or more genes from another kind of animal. In this case, a human gene was placed in a mouse, making the mouse a **transgenic animal**.

Case Studies

❖ The research at Harvard that produced the transgenic mouse was funded by a corporation. Therefore, this company owns the commercial rights to the results of this research. The transgenic mice are now being bred and offered for sale for research. These patented mice have a genetic tendency to develop certain tumors. The patent that was issued covers all transgenic nonhuman mammals engineered with the human genes that cause tumors. Since there are about 40 known genes that cause tumors, the company can now own other transgenic mammals that it develops for cancer research.

❖ Farm organizations and animal rights groups oppose the development and subsequent patenting of transgenic animals. The farm organizations believe that the present distribution of genes in farm animals should be known before transgenic animals are developed. Religious leaders have expressed concerns about the moral and social issues regarding developing new animals by transgenic means.

❖ In 1973, a convention of European nations agreed to prohibit patents on transgenic organisms in the European Economic Community. Now economic leaders are beginning to recognize that patents for transgenic organisms may be necessary for the growth of the biotechnology industries in Europe. A European commission is now working to establish a legal framework for patenting transgenic organisms.

❖ A group of companies has been studying human genes to map the gene that are on specific chromosomes. Some people believe that the companies withhold the Information they acquire in order to make a profit. The Human Genome Project, sponsored by the National Institutes of Health and the Department of Energy, would like to use the information that the companies have acquired. However, the companies have spent large amounts of money and time developing the information and techniques. These companies can prosper and continue only if they are guaranteed ownership rights through copyrights and patents. Some researchers think that the scientific tradition of sharing information should be followed.

❖ Tissue that people pay to have removed such as an appendix or tumor are sometimes used by researchers to make products such as unusual proteins or scarce human hormones. The products, cells or processes are then patented. However, who owns the tissues after they are removed and who should profit from them? Hormones and cancer drugs have been produced from such tissues.

In 1987, about 350 companies produced about $600 million worth of such products from human tissues that number has continued to increase to this day. Lawsuits over the ownership of the removed tissues are slowing research. A spokesperson for the industry states that only 1 In 10,000 tissue samples may have any commercial value. Their argument against paying for the tissues is that the tissues are worthless without the skills of the researchers. Carefully consider these case studies and then express your viewpoint and position on this matter as you answer the questions that follow.

A. Why would transgenic mice be useful to the researchers who buy them?

B. Should the concerns of farm organizations, animal rights groups, and religious leaders be taken into account before transgenic animals are developed and patented?

C. What are the main arguments for and against patenting of transgenic organisms and human tissue products?

D. What do you think could be accomplished now that researchers have "mapped" (sequenced) all the human genes?

E. Suppose a hormone or cancer drug is produced from tissues taken from your body. Do you think you are entitled to profit from the sales of that material? Explain

F. Discuss whether or not all scientific information should be freely available to all scientists. Is there any instance when this information should be withheld and kept secret?

2. *Embryonic Stem Cells.* The use of embryonic stems cells is controversial on several different levels—socially, medically, religiously, politically, and legally. Delve into this controversy and then write a position paper on the use of embryonic stem cells in medical research.

Guidelines

A. Compose a position paper, not an <u>opinion</u> paper. Defend your position with as many facts, figures, quotes, and pertinent information as possible.

B. Your work will be evaluated not on the "correctness" of your position but the quality of the defense of your position.

C. Your instructor may provide additional details or further instructions.

3. *Model a Cell* Imagine you have been challenged to make a model of a eukaryotic plant cell using common items found around the home, in the garage, or in the basement perhaps. You can also buy small items at the grocery store or hardware store. Explain what object you would use and why to represent the following parts of a plant cell:

- Cell wall
- Plasma membrane
- Cytoplasm
- Nucleus
- Central vacuole
- Chloroplasts
- Mitochondria
- Endoplasmic reticulum

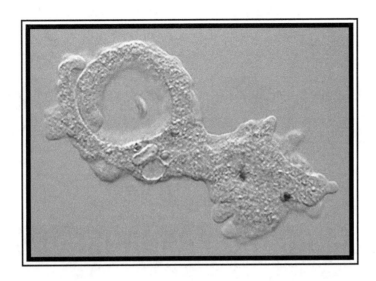

CELL PROCESSES: TRANSPORT AND METABOLISM

What drives life is thus a little electric current, kept up by the sunshine. All the complexities
of intermediary metabolism are but the lacework around this basic fact.

—Albert Szent-Györgyi

Introduction

For every instant of its existence, the internal components of a cell are involved in a frenzy of coordinated activities and processes that occur with an accuracy and intricacy that boggles the mind, and all of it occurring in a tiny fleck no bigger than the period at the end of this sentence.

A cell achieves its living individuality by being compartmentalized from its environment and other cells by a permeable barrier—the plasma membrane (Chapter 5). To survive, however, a cell must transport materials both in and out through this permeable membrane.

Cellular Transport

Cells employ several processes for moving molecules and macromolecules in and out. As we will see the same kinds of processes that allow entry can be used to achieve exit. The different kinds of cellular transport can be divided into two categories: those that do not require energy (passive) and those that do require energy (active).

Passive Transport

Cells passively transport materials in and out three ways: simple diffusion, facilitated transport, and bulk transport.

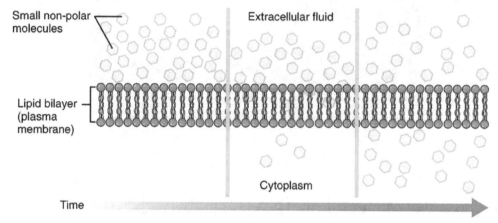

Figure 6.1 Simple Diffusion. Over time molecules will move across permeable plasma membranes from a place of high concentration (extracellular) to a place of low concentration (cytoplasm) until equilibrium is reached.

Simple diffusion The plasma membrane is selectively permeable in that it allows only a few molecules in and even fewer out. This movement of molecules in or out is often accomplished passively through a process known as **diffusion** (**Figure 6.1**). Diffusion is considered a passive process because it will occur spontaneously without the need of energy to power it forward. Diffusion occurs because molecules have the tendency to move from where there are many to where there are few until equilibrium is reached. Therefore, if there are more oxygen molecules outside a cell (higher concentration gradient) than within (lower concentration gradient), oxygen molecules will move through the plasma membrane into the cell until there are an equal number of oxygen molecules on both sides of the plasma membrane. In diffusion, molecules always move with the concentration gradient from high to low concentration, and once a dynamic equilibrium is reached, diffusion stops. The greater the concentration gradient, the faster diffusion proceeds.

Carbon dioxide, water, and a few other small, simple molecules also flow passively in and out of a cell through diffusion due to differences in concentration gradient. Examples of diffusion are all around us such as when we add sugar to our tea (Stirring the tea speeds up the diffusion process) or the smell of percolating coffee or cooking food wafting throughout the whole house.

Osmosis is a form of diffusion in which water specifically moves across a *differentially permeable membrane* toward a higher concentration of solute. In differentially permeable membranes such as the plasma membrane, small solvent molecules like water move easily across the membrane, but large macromolecules cannot. This keeps the insides of a cell from oozing out through the plasma membrane.

In biological systems, the solvent is typically water, but osmosis can occur in other liquids and even gases. There is water outside a cell and inside a cell. If the concentration of water molecules is the same on both sides of the plasma membrane, there will no net flow of water molecules in or out. However, should the cell be placed in salt water (solute), the inside of the cell has a greater concentration of water molecules (solvent) than does the salt water on the outside and thus water would move <u>out</u> of the cell. (Think of it in terms of the water trying to dilute the salt). If the cell should be placed in fresh water (solvent), the concentration of water is greater outside than inside in the contents of cytosol (solute) and water would thus move <u>into</u> the cell (**Figure 6.2**).

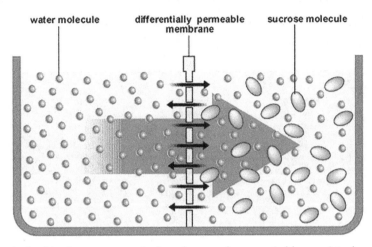

Figure 6.2 Osmosis. In this diagram, water is the solvent and sucrose (table sugar) is the solute. Because the concentration of solvent is greater on the left side of the differentially permeable membrane than on the right and because sucrose cannot pass through the membrane, water molecules move into the sugar solution.

The movement of water due to osmosis can generate pressure known as **osmotic pressure**. The greater the osmotic gradient, the more water diffuses, and the more water diffuses, the higher the osmotic pressure (**Figure 6.3**). Due to osmotic pressure, water is absorbed by the kidneys and taken up by the capillaries in tissues. In plants, osmotic pressure helps maintain the turgor pressure of each cell and opens the stomata in plant leaves to provide uptake of carbon dioxide for photosynthesis.

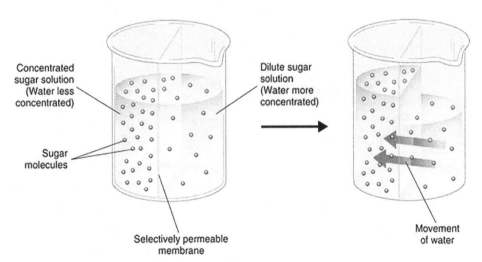

Figure 6.3 Osmotic Pressure. As water diffuses into the sugar solution, osmotic pressure is generated pushing the water higher on solution side (left) of the beaker than on the solute side (right). The increase in height has been greatly exaggerated to illustrate the point.

The variation in osmotic concentration gradient between the inside and the outside of a cell differ. If the solvent (water) concentration is greater inside the cell than out, the solution is said to by **hypotonic**, but if the solvent (water) concentration is greater outside the cell than in, the solution is said to be **hypertonic**. A solution in which the water concentration is the same on both sides of the plasma membrane is said to be **isotonic**. A 0.9% solution of the salt sodium chloride (NaCL) is known to be isotonic to human red blood cells. Therefore, intravenous solutions medically administered usually have this tonicity.

The tonicity of a solution can affect the chemistry of cells and thus the entire organism. Cells in hypotonic conditions can swell and burst (**cytolysis**) whereas those in hypertonic situations can shrivel and shrink (**plasmolysis**) (**Figure 6.4**).

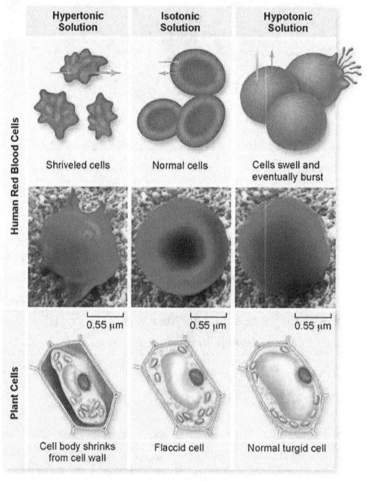

Figure 6.4 To survive, cells must make constant osmotic adjustments to remain on the razor's edge that is an isotonic environment.

Many animals living in an estuary or saltwater (hypertonic) such as oysters, crabs, some fish, seabirds, and sea turtles are able to cope with changes in the salinity (salt concentration) of their environment using specialized gills, kidneys, or salt glands. Sharks increase or decrease urea in their blood until their blood is isotonic with the surrounding salt water. Marine fish drink no water but excrete salts across their gills. The sight of a sea turtle shedding tears might lead one to think that animal to be sad when in actuality the turtle is ridding itself of excess salt by means of glands near the eye.

Facilitated transport Another form of passive transport is *facilitated transport* (also known as *passive-mediated transport*). Due to the hydrophobic nature of the phospholipids that make up the plasma membrane, only small nonpolar molecules, such as oxygen can easily diffuse across the membrane. However, large polar molecules (those with a positive charge [+] on one end and a negative charge [-] on the other) and charged ions cannot diffuse freely across the membrane. Because simple diffusion will not suffice to move these molecules across the plasma membrane, they must be carried across by special membrane transport proteins. There are two types of membrane transport proteins: carrier proteins and ion channel proteins. Small polar molecules are transported across the membranes by **carrier proteins** (**Figure 6.5**). After a carrier combines with a molecule, the carrier is believed to undergo a conformational change in shape that moves the molecule across the membrane.

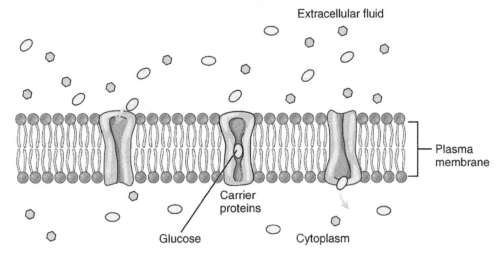

Figure 6.5 Although the functions of transport carrier proteins are not fully understood, we do know they are specific and can only transport a certain type of molecule across the membrane.

Ions move in or out through **ion channel proteins**. Ion channels are protein pores in the membrane where ions flow according to their electrochemical gradient. Ion channels may allow ion diffusion at all times, or they may be gated channels, requiring a signal to open or close them (**Figure 6.6**). Neither diffusion nor facilitated transport requires an expenditure of energy because the molecules are moving down their concentration gradient.

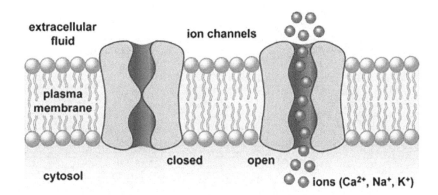

Figure 6.6 When an ion channel is open, ions move into or out of the cell in single-file fashion. Individual ion channels are specific to particular ions, meaning that they usually allow only a single type of ion to pass through them.

Bulk Transport Very large molecules cannot move through the plasma membrane and thus must be transported in, out or through the cell via the formation of vesicles. As we learned earlier in this chapter, vesicles are small membrane-bound sacs within the cell that store, transport, or digest cellular products. The membrane of a vesicle is structured in such a way that it can fuse not only with the plasma membrane but also with the membrane of organelles within the cell. A cell may capture material from its external surroundings through a process known as **endocytosis**. Endocytosis occurs in three forms: phagocytosis, pinocytosis, and receptor-mediated endocytosis (**Figure 6.7**). On the other hand, vesicles formed inside the cell may be expelled through a process known as **exocytosis**. There are two types of exocytosis: Non Ca^{2+} triggered or constitutive exocytosis and Ca^{2+} triggered or regulated exocytosis (**Figure 6.8**). In *constitutive exocytosis*, proteins constantly being packaged into vesicles by the Golgi complex are carried promptly to the plasma membrane where they released to the extracellular environment or incorporated into the plasma membrane.

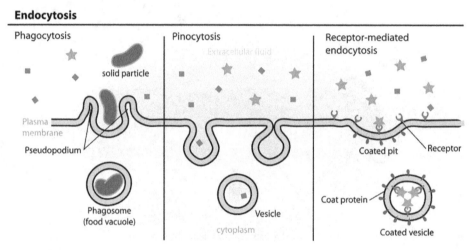

Figure 6.7 Types of Endocytosis. In phagocytosis, the cell membrane binds to a large particle and extends around it. In pinocytosis, small areas of the cell membrane . bearing specific receptors for small molecules or ions, invaginate to form caveolae (vesicles). Receptor-mediated endocytosis is a mechanism for selective uptake of large molecules in clathrin-coated pits. Binding of the ligand to the receptor on the surface membrane triggers invagination of pits.

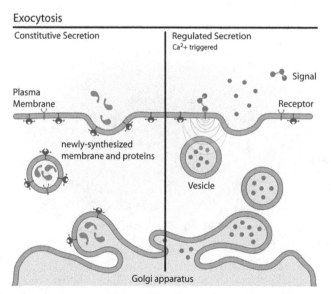

Figure 6.8 Regulated secretion is triggered by Ca2+ whereas constitutive secretion is not.

In *regulated exocytosis*, proteins and small molecules stored in vesicles only fuse with the plasma membrane after an extracellular signal triggers the cell. Unlike the constitutive pathway, the regulated pathway is found primarily in cells that are specialized for secreting their products (hormones, neurotransmitters, digestive enzymes) rapidly on demand. In these secretory cells, the extracellular signal is often a chemical messenger (such as a hormone), and the activation of receptors generates various intracellular signals, including a transient increase in the amount of free Ca^{2+} in the cytosol. These signals initiate exocytosis, leading to fusion of secretory vesicles with the plasma membrane and release of the products to the extracellular space.

Active Transport

If possible, cells want to carry on as much of their transport business as possible using diffusion and facilitated transport because it requires no energy expenditure. All the cell needs to do is let molecules move down their concentration gradient from high to low concentration. However, there are times when cells must transport materials against the concentration gradient (from low to high concentration), especially when the cell needs to accumulate high concentrations of vital materials, such as ions, glucose, and amino acids or remove unwanted ions that are diffusing freely into the cell. The process by which cells move materials in or out against a concentration gradient is known as **active transport** because the cell is required to expend energy to accomplish the task. The energy source that drives active transport is contained primarily in adenosine triphosphate (ATP) molecule. ATP powers specialized carrier proteins that "pump" materials against a concentration gradient. In fact, up to 40% of the ATP formed within a cell may be used up in the process of powering active transport (**Figure 6.9**).

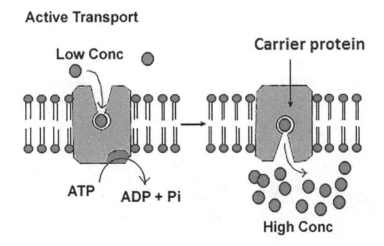

Figure 6.9 In active transport, ATP activates carrier proteins to change shape and carry molecules across the plasma membrane up the concentration gradient."

Box 6.1
Cell Signaling

In order to respond to changes in their immediate environment, cells must be able to receive and process signals that originate outside their borders. Individual cells often receive many signals simultaneously, and they then integrate the information they receive into a unified action plan. But cells aren't just targets. They also send out chemical signaling messages to other cells both near and far. For example, the pancreas releases a hormone called insulin that is transported in blood vessels to the liver, and this signal causes the liver to store glucose as glycogen. Failure of the liver cells to respond appropriately is one of the contributing factors in Type II diabetes.

Most cell signals are chemical in nature. For example, prokaryotic organisms have sensors that detect nutrients and help them navigate toward food sources. In multicellular organisms, growth factors, hormones, neurotransmitters, and extracellular matrix components are some of the many types of chemical signals cells use. These substances can exert their effects locally, or they might travel over long distances. For instance, *neurotransmitters* are a class of short-range signaling molecules that travel across the tiny spaces between adjacent neurons or between neurons and muscle cells. Other signaling molecules must move much farther to reach their targets. One example is follicle-stimulating hormone, which travels from the mammalian brain to the ovary, where it triggers egg release.

How do cells recognize incoming signals? Cells have proteins called *receptors* that bind to signaling molecules and initiate a physiological response. Different receptors are specific for different molecules. Dopamine receptors bind dopamine, insulin receptors bind insulin, nerve growth factor receptors bind nerve growth factor, and so on. In fact, there are hundreds of receptor types found in cells, and varying cell types have different populations of receptors. Receptors can also respond directly to light or pressure, which makes cells mechanically sensitive to events in their external environment.

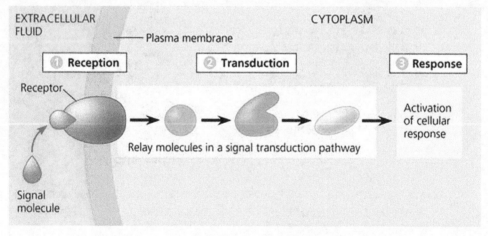

Figure 6.10

The first step in cell signaling begins with signaling molecules interacting with their receptors (**Figure 6.10**). Receptors are generally transmembrane proteins that bind to signaling molecules outside the cell and subsequently transmit the signal through a sequence of molecular switches to internal signaling pathways. Membrane receptors fall into three major classes: G-protein-coupled receptors, ion channel receptors, and enzyme-linked receptors. The names of these receptor classes refer to the mechanism by which the receptors transform external signals into internal ones via protein action, ion channel opening, or enzyme activation, respectively. Because membrane receptors interact with both extracellular signals and molecules within the cell, they permit signaling molecules to affect cell function without actually entering the cell. This is important because most signaling molecules are either too big or too charged to cross a cell's plasma membrane

Once a receptor protein receives a signal, it undergoes a conformational change, which, in turn, launches a series of biochemical reactions within the cell. These intracellular signaling pathways, also called *signal transduction cascades*, typically amplify the message, producing multiple intracellular signals for every one receptor that is bound. The cell response to a transduction pathway can be a change in the shape or movement of a cell, the activation of a particular enzyme, or the activation of a specific gene.

Cellular Metabolism

To maintain their structural organization and carry out necessary life functions, every cell needs a constant supply of energy. Cells produce chemical energy from the food molecules being transported into them across the plasma membrane.

Cellular metabolism may be thought of as the sum total of all the biochemical processes that occur within a cell. Such biochemical processes are categorized as either anabolistic or catabolistic. Anabolic processes (**anabolism**) are those in which a cell constructs large macromolecular assemblages (polymers) from smaller units, such as building proteins from various amino acids. Anabolic processes are powered by catabolic processes (**catabolism**) in which large macromolecules are broken down into smaller units (monomers).

Proteins (polymer) ⊠ amino acids (monomer)
Polysaccharides ⊠ monosaccharides
Fats ⊠ fatty acids

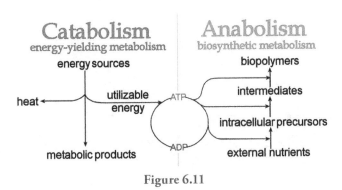

Figure 6.11

Cells use the monomers released from breaking down polymers to either construct new polymer molecules or degrade the monomers further forming the energy-carrying molecule adenosine triphosphate (ATP) and simple waste products (**Figure 6.11**). Metabolism includes both spontaneous reactions and energy-requiring reactions. *Exergonic reactions* are spontaneous and release energy whereas *endergonic reactions* require an input of energy to occur.

ATP is considered by biologists to be the "energy currency" of metabolism and thus life itself. Formed in the mitochondria, ATP is present in the cytoplasm of the cell and provides the energy required for every biochemical process within the cell. The importance of ATP can be seen in the fact that the human body produces 2×10^{26} molecules of ATP daily!

ATP may be thought of as a molecular "battery." It is discharged (energy released) when one of the phosphate groups on the "tail" of the molecule is lost. The ADP molecule that remains can be recharged through oxidative phosphorylation back into ATP (**Figure 6.12**). In living systems, ATP performs a number of different tasks:

- **Chemical work**. ATP supplies the energy needed to synthesize macromolecules (anabolism) that make up the cell and entire organism.
- **Transport work**. ATP supplies the energy needed to pump substances across the plasma membrane up the concentration gradient.
- **Mechanical work**. ATP supplies the energy needed for movement of any kind.

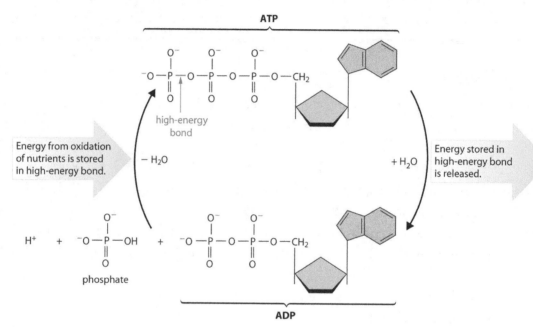

Figure 6.12 Many biological processes use ATP as an energy carrier between energonic reactions (left) and exogonic reactions (right).

Metabolic Pathways and Enzymes

Cellular metabolism encompasses:

- Degrading food molecules (catabolism)
- Synthesizing macromolecules needed by the cell (anabolism)
- Generating small precursor molecules needed by the cell such as some amino acids
- All reactions involving electron transfers (ATP $\leftrightarrow$ ADP)

Metabolism occurs in a sequence of biochemical reactions known as *metabolic pathways*. Metabolic pathways may be simple linear sequences consisting of only a few steps, or they may be highly branched and convoluted derivations from a central main pathway (**Figure 6.13**). Some pathways can serve multiple functions. For example, the citric acid cycle of aerobic respiration functions mainly to produce adenosine triphosphate (ATP), the molecule from which a cell derives most of its chemical energy, but it can also yield small precursor molecules necessary for many biochemical reactions within the cell.

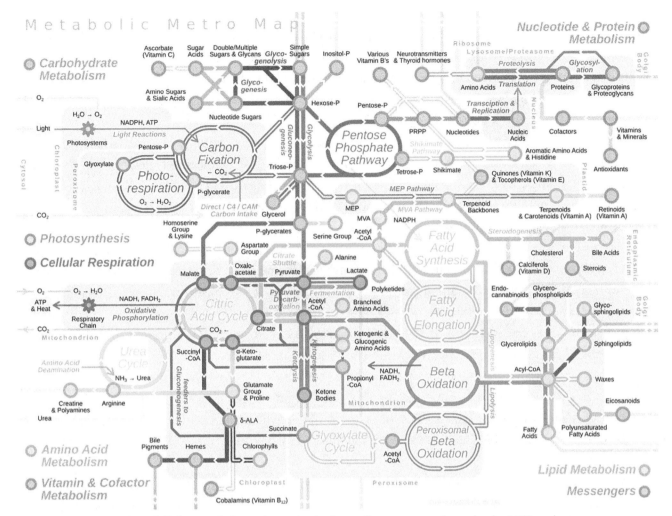

Figure 6.13 Cellular metabolic pathways in the form of a metro map. Just take the MEP pathway to Vitamin A and be careful not to get trapped in the roundabout that is the Citric Acid Cycle.

The chemical reactions that constitute metabolism would not easily occur or might not occur at all without the use of organic catalysts called enzymes. All biochemical reactions within metabolic pathways are catalyzed and controlled by proteins known as enzymes. An **enzyme** is a protein molecule that speeds up a chemical reaction without itself being affected by the reaction. Enzymes allow reactions to occur under certain conditions, and they regulate metabolism, partly by eliminating nonspecific side reactions.

Not all enzymes are proteins. **Ribozymes** are made of RNA instead of protein and can also serve as biological catalysts. Ribozymes are involved in the synthesis of RNA and the synthesis of proteins at ribosomes.

Many specific steps can be involved in a metabolic pathway, and each step is a chemical reaction catalyzed by an enzyme. The reactants in an enzymatic reaction are called the *substrates* for that enzyme. The substrates for the first reaction are converted into products, and those products then serve as the substrates for the next enzyme-catalyzed reaction. One reaction leads to the next in a highly regulated manner.

$$A \xrightarrow{R_1} B \xrightarrow{R_2} C \xrightarrow{R_3} D$$

In this diagram, A is the substrate for R_1, and B is the product. B then becomes the substrate for R_2, and C is the product. This process continues until the final product, D forms. Also, any one of the molecules (A-D) in this metabolic pathway could also be a reactant in another pathway. Many of the metabolic pathways in living organisms are highly branched, and interactions between metabolic pathways are very common as you can see in the Metabolic Metro Map (Figure 6.13).

The specificity of enzymes allows the regulation of metabolism. The presence of particular enzymes helps determine which metabolic pathways are operative. In addition, some substrates can produce more than one type of product, depending on which pathway is open to them. Therefore, which enzyme is present determines which product is produced, as well as determining the direction of metabolism, without several alternative pathways being activated.

Enzymes speed up reactions dramatically, and in some instances they can increase the reaction rate more than 10 million times. The rate depends on the amount of substrate available to associate at the active sites of enzymes. Therefore, increasing the amount of substrate and the amount of enzyme can increase the rate of reaction. Enzymes require specific conditions to be met in order to be fully operational. Some of these conditions are:

- **Substrate concentration**. Enzyme activity increases as substrate concentrations increases, because there are more collisions between substrate molecules and the enzyme. As more substrate molecules fill active sites, more product results per unit of time. When the active sites are filled almost continuously with substrate, the rate of reaction can no longer increase.

- **Optimal pH**. Each enzyme has an optimal pH at which the enzyme rate is highest. At their respective pH values, each enzyme can maintain its structural configuration, which enables optimal function.

- **Temperature**. As temperature rises, enzyme activity increases to a point because of greater collisions between substrate and enzyme. If the temperature rises beyond a certain point, enzyme activity eventually levels out and then declines rapidly, because the enzyme is *denatured*. There are exceptions to this generalization. For example, some prokaryotes can live in boiling hot springs because their enzymes do not denature.

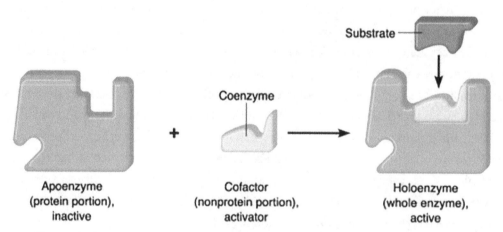

Figure 6.14 An inactive enzyme without the cofactor is called an apoenzyme whereas the complete activated enzyme with cofactor is called a holoenzyme.

Many enzymes require the presence of an inorganic ion or nonprotein organic molecule at the active site in order to work properly. These necessary ions or molecules are called *cofactors* (**Figure 6.14**). **Vitamins**, small, organic molecules required in trace amounts in our diets, are required for the synthesis of some coenzymes and become part of a coenzyme's molecular structure. For example, the vitamin niacin is part of the coenzyme NAD^+ and riboflavin (B_2) is a component of the coenzyme FAD. If a vitamin is not available, enzymatic activity will decrease, and eventually lead to a vitamin-deficiency disorder. In humans, a niacin deficiency results in pellagra, and riboflavin deficiency results in cracks at the corners of the mouth.

Sometimes it is necessary to limit or inhibit the activity of an enzyme. Enzyme inhibition occurs when a molecule (the inhibitor) binds to an enzyme and decreases its activity. There are two basic types of enzyme inhibitors: competitive and noncompetitive. *Competitive inhibition* occurs when an inhibitor and a substrate compete for the active site on an enzyme. *Noncompetitive inhibition* occurs when an inhibitor binds to an enzyme at a location other than the active site. This changes the shape of the enzyme, which, in turn, changes the functionality of the enzyme (**Figure 6.15**).

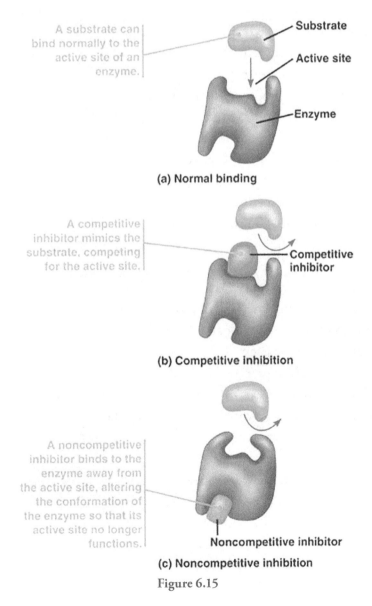

A substrate can bind normally to the active site of an enzyme.

Substrate
Active site
Enzyme

(a) Normal binding

A competitive inhibitor mimics the substrate, competing for the active site.

Competitive inhibitor

(b) Competitive inhibition

A noncompetitive inhibitor binds to the enzyme away from the active site, altering the conformation of the enzyme so that its active site no longer functions.

Noncompetitive inhibitor

(c) Noncompetitive inhibition

Figure 6.15

In Summary

- The different kinds of cellular transport can be divided into two categories: those that do not require energy (passive) and those that do require energy (active).

- Cells passively transport materials in and out three ways: simple diffusion, facilitated transport, and bulk transport.

- Diffusion occurs because molecules have the tendency to move from where there are many to where there are few until equilibrium is reached.

- Osmosis is a form of diffusion in which a solvent moves across a differentially permeable membrane toward a higher concentration of solute. In living systems, water is usually the solvent.

- The movement of water due to osmosis can generate pressure known as osmotic pressure. The greater the osmotic gradient, the more water diffuses, and the more water diffuses, the higher the osmotic pressure.

- If the solvent (water) concentration is greater inside the cell than out, the solution is said to by hypotonic, but if the solvent (water) concentration is greater outside the cell than in, the solution is said to be hypertonic. A solution in which the water concentration is the same on both sides of the plasma membrane is said to be isotonic.

- Cells in hypotonic conditions can swell and burst (cytolysis) whereas those in hypertonic situations can shrivel and shrink (plasmolysis).

- Facilitated transport is the process of spontaneous passive transport of molecules or ions across a plasma membrane via specific carrier proteins.

- The movement of macromolecules such as proteins or polysaccharides into or out of the cell is called bulk transport. There are two types of bulk transport: exocytosis and endocytosis and both require the expenditure of energy (ATP).

- The process by which cells move materials in or out against a concentration gradient is known as active transport because the cell is required to expend energy to accomplish the task.

- Cellular metabolism may be thought of as the sum total of all the biochemical processes that occur within a cell.

- ATP may be thought of as a molecular "battery." It is discharged (energy released) when one of the phosphate groups on the "tail" of the molecule is lost. The ADP molecule that remains can be recharged through oxidative phosphorylation back into ATP.

- Cellular metabolism encompasses:
  - Degrading food molecules (catabolism)
  - Synthesizing macromolecules needed by the cell (anabolism)
  - Generating small precursor molecules needed by the cell such as some amino acids
  - All reactions involving electron transfers (ATP $\leftrightarrow$ ADP)

- Metabolism occurs in a sequence of biochemical reactions known as metabolic pathways.

- The chemical reactions that constitute metabolism would not easily occur or might not occur at all without the use of organic catalysts called enzymes.

- Enzymes speed up reactions dramatically, and in some instances they can increase the reaction rate more than 10 million times.

- Many enzymes require the presence of an inorganic ion or nonprotein organic molecule at the active site in order to work properly. These necessary ions or molecules are called cofactors.

Review and Reflect

1. **Roll the Rock** Diffusion may be thought of as an immense boulder at the top of the concentration gradient. A slight tap and the boulder spontaneously rolls downhill (concentration gradient). Active transport may be thought of as an immense boulder at the bottom of concentration gradient that molecular Sisyphus must roll uphill (concentration gradient). Explain this analogy from the perspective of cellular energy expended.

2. **Wrong Way Water** Diagram and explain the effect on the cells of a freshwater fish placed in saltwater and on the cells of a saltwater fish place in fresh water.

3. **A Deadly Drink** In his poem, *Rhyme of the Ancient Mariner*, Samuel Taylor Coleridge describes the plight of anyone cast adrift on the ocean.

   Water, water everywhere, and all the boards did shrink; water, water everywhere, nor any drop to drink.

   Explain why it is osmotic suicide for someone adrift on the ocean and dying of thirst to drink salt water.

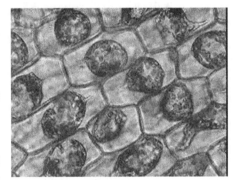

Figure 6.16

4. **Tone Up Your Tonicity** Examine the plant cells in **Figure 6.16**. Are these cells in a hypotonic, hypertonic, or isotonic solution? How can you tell?

5. **Too Bulky** Some macromolecules that cells need to get in or get out are just too large to pass through the plasma membrane. What's a cell to do?

6. **Packs a Punch** ATP has been described as the "spark-plug of life." Is this analogy biologically accurate? Explain.

7. **Explaining Enzymes** Your father is reading a story about cancer treatments that use enzyme inhibitors. He knows you are a biology student so he asks, "What are enzymes and what do they do?" How would you respond?

8. **Temperature Control** Examine **Figure 6.17**. What conclusions might be drawn from this graph?

9. **Vitamins in Action** What is the connection between vitamins, coenzymes, and vitamin-deficiency disorders?

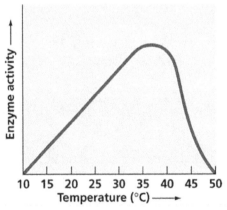

Figure 6.17

Create and Connect

1. *Under Pressure* Imagine your instructor has handed you a carrot and challenged you to use the carrot to demonstrate osmotic pressure. Diagram and explain how you would go about meeting the challenge.

2. *Target the Scourge* In targeted therapy to used to treat cancer, drugs target certain parts of the cell and the signals that are needed for cancer to develop and grow. Some targeted therapies make use of enzyme inhibitors that target enzymes such as DNA polymerase, an enzyme required for DNA replication. Write a short essay in which you describe how enzyme inhibitors function and explain why this method of cancer treatment is a viable option.

3. *Experimental Analysis* An experiment is designed to study the mechanism of sucrose uptake by plant cells. Cells are immersed in a sucrose solution, and the pH of the solution is monitored. Samples of the cells are taken at intervals, and their sucrose concentration is measured. After a decrease in pH of the solution to a steady, slightly acidic level, sucrose uptake begins.

   Propose a hypothesis for these results. What do you think would happen if an inhibitor of ATP reformation by the cell were added to the beaker once the pH was at a steady state?

CELL PROCESSES: RESPIRATION

The food we eat goes beyond its macronutrients of carbohydrates, fat, and protein. It's infor-mation. It interacts with and instructs our genome with every mouthful, changing genetic expression

—David Perlmutter

Introduction

Life is work. Living cells require transfusions of energy from outside sources to perform the many tasks required for their survival—pumping molecules across the plasma membrane, assembling polymers, replicating, and so on. The energy stored in organic molecules of food required by cells and thus organisms ultimately comes from the sun.

Photosynthesis generates oxygen and organic molecules that are used by the mitochondria of eukaryotic cells as fuel for cellular respiration. Respiration breaks the fuel down, generating ATP. The waste products of this type of respiration, carbon dioxide, and water, are then again available as the raw products for photosynthesis. Harvesting energy is the fundamental function of cellular respiration.

Cellular Respiration

Cellular respiration should not be confused with the term "respiration" as it is applied to the process of breathing. *Cellular respiration* is the process by which cells harvest energy by breaking down nutrient molecules produced by photosynthesis. The process of cellular respiration extracts energy from chemical bonds through the oxidation of organic compounds.

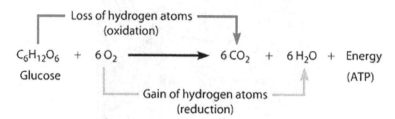

Figure 7.1 This equation indicates that the reactant molecules glucose and O2 disassociate, and their atoms regroup to form the products CO2 and H2O. In the process, glucose releases energy stored in the electrons (hydrogen) forming the covalent bonds. The cell stores this energy in the chemical bonds of ATP.

Cellular respiration involves the complete breakdown of glucose to carbon dioxide and water (**Figure 7.1**). This equation reveals that cellular respiration is an oxidation-reduction reaction. An atom that loses electrons is said to be *oxidized*, and an atom accepting electrons is said to be *reduced*. Oxidation reactions are often coupled with reduction reactions in living systems, and these paired reactions are called *redox reactions*. In the equation, glucose has been oxidized because it lost hydrogen atoms (electrons) and O_2 has been reduced because it gained hydrogen atoms (electrons). In general, organic molecules that have an abundance of hydrogen are excellent fuels because their bonds are a source of "hilltop" electrons, whose energy may be released as these electrons "roll" down an energy gradient when they are transferred to oxygen. The summary equation for respiration indicates that hydrogen is transferred from glucose to oxygen. What is not apparent, however, is that the energy state of the electron changes as hydrogen (with its electron) is transferred to oxygen. In respiration, the oxidation of glucose transfers electrons to a lower energy state, liberating energy that becomes available for ATP synthesis.

The main energy-yielding foods—carbohydrates and fats—are reservoirs of electrons associated with hydrogen. Only the barrier of activation energy holds back the flood of electrons to a lower energy state. Without this barrier, a food substance like glucose would combine almost instantaneously with O_2. Body temperature is not high enough to lower the barrier. Instead, when you swallow glucose, enzymes in your cells will lower the barrier of activation energy, allowing the sugar to be oxidized in a series of steps.

In the equation, the energy stored as the result of the redox of glucose is in the form of adenosine triphosphate (ATP). Each ATP contains only 1% of energy present in 1 glucose molecule. Cellular respiration cannot harvest all the energy present in glucose in a usable form. Usually, only about 40% of total energy in glucose is converted to ATP. The rest is lost as heat. By comparison, only between 20% and 30% of the energy within gasoline is converted to the motion of a car. The rest is lost as heat. When the chicks pictured in the chapter-opening photo breathe, they acquire oxygen, and when they feed on the insect, they acquire glucose. Both molecules enter the bloodstream and are carried to body cells, where cellular respiration occurs. Carbon dioxide and water are released as glucose breaks down in the mitochondria providing the energy for the formation of ATP molecules.

If the gasoline tank of a car explodes, it cannot drive a car very far. Cellular respiration does not oxidize glucose in a single explosive step either. Rather, glucose is broken down in a series of steps, each one catalyzed by an enzyme. In the equation, hydrogen atoms are not transferred directly to oxygen but instead are usually passed first to an electron carrier, a coenzyme called NAD^+ (nicotinamide adenine dinucleotide, a derivative of the vitamin niacin). NAD is well suited as an electron carrier because it easily cycles between being oxidized (NAD^+) and reduced (NADH). FAD (flavin adenine dinucleotide) is another coenzyme that serves as an electron carrier being reduced to become $FADH_2$.

How are the electrons extracted from glucose and stored as potential energy in NADH finally reach oxygen? Instead of occurring in one explosive reaction, cellular respiration uses an electron transport chain to break the fall of electrons to oxygen into several energy-releasing steps. An **electron transport chain** consists of a number of molecules, mostly proteins, built into the inner membrane of the mitochondria (cristae) of eukaryotic cells (**Figure 7.2**). During electron transfer, a build-up of hydrogen ions occurs. If these hydrogen ions are not removed, they will cause a large decrease in pH. Acid accumulation is avoided because oxygen reacts with the hydrogen ions to form water and in the process, oxygen is reduced. The oxygen in this situation comes from the air you breathe in. In summary, during cellular respiration, most electrons travel a "downhill" route: glucose→NADH and FAD→ electron transport chain→ oxygen.

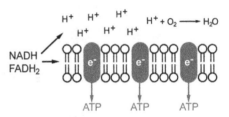

Figure 7.2 The carriers FAD and NAD+ bring in the hydrogen which separates to H+ and electrons (e-). The electrons pass from carrier to carrier and loose energy. These small amounts of released energy are used to synthesize ATP

Phases of Cellular Respiration

The harvesting of energy from glucose by cellular respiration is accomplished in four phases: glycolysis, Aceytl CoA formation, the citric acid cycle, and oxidative phosphorylation (**Figure 7.3**).

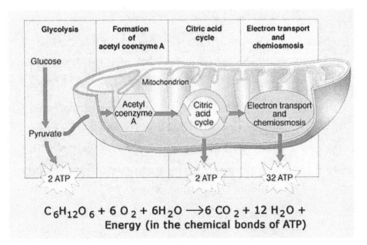

Figure 7.3 Overview of Cellular Respiration. Glycolysis in the cytoplasm produces pyruvate that enters mitochondria if oxygen is available. The formation of Aceytl CoA and the citric acid cycle that follow occur inside the mitochondria. Also, inside the mitochondria, the electron transport chain receives the electrons that were removed from glucose breakdown products. Each stage generates electrons from chemical breakdown and oxidation reactions. The theoretical yield per glucose is 36 to 38 ATP.

Glycolysis Glycolysis takes place in the cytosol and is the breakdown of C_6 (6-carbon) glucose to two C_3 (3-carbon) pyruvate molecules. Glycolysis is a series of ten steps. From an energy standpoint, glycolysis moves through two phases: the energy-investment phase and the energy-generation phase.

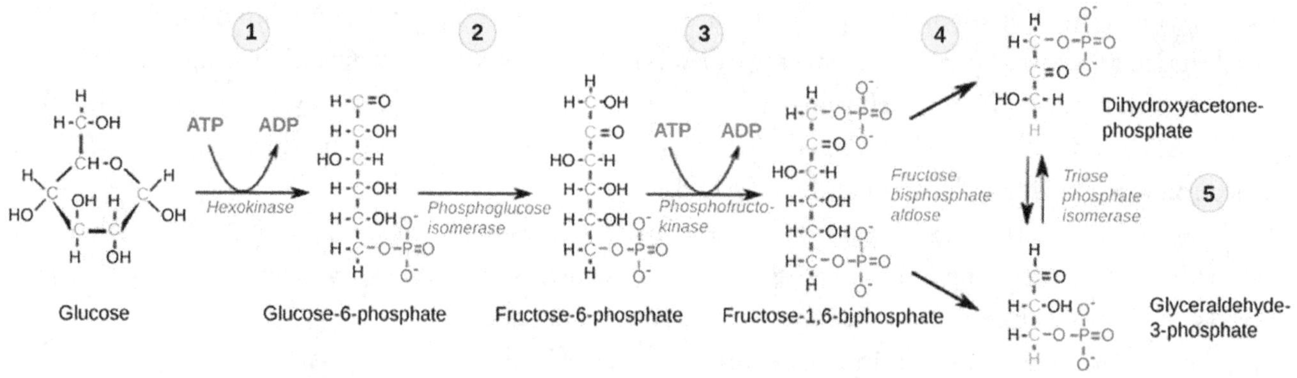

Energy-Investment Phase (**Figure 7.4**)

Step 1 The first step in glycolysis is catalyzed by hexokinase, an enzyme with broad specificity that catalyzes the phosphorylation of six-carbon sugars. Hexokinase phosphorylates glucose using ATP as the source of the phosphate, producing glucose-6-phosphate, a more reactive form of glucose.

Step 2 In the second step of glycolysis, an enzyme converts glucose-6-phosphate into one of its isomers, fructose-6-phosphate. This change from phosphoglucose to phosphofructose allows the eventual split of the sugar into two three-carbon molecules.

Step 3 The third step is the phosphorylation of fructose-6-phosphate, catalyzed by the enzyme phosphofructokinase. A second ATP molecule donates a high-energy phosphate to fructose-6-phosphate, producing fructose-1,6-bisphosphate.

Step 4 The newly-added high-energy phosphates further destabilize fructose-1,6-bisphosphate. The fourth step in glycolysis employs an enzyme, aldolase, to cleave 1,6-bisphosphate into two three-carbon isomers: dihydroxyacetone-phosphate and glyceraldehyde-3-phosphate.

Step 5 In the fifth step, an isomerase transforms the dihydroxyacetone-phosphate into its isomer, glyceraldehyde-3-phosphate. Thus, the pathway will continue with two C3 molecules of a single isomer. At this point in the pathway, there is a net investment of energy from two ATP molecules in the breakdown of one glucose molecule.

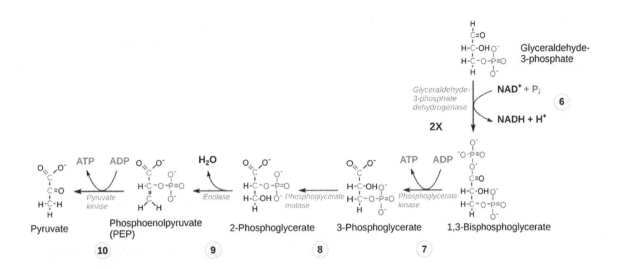

Energy-Generation Phase (**Figure 7.5**).So far, glycolysis has cost the cell two ATP molecules and produced two small, three-carbon sugar molecules. Both of these molecules will proceed through the second half of the pathway where sufficient energy will be extracted to pay back the two ATP molecules used as an initial investment while also producing a profit for the cell of two additional ATP molecules and two even higher-energy NADH molecules.

Step 6 The sixth step in glycolysis oxidizes the sugar (glyceraldehyde-3-phosphate), extracting high-energy electrons that are picked up by the electron carrier NAD⁺, producing NADH. The sugar is then phosphorylated by the addition of a second phosphate group, producing 1,3-bisphosphoglycerate. Note that the second phosphate group does not require another ATP molecule.

Step 7 In the seventh step, catalyzed by phosphoglycerate kinase, 1,3-bisphosphoglycerate donates a high-energy phosphate to ADP, forming one molecule of ATP. A carbonyl group on the 1,3-bisphosphoglycerate is oxidized to a carboxyl group, and 3-phosphoglycerate is formed.

Step 8 In the eighth step, the remaining phosphate group in 3-phosphoglycerate moves from the third carbon to the second carbon, producing 2-phosphoglycerate. The enzyme catalyzing this step is a mutase (isomerase).

Step 9 Enolase catalyzes the ninth step. This enzyme causes 2-phosphoglycerate to lose water from its structure through a dehydration synthesis reaction, resulting in the formation of a double bond that increases the potential energy in the remaining phosphate bond and produces phosphoenolpyruvate (PEP).

Step 10 The last step in glycolysis is catalyzed by the enzyme pyruvate kinase and results in the production of a second ATP molecule by phosphorylation and pyruvate.

Glucose breakdown requires an input of oxygen (**aerobic**) in the cytosol to keep the electron transport chain operating. However, some cells and organisms often find themselves in an oxygen-depleted environment (**anaerobic**). In anaerobic conditions, the pyruvate that forms from glycolysis is reduced to lactate (lactic acid) in animals and bacteria or alcohol in plants and yeasts in a process known as **fermentation** (**Figure 7.6**). Lactate is produced in human muscle cells when they become starved for oxygen during strenuous activity, and yeasts are examples of organisms that generate ethyl alcohol as a result of fermentation. Some bacteria produce chemicals of industrial importance—isopropanol, butyric acid, propionic acid, and acetic acid—through the process of fermentation.

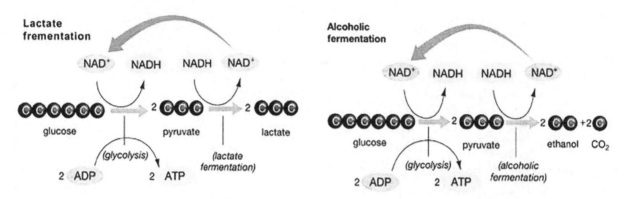

Figure 7.6 The Process of Fermentation. Fermentation consists of glycolysis followed by a reduction of pyruvate into lactate or alcohol.

Despite its low yield of only two ATP made by substrate-level ATP synthesis, lactate fermentation is essential in certain situations to some animals and animal tissue. Animals such as a hunting cheetah or a human sprinting to class because they are late use lactate fermentation for a rapid burst of energy. Also, when muscles are working vigorously over a short period of time, lactate fermentation provides them with ATP, even though oxygen is temporarily in limited supply.

Box 7.1
Fermentation and Food

Strolling down the aisles of any grocery store will reveal many products that produced through the action of fermentation such as bread, cheese and yogurt, soy sauce, pickles, and alcoholic products such as beer and wine. The use of fermentation by humans to produce food and beverages has been going on for centuries, long before we understood the biochemistry of it all (**Figure 7.7**).

Baker's yeast (*Saccharomyces cerevisiae*) is added to bread dough to cause it to rise. As the yeast ferment the sugar in the dough, they give off carbon dioxide that causes the dough to swell and ethyl alcohol (**Figure 7.8**). The ethyl alcohol evaporates as the bread bakes. The large holes or pockets you see in some bread, especially home-made bread were bubbles of carbon dioxide that formed as the bread was rising.

Figure 7.7 Paintings on the walls in the Tomb of Nakht, 18th dynasty, Thebes, Ancient Egypt depict grapes being crushed into juice to produce wine.

Yogurt, sour cream, and cheese are produced through the action of various lactate bacteria that cause milk to sour. Milk contains lactose that these bacteria use as a carbohydrate source for fermentation. Yogurt is made by adding lactate bacteria, such as *Streptococcus thermophilus* and *Lactobacillus bulgaricus* to milk and then incubating (warming) it to encourage the bacteria to convert the lactose. To make cheese, one starts with milk from a cow, goat, or sheep. The milk may be heated (pasteurized) or not heated (unpasteurized). Starter cultures of bacteria are added to the milk to begin the fermentation process. Specific bacteria are used to produce specific types of cheese. As the process proceeds, rennet containing the enzyme chymosin is added to the fermenting milk to coagulate (thicken) it into the curd that eventually becomes cheese (**Figure 7.9**).

Figure 7.8 Bread dough is often left to rise for some time before baking. Bread with yeast added is said to be leavened whereas bread baked with no yeast is said to be unleavened.

Figure 7.9 Adding rennet to fermenting milk causes the milk to thicken into cheese. Studies show humans have been eating cheese for over 8,000 years.

Brine cucumber pickles, sauerkraut, and kimchi are pickled vegetables produced by the action of acid-producing fermenting bacteria that can survive in high-salt environments. Soy sauce is traditionally made by adding mold to a combination of yeasts and fermenting bacteria to soybeans and wheat. The mold breaks down the starch in the soybeans supplying the fermenting microorganisms with sugar that they ferment into alcohol and organic acids.

Beer and wine are produced when yeasts ferment carbohydrates. Wine is produced when grape juice ferments and beer is produced when malt and hops ferments (**Figure 7.10**). A few specialized varieties of beer have a distinct taste because of the particular ingredients used in the fermentation process. Stronger alcoholic drinks, such as whiskey and vodka, require distillation to concentrate the alcohol content.

As we can see from these examples, fermentation is a biologically and economically important process that makes our lives and our diets tangy and tasty.

Figure 7.10 The frothy head on a glass of beer is escaping carbon dioxide bubbles that were formed during fermentation.

Aceytl CoA Formation Once pyruvate is produced through gly- colysis, it moves into a cell and the mitochondria in the cytosol. In the mitochondria, the C_3 pyruvate is converted into C_2 acetyl-CoA and CO_2 is given off (**Figure 7.11**). This is an oxidation reaction in which electrons are removed from pyruvate by NAD^+ and NADH is formed. One reaction occurs per pyruvate, so this reaction occurs twice per glucose molecule.

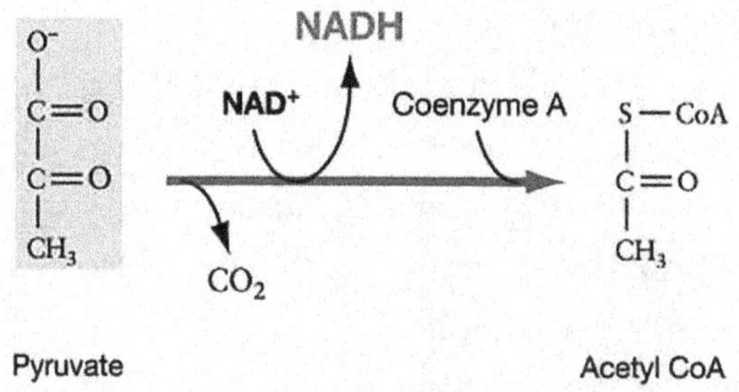

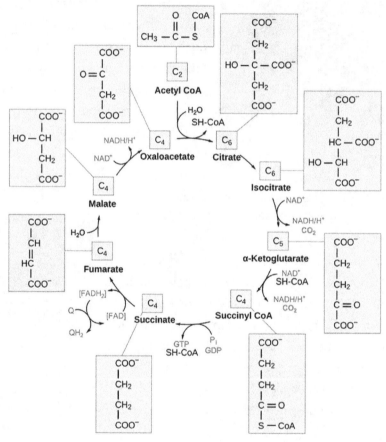

Figure 7.11

The CoA molecule will carry the acetyl group into the mitochondrial matrix and the citric acid cycle. In vertebrates, the CO_2 formed by this reaction freely diffuses out of cells and into the blood where it is carried to the lungs and exhaled.

Figure 7.12

Citric Acid Cycle The *citric acid cycle* (or *Kreb's cycle*) is a cyclical metabolic pathway located in the matrix of mitochondria (**Figure 7.12**). As with the conversion of pyruvate to acetyl-CoA, the citric acid cycle takes place in the matrix of the mitochondria. Unlike glycolysis, the citric acid cycle is a closed loop: the last part of the pathway regenerates the compound used in the first step. The eight steps of the cycle are a series of redox, dehydration, hydration, and decarboxylation reactions that produce two carbon dioxide molecules, one GTP/ATP, and reduced forms of NADH and FADH2. This is considered an aerobic pathway because the NADH and FADH2 produced must transfer their electrons to the next pathway in the system, which will use oxygen. If this transfer does not occur, the oxidation steps of the citric acid cycle also do not occur. Note that the citric acid cycle produces very little ATP directly and does not directly consume oxygen.

Step 1 In the first step of the citric acid cycle, acetyl-CoA joins with a four-carbon molecule, oxaloacetate, releasing the CoA group and forming a six-carbon molecule called citrate.

Step 2 In the second step, citrate is converted into its isomer, isocitrate. This is a two-step process, involving first the removal and then the addition of a water molecule.

Step 3 In the third step, isocitrate is oxidized and releases a molecule of carbon dioxide, leaving behind a five-carbon molecule—α-ketoglutarate. During this step, NAD^+ is reduced to form NADH. The enzyme catalyzing this step, **isocitrate dehydrogenase**, is important in regulating the speed of the citric acid cycle.

Step 4 The fourth step is similar to the third. In this case, it's α-ketoglutarate that's oxidized, reducing NAD^+ to NADH and releasing a molecule of carbon dioxide in the process. The remaining four-carbon molecule picks up Coenzyme A, forming the unstable compound succinyl-CoA. The enzyme catalyzing this step, **α-ketoglutarate dehydrogenase,** is also important in the regulation of the citric acid cycle.

Step 5 In step five, the CoA of succinyl-CoA is replaced by a phosphate group, which is then transferred to ADP to make ATP. In some cells, GPD (guanine diphosphate) is used instead of ATP forming GTP (guanine triphosphate) as a product. The four-carbon molecule produced in this step is called succinate.

Step 6 In step six, succinate is oxidized, forming another four-carbon molecule called fumarate. In this reaction, two hydrogen atoms—with their electrons—are transferred to FAD, producing $FADH_2$. The enzyme that carries out this step is embedded in the inner membrane of the mitochondrion, so that $FADH_2$ can transfer its electrons directly into the electron transport chain.

Step 7 In step seven, water is added to the four-carbon molecule fumarate, converting it into another four-carbon molecule called malate.

Step 8 In the last step of the citric acid cycle, oxaloacetate—the starting four-carbon compound—is regenerated by oxidation of malate. Another molecule of NAD^+ is reduced to NADH in the process.

Let's take a step back and do some accounting and trace the fate of the carbons that entered the citric acid cycle and counting the reduced electron carriers—NADH and $FADH_2$ produced. In a single turn of the cycle,

- Two carbons enter from acetyl-CoA, and two molecules of carbon dioxide are released.
- Three molecules of NADH and one molecule of $FADH_2$ are generated.
- One molecule of ATP or GDP is generated.

These figures are for one turn of the cycle, corresponding to one molecule of aceytl CoA. Each glucose produces two acetyl-CoA molecules, so we need to multiply these numbers by 2 if we want the per-glucose yield.

Two carbons from aceytl-CoA enter the citric acid cycle in each turn, and two carbon dioxide molecules are released. However, the carbon dioxide molecules don't contain carbon atoms from the aceytl-CoA that just entered the cycle. Instead, the carbons from aceytl-CoA are initially incorporated into the intermediates of aceytl-CoA and are released as carbon dioxide only during later turns. After enough turns, all the carbon atoms from the acetyl group of aceytl-CoA will be released as carbon dioxide.

The output of the citric acid cycle seems unimpressive. All that work for just one ATP or GTP? Although true that the citric acid cycle doesn't produce much ATP directly, it can make a lot of ATP indirectly by way of the NADH and FADH2 it generates. These electron carriers will connect with the last portion of cellular respiration, depositing their electrons into the electron transport chain to drive synthesis of ATP molecules through oxidative phosphorylation.

The electron transport chain is the final component of aerobic respiration and is the only part of glucose metabolism that uses atmospheric oxygen. Electron transport is a series of redox reactions that resemble a relay race. Electrons are passed rapidly from one component to the next to the endpoint of the chain, where the electrons reduce molecular oxygen, producing water. The electron transport chain is an aggregation of four complexes together with associated electron carriers that is present in multiple sites in the inner mitochondrial membrane (**Figure 7.13**).

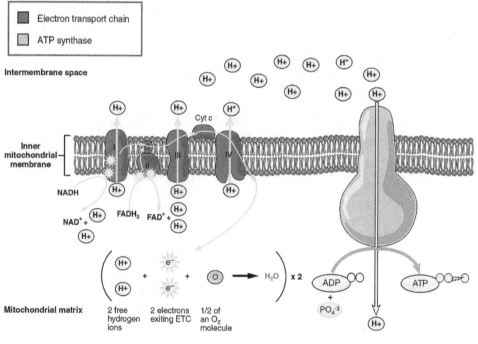

Figure 7.13

NADH and FADH2 move electrons to the transport chain. As electrons move from one protein complex to the other via redox reactions, energy is used to pump hydrogen ions (H+) from the matrix into the intermembrane space. As hydrogen ions flow down a concentration gradient from the intermembrane space into the mitochondrial matrix, ATP is synthesized by the enzyme ATP synthase. For every pair of electrons that enters by way of NADH, three ATP result. For every pair of electrons that enters by way of FADH2, two ATP result. ATP leaves the matrix through a channel protein.

The theoretical yield of energy per glucose molecule is 36 to 38 ATP molecules. However, we know that cells rarely ever achieve these theoretical values. Several factors can lower the ATP yield for each molecule of glucose entering the pathway:

- In some cells, NADH cannot cross mitochondrial membranes, but a "shuttle" mechanism allows its electrons to be delivered to the electron transport chain inside the mitochondria. The cost to the cell is one ATP for each NADH that is shuttled to the electron transport chain. This reduces the overall count of ATP produced as a result of glycolysis, in some cells, to four instead of six ATP.
- At times, cells need to expend energy to move ADP molecules and pyruvate into the cell and to establish gradients in the mitochondria.

Most estimates place the yield at around 30 ATP per glucose molecule, about 32 to 39 percent of the available energy in a glucose molecule. The rest of the energy is lost as heat.

Metabolic Pool

The total amount of a substance or group of substances that is available for participation in a specified metabolic reaction or pathway is referred to as the *metabolic pool* for that substance or group of substances.

In certain ways, a metabolic pathway is similar to a factory assembly line. Products are assembled from parts by workers who each perform a specific step in the manufacturing process. Enzymes of a cell are like workers on an assembly line; each is only responsible for a particular step in the assembly process. A lag period also occurs when a new factory is constructed, a time period before finished products begin to roll off the assembly line at a steady rate. This lag period partially results from the time needed to fill supply bins with the necessary parts. As you might imagine, when parts are not readily available, production is slow or stops. Molecular pools are somewhat analogous to the parts bins of a factory.

Key metabolic pathways routinely draw from pools of particular substates needed to synthesize or degrade larger molecules. Substrates like the end product of glycolysis, pyruvate, exist as a pool that is continuously affected by changes in cellular and environmental conditions (**Figure 7.14**).

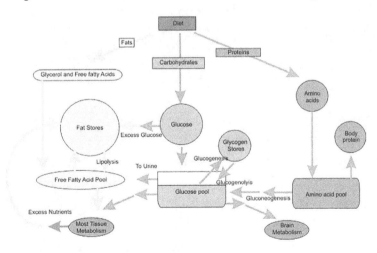

Metabolism Summary

Figure 7.14 Carbohydrates, fats, and proteins can be used as energy sources, and their monomers can enter degradative pathways at specific points. Catabolism produces molecules that can also be used for anabolism of other compounds.

In Summary

- Cellular respiration is the process by which cells harvest energy by breaking down nutrient molecules produced by photosynthesis.

- Oxidation reactions are often coupled with reduction reactions in living systems, and these paired reactions are called redox reactions.

- In respiration, the oxidation of glucose transfers electrons to a lower energy state, liberating energy that becomes available for ATP synthesis.

- The energy stored as the result of the redox of glucose is in the form of adenosine triphosphate (ATP).

- The harvesting of energy from glucose by cellular respiration is accomplished in four phases: glycolysis, Aceytl-CoA formation, the citric acid cycle, and oxidative phosphorylation.

- Glycolysis takes place in the cytosol and is the breakdown of C_6 (6-carbon) glucose to two C_3 (3-carbon) pyruvate molecules.

- Glycolysis is a series of ten steps. From an energy standpoint, glycolysis moves through two phases: the energy-investment phase and the energy-generation phase.

- In anaerobic conditions, the pyruvate that forms from glycolysis is reduced to lactate (lactic acid) in animals and bacteria or alcohol in plants and yeasts in a process known as fermentation.

- The eight steps of the citric acid cycle are a series of redox, dehydration, hydration, and decarboxylation reactions that produce two carbon dioxide molecules, one GTP/ATP, and reduced forms of NADH and FADH2.

- The electron transport chain is an aggregation of four complexes together with associated electron carriers that is present in multiple sites in the inner mitochondrial membrane (cristae).

- The theoretical yield of energy per glucose molecule is 36 to 38 ATP molecules. However, we know that cells rarely ever achieve these theoretical values.

- The total amount of a substance or group of substances that is available for participation in a specified metabolic reaction or pathway is referred to as the metabolic pool for that substance or group of substances.

- Key metabolic pathways routinely draw from pools of particular substates needed to synthesize or degrade larger molecules.

Review and Reflect

1. *Memory Crutch* Many students use mnemonics to help them remember facts and concept. A common mnemonic for remembering lists is to create an easily remembered acronym, or, taking each of the initial letters of the list members. For example, to remember the Great Lakes, think HOMES (Huron, Ontario, Michigan, Erie, and Superior). One mnemonic used when it comes to redox reactions is: L.E.O. the lion goes G.E.R. What does this mnemonic help you remember about redox reactions?

2. *Your Electrons or Mine?* If the following redox reaction occurred, which compound would be oxidized? Reduced?

$$C_4H_6O_5 + NAD^+ \rightarrow C_4H_4O_5 + NADH + H^+$$

3. ***Sugar Boom*** Oxygen is a very reactive gas that can explode under the right conditions. Why doesn't oxygen explode when it comes in contact with glucose? In other words, why don't the oxygen and glucose combine instantaneously?

4. ***Energy Conversion*** Your study buddy turns you and asks, "How can one molecule of glucose end up becoming 30+ ATP molecules?" Explain cellular respiration in general terms.

5. ***Invest or Generate?*** Compare and contrast the energy investment phase of glycolysis with the energy generation phase.

6. ***I Need Air*** Compare and contrast aerobic and anaerobic respiration.

7. ***Fermentation Fun*** Fermentation is important in many of the foods we eat and the beverages we drink. How does the production of fermentation products differ from that of other food products? What products of fermentation do you personally use on a regular basis?

8. ***Fermentation Fail*** Your little brother plans to take grape juice, stir in some table sugar, and then hide the concoction in the basement for a few months hoping it will turn into wine. Will this work? Explain

9. ***Off the Deep End*** Explain how the metabolic pool is like an assembly line in a factory.

Creative and Connect

1. ***Head of the Class*** You have been asked by the high school biology teacher of the local high school to present a lecture on cellular respiration. Prepare the lecture you would give including any diagrams you might present.

2. ***Truth in Advertising*** (**Figure 7.15**) Coenzyme Q (CoQ) is sold as a nutritional supplement. Many companies use a marketing slogan for CoQ that claim this product "promotes heart health." Considering the role of coenzyme Q, is this slogan accurate? Write your own slogan for this product using sound biochemical facts.

3. ***Mitochondria, Weight Loss, and Death*** In the 1930s, some physicians prescribed low doses of a compound called dinitrophenol (DNP) to help patients lose weight. After some patients died, this compound was considered unsafe to use for that purpose. DNP uncouples the chemiosmotic machinery by making the lipid bilayer of the inner mitochondria leaky to H^+. Explain how this could cause weight loss and death.

Figure 7.15

CELL CYCLE AND CELL DIVISION

Any living cell carries with it the experience of a billion years of experimentation by its ancestors.

—Max Ludwig Henning Delbruck

Introduction

Cells are packets of life. And because every living thing is composed of cells, the entities we call organisms are alive. The life force of an organism resides in its cells and these cells pass their life force on to other cells. Our understanding of this was first voiced by the German physician Rudolph Virchow in 1855 in a powerful axiom, *Omnis cellula e cellula* meaning "Every cell from a cell." Virchow further stated: "Where a cell exists, there must have been a preexisting cell, just as the animal arises from an animal and a plant only from a plant." Thus, the continuity of life is based on the replication of cells or cell division.

Cell Cycle

Humans (or any other complex animal that undergoes sexual reproduction) start life with a cellular bang. Amazingly it all begins with a single fertilized egg cell. Some 42 turns of the cell cycle later (a generational time of six to seven days per turn of the cycle) a human infant composed of approximately 2 trillion cells is born. Cell division is rapid during early embryonic development then slows with age eventually resulting in a human adult with approximately 200 trillion cells differentiated into over 200 different types of specialized cells working together in harmony. Approximately 40-50 billion replacement cells—about 2 million cells per second—are produced by an adult human body every 24 hours.

How is such a near-miraculous feat possible? The answer lies in the process of cell division (replication). Cell division is a method by which the genetic material (DNA) of a single cell (parent cell) is distributed equally and identically into two daughter cells. The replication of a cell with all its complexity, cannot occur merely by pinching in half. The most remarkable aspect of cell division is the precise fidelity with which the DNA is passed from one generation of cells to the next.

Cell division, however, is only one milestone in a cell's life. The wholeness of a cell's existence is not a linear expression of its internal activities, but rather a cyclic one—the cell cycle (**Figure 8.1**). The cell cycle is a progressive and repeating series of events in the life of a cell. The cell cycle is delineated into four distinct phases: G_1 phase, S phase, G_2 phase (collectively known as *Interphase*), and *M phase*. The end result of the turning of the cell cycle is the production of two daughter cells genetically identical to the original parent cell.

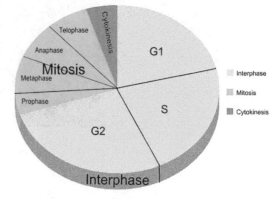

Figure 8.1

Table 8.1		
The Cell Cycle		
State	**Phase**	**Activity**
Interphase	*Gap 1* (G$_1$)	Also known as the growth phase because biosynthetic activity proceeds at an accelerated rate. This phase is also marked by the synthesis of certain enzymes and materials required during the S phase.
	Synthesis (S)	DNA replication occurs during this phase. During this phase, the amount of DNA in the cell has effectively doubled.
	Gap 2 (G$_2$)	During the gap between DNA synthesis and mitosis, increased biosynthetic activity again occurs, mainly involving the production of microtubules, which are required during the process of mitosis.
Cell division	*Mitosis* (M)	Cell growth stops at this stage, and cellular energy is focused on the orderly division into two identical daughter cells. M phase proceeds sequentially: *Prophase* *Metaphase* *Anaphase* *Telophase* *Cytokinesis*
Quiescent/ senescent	*Gap 0* (G$_0$)	A phase where the cell has left the cycle and is no longer dividing. Examples are neurons (nerve cells) and erythrocytes (red blood cells). G$_0$ cells usually do not reenter the cell cycle but instead, will carry out their function in the organism until they die.

Command and Control of Cell Division

Interphase is the phase of the cell cycle in which a typical cell spends most of its life (**Figure 8.2**). During this phase, the cell copies its DNA in preparation for mitosis (M). Interphase is the "daily living" or metabolic phase of the cell, in which the cell obtains nutrients and metabolizes them, grows, reads its DNA, and conducts other "normal" cell functions. This phase was formerly called the resting phase. However, interphase does not describe a cell that is merely resting; rather, the cell is actively living and preparing for later cell division, so the name was changed. A common misconception is that interphase is the first stage of mitosis. However, since mitosis is the division of the nucleus, prophase is actually the first stage.

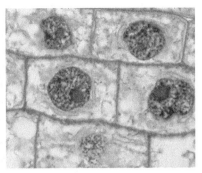

Figure 8.2 Plant Cells in Interphase. The nucleolus is clearly visible as is the thickening chromatin.

In interphase, the cell gets itself ready for mitosis or meiosis. Somatic cells, or normal diploid cells of the body, go through mitosis in order to reproduce themselves through cell division, whereas diploid germ cells (primary spermatocytes and primary oocytes) go through meiosis in order to create haploid gametes (sperm and ova) for the purpose of sexual reproduction. Chromosomes are copied.

What prompts a cell to leave interphase and begin the mitotic process? In eukaryotic cells, the passage of a cell through the cell cycle is controlled by genetically coded proteins in the cytoplasm, mainly cyclins and cyclin-dependent kinases.

Functioning much like chemical traffic signals, a series of checkpoints is built into the cell cycle. Each phase ends when a cellular checkpoint "approves" the accuracy of the stage's completion before proceeding to the next. The stages of interphase are:

G_1/S checkpoint This is the main checkpoint in the cell cycle. If cellular DNA is undamaged and healthy and necessary nutrients are present, proteins will initiate the S phase (DNA synthesis and replication). If conditions for DNA synthesis and replication are not favorable, the cell cycle will stop at this point.

G_2 checkpoint If DNA has properly replicated, mitosis will proceed. If the DNA is damaged and cannot be repaired, apoptosis (cell death) will occur.

M checkpoint Mitosis will not continue if the chromosomes are not properly aligned on the mitotic spindle.

There are great differences in the generational time of cells that compose different types of tissues. In some, one turn of the cell cycle may be measured in hours whereas in others it is measured in days, months, or even years. A particular human cells might undergo one division in 24 hours. Of this time, the M phase would occupy less than 1 hour whereas the S phase might occupy 10-12 hours, about half the cycle. The remainder of the time would be apportioned between the G_1 and G_2 phases.

In some cases, cells divide only in the early stages of development. Muscle cells stop dividing during the third month of fetal development with further growth dependent on the enlargement of fibers already present. Cells of the nervous system stop dividing early on in fetal development but persist throughout the life of the individual.

Collectively, cell replication in an animal is critical for growth, replacement of dying cells, and wound repair. However, each cell eventually leaves the cell cycle and enters into what cytologists call quiescent/senescent (Gap 0) leading to the death of the cell. There are thought to be several ways in which a cell that is perfectly healthy and supplied with nutrients can die. One process is known as *programmed apoptosis*. As contradictory as it sounds, there are times when cell death is necessary for the continued well-being of the individual. This system works primarily to kill sick cells. For example, cells that have become infected with a virus behave differently biochemically. Nearby cells can "sense" this unusual activity and signal for that infected cell to undergo apoptosis. Apoptosis in select tissues is also critical in the formation of the fingers and toes of vertebrates.

Cells die when their DNA accumulates enough mutations and damage that apoptosis is triggered. Finally, cell death can result because of the way DNA replicate. With every round of replication, the ends of the DNA molecules (**telomeres**) get shorter and shorter. As a result, critical genes will eventually be lost, and the cell will die.

Mitosis In eukaryotic organisms, cell division is a vital process for both the individual and species level. A type of cell division known as **mitosis** allows for the development (and in some cases regeneration), growth, and maintenance (cell replacement) of an individual animal whereas another type of cell division known as **meiosis** produces the gametes (sperm and egg) that allow for reproduction and the continuation of the species. Mitosis occurs in the somatic (body) cells of an animal, but meiosis occurs only in the gonads (testes and ovaries). Meiosis will be detailed in a later chapter.

In the 1880s, chromosomes were discovered. A few years later chromosomes were found to segregate by an orderly process into the daughter cells formed by cell division as well as into the gametes formed by the division of reproductive cells (germ cells). Three important regularities were observed about the chromosome complement (complete set of chromosomes) of plants and animals:

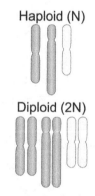

Haploid (N)

Diploid (2N)

1. During cell division, the chromosomes become visible and present as homologous pairs.
2. The number of chromosomes in somatic cells differs from the chromosome number of gametes.
3. Somatic cells, containing a full complement of chromosomes, are said to be **diploid** (2n) whereas gametes, containing a half set of chromosomes, are said to be **haploid** (n) (**Figure 8.3**). A diploid individual carries two allelic copies of each gene present in each pair of chromosomes. The chromosomes occur in pairs because one chromosome of each pair derives from the maternal parent and the other from the paternal parent of the organism.

Figure 8.3 Diploid cells have two homologous copies of each chromosome. Haploid cells have one of each chromosome

In multicellular organisms that develop from single cells, the presence of the diploid chromosome number in somatic cells and the haploid chromosome number of germ cells indicates that there are *two* processes of nuclear division that differ in their outcome. One these (mitosis) maintains the chromosome number; the other (meiosis) reduces the number by half.

Table 8.2
Diploid and Haploid Numbers for Select Organisms

Common Name	Scientific Name	Haploid Number	Diploid Number
Amoeba	*Amoeba proteus*	25	50
Fern	Phylum Pteridophyta	505	1010
Onion	*Allium cepa*	8	16
Fruit fly	*Drosophila melanogaster*	4	8
Earthworm	Class Oligochaeta	18	36
Cow (domestic)	*Bos taurus*	30	60
Chimpanzee	*Pan troglodytes*	24	48
Dog (domestic)	*Canis familiaris*	39	78
Horse	*Equus caballus*	32	64
Human	*Homo sapiens*	23	46

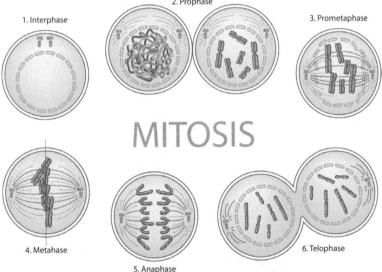

Figure 8.4 Phases of Mitosis

Through the process of mitosis, a single somatic parent cell duplicates its diploid genome exactly and precisely then partitions a complete and identical diploid genome into each of two new daughter cells (**Figure 8.4**). Mitosis is conventionally divided into four stages: prophase, metaphase, anaphase, and telophase.

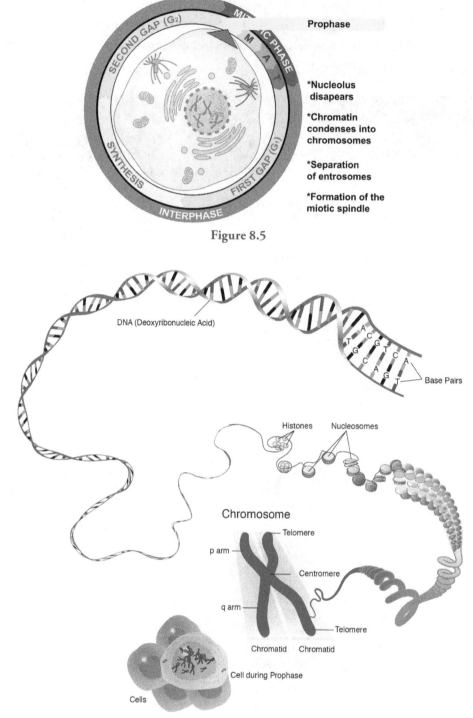

Prophase

*Nucleolus disapears

*Chromatin condenses into chromosomes

*Separation of entrosomes

*Formation of the miotic spindle

Figure 8.5

Figure 8.6 Construction of a Chromosome.

(**Figure 8.5**) **Prophase** In the early stages of prophase the chromatin is thread-like and not visible with a light microscope. As prophase proceeds, the chromatin begins to coil and then supercoil until it becomes visible as chromosomes (**Figure 8.6**). At this juncture, each chromosome is longitudinally double consisting of two closely associated subunits called chromatids. The chromatids in a pair are held together in a specific region of the chromosome called the **centromere**. At the end of prophase, the nucleoli disappear, and the nuclear membrane abruptly disintegrates, and the chromatids move freely in the cytoplasm.

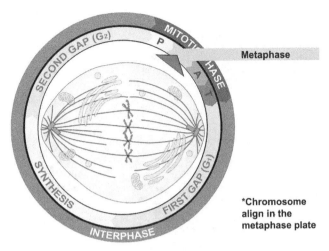

Figure 8.7

(**Figure 8.7**) **Metaphase** As the cell transitions into metaphase, the mitotic spindle forms as the centrioles move to opposite poles of the cell (*prometaphase*). The spindle is a football-shaped array of fibers consisting mainly of microtubules. A structure known as a **kinetochore** forms on the centromere of each chromatid pair. Spindle fibers then attach to each kinetochore and the chromatids align on the metaphase plate, an imaginary plane equidistance from each spindle pole. Proper chromatid alignment is an important cell cycle control checkpoint at metaphase in both mitosis and meiosis. In fact, proper chromatid alignment is so critical that mitosis will stop moving forward if the alignment is not perfect.

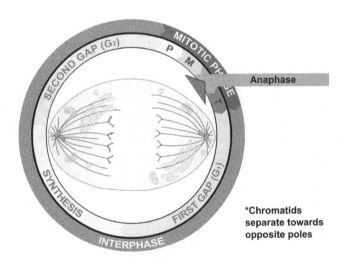

Figure 8.8

(**Figure 8.8**) **Anaphase** The centromeres of each chromatid pair separate and the two sister chromatids move toward opposite poles of the spindle as the fibers shorten. Once the centromeres separate, each sister chromatid is regarded as a separate chromosome in its own right. At the completion of anaphase, two identical sets of chromosomes lie near opposite poles of the spindle. Each chromosome set contains the same number of chromosomes and the same genes that were present in the pre*Synthesis* (S) interphase nucleus.

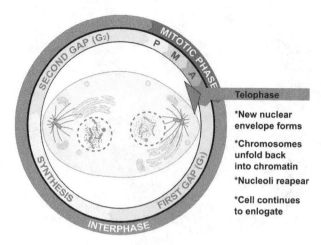

Figure 8.9

(**Figure 8.9**) **Telophase** As mitosis winds down, the nuclear membrane reforms, the chromosomes begin to uncoil, and the spindle disappears. Finally, in a process called *cytokinesis*, the original cell is cleaved into two new cells. The process of cytokinesis differs somewhat between plant and animal cells (**Figure 8.10**).

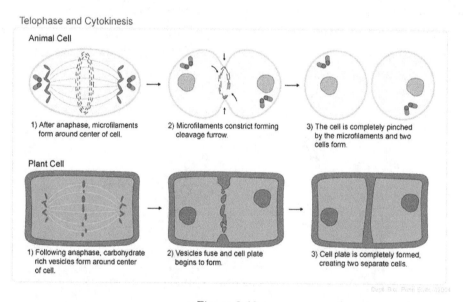

Figure 8.10

Functions of Mitosis

Growth and repair in both plants and animals are dependent on mitosis. Mitosis is responsible for the growth from a single fertilized egg cell to an adult human containing hundreds of billions of cells or a giant sequoia tree nearly 300 feet high and over 25 feet in diameter.

In some parts of body (skin and digestive tract), cells are constantly sloughed off and replaced by new ones. New cells are formed by mitosis and so are exact copies of the cells being replaced. In like manner, red blood cells have short lifespan (only about four months) and new red blood cells are formed by mitosis. Some organisms can use mitosis to regenerate lost parts. For example, starfish can regenerate lost arms through mitosis. Other organisms can replicate through mitosis. *Hydra* form asexual buds, and in the case of single-celled microorganisms such as bacteria, asexual reproduction via mitosis reproduces the entire organism.

Box 8.1
Stem Cells

In the beginning, there is the stem cell; it is the origin of an organism's life.

—Stewart Sell

Stem cells are mother cells that have the potential to become any type of cell in the body. One of the main characteristics of stem cells is their ability to self-renew or multiply while maintaining the potential to develop into other types of cells. Stem cells can become cells of the blood, heart, bones, skin, muscles, brain, etc. There are different sources of stem cells, but all types of stem cells have the same capacity to develop into multiple types of cells.

Types of stem cells

Pluripotent Stem Cells (PS cells) These possess the capacity to divide for long periods and retain their ability to make all cell types within the organism. The best-known type of pluripotent stem cell is the one present in embryos that help babies grow within the womb. These are termed embryonic stem cells. These cells form at the blastocyst stage of development. A blastocyst is a hollow ball of cells that is smaller than a pinhead. The embryonic stem cells lie within this ball of cells. Recent research has enabled scientists to derive pluripotent cells from adult human skin cells. These are termed induced pluripotent stem cells or iPS cells.

Fetal (embryonic)stem cells These are obtained from tissues of a developing human fetus. These cells have some characteristics of the tissues they are taken from. For example, those taken from fetal muscles can make only muscle cells. These are also called progenitor cells.

Adult stem cells These are obtained from some tissues of the adult body. The most commonly used example is the bone marrow. Bone marrow is a rich source of stem cells that can be used to treat some blood diseases and cancers.

Scientists first studied the potential of stem cells in mouse embryos over two decades ago. Over years of research, they discovered the properties of these stem cells in 1998. They found methods to isolate stem cells from human embryos and grow the cells in the laboratory. Early studies utilized embryos created for infertility purposes through in vitro fertilization procedures and when they were no longer needed for that purpose. The use required the voluntary donation of the embryos by the owners.

Stem cell research is improving by leaps and bounds. These may soon become the basis for treating diseases such as Parkinson's disease, diabetes, heart failure, cerebral palsy, heart disease and host of other chronic ailments. Stem cells may also be used for screening new drugs and toxins and understanding birth defects without subjecting human volunteers to the toxins and drugs.

Cell cycle checkpoints normally ensure that DNA replication and mitosis occur only when conditions are favorable, and all processes are proceeding correctly. However, unregulated cell division can lead to cancer.

Mutations in genes that encode cell cycle proteins can lead to unregulated growth resulting in tumor formation and ultimately the invasion of surrounding tissue by cancerous cells.

Cancer cells do not heed the normal signals that regulate the cell cycle. Cancer is a cellular growth that occurs when cells divide uncontrollably. In culture, cancer cells do not stop dividing when growth factors are depleted. A logical hypothesis is that cancer cells do not need growth factors in their culture medium to grow and divide. They may make a required growth factor themselves, or they may have an abnormality in the signaling pathway that conveys the growth factor's signal to the cell cycle control system even in the absence of that factor. Another possibility is that the cell cycle control system itself is faulty.

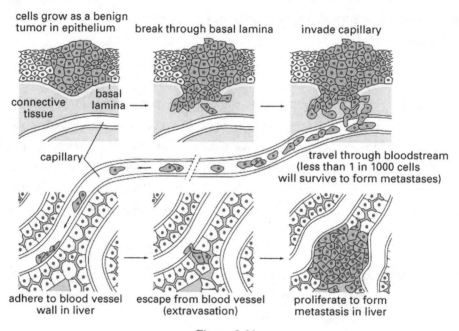

Figure 8.11

Although cancers vary greatly, they usually follow a common multistep progression (**Figure 8.11**). Most cancers begin as an abnormal cell growth that is *benign* or not cancerous and does not grow larger. However, additional mutations may occur causing the abnormal cells to fail to respond to inhibiting signals that control the cell cycle. When this occurs, the growth becomes *malignant,* meaning that it is cancerous and possesses the ability to spread (**metastasize**).

Cancer can take years or decades to develop. A mutation in a cell may cause it to become precancerous, but regulatory and inhibiting processes within the body often prevent the cell from becoming cancerous. Although cancers vary greatly, cancer cells share common characteristics:

- *Cancer cells lack differentiation.* Cancer cells look distinctly abnormal. Cancer cells are not specialized nor do they contribute to the functioning of tissue.
- *Cancer cells have abnormal nuclei.* The nuclei of cancer cells are enlarged and may contain an abnormal number of chromosomes and/or some of the chromosomes have deleted portions.
- *Cancer cells do not undergo apoptosis.* Normally, cells with damaged DNA undergo apoptosis or programmed cell death. Cancer cells do not respond to the signals that trigger apoptosis and thus refuse to die.

- *Cancer cells form tumors.* Normal cells adhere to a substratum or to each other and when they do so they stop dividing. Cancer cells, however, do not exhibit contact inhibition and pile on top of one another growing into a multi-layer lump known as a *tumor.*
- *Cancer cells undergo metastasis.* Benign tumors are usually contained within a capsule and cannot invade adjacent tissue. Additional mutations may result in the tumor becoming malignant and breaking through the capsule to spread (metastisize) cancer cells through the blood and lymph. This process is known as *metastasis.*

Each cancer is thought to start from one abnormal mutated cell. What seems to happen is that certain vital genes which control how cells divide and multiply are damaged or altered. This makes the cell abnormal. If the abnormal cell survives, through a series of additional mutations it may eventually begin to multiply out of control into a cancerous (malignant) tumor.

We all have a risk of developing cancer. Many cancers seem to develop for no apparent reason. However, certain risk factors are known to increase the chance that one or more of your cells will become abnormal and lead to cancer. Risk factors include the following:

- **Chemical carcinogens** A *carcinogen* is something that can damage a cell and make it more likely to turn into a cancerous cell. As a general rule, the more the exposure to a carcinogen, the greater the risk. Well-known examples include:
  - *Tobacco* If you smoke, you are more likely to develop cancer of the lung, mouth, throat, esophagus, bladder, and pancreas. Smoking is thought to cause about 1 in 4 of all cancers. About 1 in 10 smokers die from lung cancer. The heavier you smoke, the greater the risk. If you stop smoking, your risk goes down considerably.
  - *Workplace chemicals* such as asbestos, benzene, formaldehyde, etc. If you have worked with these without protection, you have an increased risk of developing certain cancers.
- **Age** The older you become, the more likely that you will develop a cancer. This is probably due to an accumulation of damage to cells over time. Also, the body's defenses and resistance against abnormal cells may become less good as you become older. For example, the ability to repair damaged cells, and the immune system which may destroy abnormal cells, may become less efficient with age. Eventually one damaged cell may manage to survive and multiply out of control into cancer. Most cancers develop in older people.
- **Lifestyle factors** Diet and other lifestyle factors (and, as mentioned, smoking) can increase or decrease the risk of developing cancer.
- **Radiation** is a carcinogen. For example, exposure to radioactive materials and nuclear fallout can increase the risk of leukemia and other cancers. Too much sun exposure and sunburn (radiation from UVA and UVB) increase your risk of developing skin cancer. The larger the dose of radiation, the greater the risk of developing cancer. But note: the risk from small doses, such as from a single X-ray test, is very small.
- **Infection** Some germs (viruses and bacteria) are linked to certain cancers. For example, people with persistent infection with the hepatitis B virus or the hepatitis C virus have an increased risk of developing cancer of the liver. Another example is the link between the human papillomavirus (HPV) and cervical cancer. Most (possibly all) women who develop cervical cancer have been

infected with a strain (subtype) of HPV at some point in their life. Another example is that a germ (bacterium) called *Helicobacter pylori* is linked to stomach cancer. However, most viruses and viral infections are not linked to cancer.

In Summary

- The continuity of life is based on the replication of cells or cell division.
- Cell division is a method by which the genetic material (DNA) of a single cell (parent cell) is distributed equally and identically into two daughter cells.
- The wholeness of a cell's existence is not a linear expression of its internal activities, but rather a cyclic one—the cell cycle.
- The cell cycle is delineated into four distinct phases: G_1 *phase*, *S phase*, G_2 *phase* (collectively known as *Interphase*), and *M phase*.
- Interphase is the "daily living" or metabolic phase of the cell, in which the cell obtains nutrients and metabolizes them, grows, reads its DNA, and conducts other "normal" cell functions.
- In eukaryotic cells, the passage of a cell through the cell cycle is controlled by genetically coded proteins in the cytosol.
- The interphase checkpoints are: G_1/S checkpoint, G_2 checkpoint, and M checkpoint.
- Cell death is known as apoptosis.
- Somatic cells, containing a full complement of chromosomes, are said to be diploid (2n) whereas gametes, containing a half set of chromosomes, are said to be haploid (n).
- There are *two* processes of nuclear division that differ in their outcome. One these (mitosis) maintains the chromosome number; the other (meiosis) reduces the number by half.
- Through the process of mitosis, a single somatic parent cell duplicates its diploid genome exactly and precisely then partitions a complete and identical diploid genome into each of two new daughter cells.
- Mitosis is conventionally divided into four stages: prophase, metaphase, anaphase, and telophase.
- Growth and repair in both plants and animals are dependent on mitosis.
- Mutations in genes that encode cell cycle proteins can lead to unregulated growth resulting in tumor formation and ultimately the invasion of surrounding tissue by cancerous cells.

Review and Reflect

1. *Cells are Life* What did Rudolph Virchow mean when he said, *Omnis cellula e cellula* meaning "Every cell from a cell."?
2. *Condense It Down* Explain the process of cell division in one sentence.
3. *Count Chromosomes* Human somatic cells have 46 chromosomes (23 pairs). Why don't cells replicate just by dividing right down the middle each time?
4. *Ride the Cycle* Why is the cell cycle a circular process rather than a linear progression?
5. *Traffic Stop* What are interphase checkpoints?
6. *Compare and Contrast* Compare and contrast cytokinesis in plant cells and animal cells.
7. *Drop Dead* How can a cell that a cell that is perfectly healthy and supplied with nutrients die?

8. *Draw It* Diagram and label the four stages of mitosis. Explain what happens at each stage.

9. *Language Barrier* Explain cancer to an alien scientist whose race is not afflicted by the cell disorder.

Create and Connect

1. *Henrietta Lacks* On February 8, 1951, cervical cancer cells were taken from Henrietta Lacks, a cancer patient who would later die of the disease. Those cells referred to as HeLa cells, were developed into an immortal cell line that became a mainstay of cell research. Write a short report on HeLa cells.

Guidelines

A. Format your report in the following manner:
   - *Title page* (including your name and lab section)
   - *Body of the Report* (include pictures, charts, tables, etc. here as appropriate). The body of the report should be a minimum of two pages long—double-spaced, 1 inch margins all around with 12 pt font.
   - *Literature Cited* A minimum of two references required. Only <u>one</u> reference may be from an online site. The *Literature Cited* page should be a separate page from the body of the report, and it should be the last page of the report. Do NOT use your textbook as a reference.

B. The instructor may provide additional details and further instructions.

2. *Stem Cells—A Position Paper* At present, research on stem cells and the funding for such research are hotly debated topics. Most commonly, this controversy focuses on embryonic stem cells and the ethics and morality of research involving the development, use, and destruction of human embryos. Delve into this controversy and then write a position paper on the use of stem cells in research.

Guidelines

A. Write a position paper, NOT an opinion paper. Defend your position with as many facts, figures, quotes, and pertinent information as possible.

B. Develop a position for the use of each type of stem cell in research.

C. Your work will be evaluated not on the "correctness" of your position but the quality of the defense of your position.

MEIOSIS AND SEXUAL REPRODUCTION

Your genetics is not your destiny.

—George M. Church

Introduction

There will never be another you, at least not genetically. Why? Because more than 70 trillion different genetic combinations are possible when sperm meets egg. The cellular process called meiosis that produces that sperm and egg determines not only inherited similarity but also variation. This is why we all somewhat resemble our parents and siblings but are not them exactly. Each of us is a unique genetic entity that has never been seen before on this planet and statistically will never be seen again.

Variations on a Theme

Have you ever been told by family friends that you have your father's nose or your mother's eyes? Family resemblance is evident in all of us, but sons and daughters are not identical copies of either their parents or their siblings (except in the case of identical twins).

Of course, parents do not literally give their children hair color, shape of nose, eye color, or any other physical trait. What then, is actually inherited? To answer that question we must turn to genetics. **Genetics** is the study of heredity, or how traits of living things are passed from one generation to the next. And what genetics tells us is that parents endow their offspring with coded information in the form of hereditary units called **genes**. The entire set of genes you inherited from your parents is referred to as your **genome** that in totality encompasses some 30,000 genes.

Your genome is written in the language of DNA, a polymer of four different nucleotides you learned about in a previous chapter. DNA and associated proteins called histones are found in the nucleus of a cell in interphase as thread-like structures known as *chromatin*. Genes are sections of DNA of the chromatin thread that code for specific proteins. The exact location of a gene within the DNA is called the gene's *locus*, and the numbers of nucleotides that make up a gene vary; some genes are long stretches of DNA whereas others are short. The DNA inside the nucleus of each cell is the blueprint for not only that cell in particular but for the entire organism in general.

Chemical messengers carry a copy of a specific gene from the DNA in the nucleus out into the cytosol where they attach to a ribosome. Chemical transfer molecules assemble free amino acids into a specific protein based on the gene code of the chemical messenger. It is the proteins formed from the genetic code of genes that determine your traits. A more specific detailing of this process will be encountered in an upcoming chapter.

Meiosis

In the human body, we find two general kinds of cells based on the number of chromosomes they contain during metaphase: somatic cells and gametes. **Somatic cells** are the cells that make up your entire physical body whereas **gametes** are specialized reproductive cells found in the testes of males (sperm) and the ovaries of females (eggs). In plants, the male gamete is pollen and the female gamete is an egg located in the ovary.

The Belgian cytologist Pierre-Joseph van Benden (1809-1894) was one of the first to observe that gametes have fewer chromosomes than do somatic cells. When studying the roundworm *Ascaris*, he noticed that the sperm and egg each contained only two chromosomes whereas the somatic cells always have four chromosomes. As we learned earlier, somatic cells are said to be *2n* or diploid because they contain a full set of chromosomes whereas gametes are said to be *n* or haploid because they contain only a half set of chromosomes. The 2n and n numbers vary between species (Table 8.2). In humans, somatic cells are diploid (2n or 46) containing 46 (23 pairs) chromosomes—22 pair of **autosomes** and 1 pair of sex chromosomes whereas gametes are haploid (n or 23) containing 23 chromosomes (**Figure 9.1**).

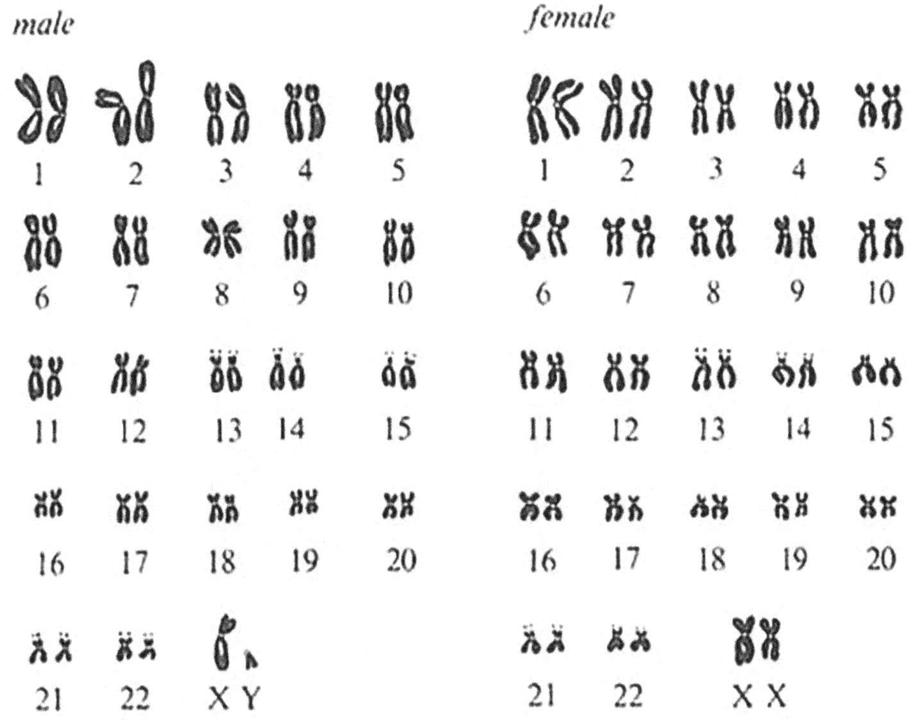

male   female

Figure 9.1 Chromosome pictures showing homologous chromosomes are called karyotypes. Pictured is the diploid karyotype of the homologous chromosomes for both a human male and female. The 23rd pair is the sex chromosomes that function in determining the gender of an individual. Because of their shape, the sex chromosomes in men are denoted as XY whereas in women they are dubbed XX. A haploid karyotype would show only 1 of each diploid chromosome pair.

Differences in chromosome numbers between somatic cells and gametes begs the question..why? As we know, the 46 somatic cell chromosomes are precisely duplicated during mitosis and each daughter cell that results from mitosis receives an exact copy of the 46 chromosomes contained by the original parent cell. To function properly, each somatic cell must contain 23 pairs (2n) of **homologous chromosomes**. If cells merely pulled apart during mitosis, each daughter cell would only get 23 (n) chromosomes. That cell would quickly die. Precisely maintaining 23 pairs of homologous chromosomes in each somatic cell is critical for the proper functioning and survival of not only each cell but of the whole organism of which those cells are a part.

One member of a homologous pair of chromosomes was inherited from the father and the other member of the pair from the mother when the haploid sperm and egg fused together. Each member of the pair carries genes controlling the same inherited characteristics. Genes work in pairs and groups so if the gene for eye color is situated at a particular locus on one of the homologous chromosomes, then the homolog of that gene will be located at the same locus on the other chromosome of the pair. Pairs of genes each on a different homologous chromosome are called **alleles**.

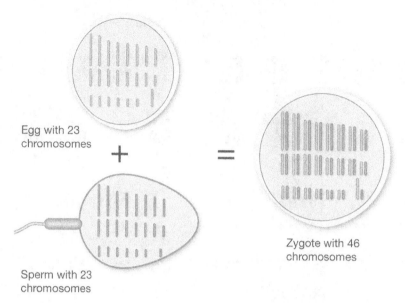

Egg with 23
chromosomes

Sperm with 23
chromosomes

Zygote with 46
chromosomes

Figure 9.2 The 23 pairs of homologous chromosomes in the zygote that results from fertilization are the blueprint for the construction of a new human body.

Gametes (sperm and egg in animals and pollen and egg in plants) each contain half (n) the chromosomes contained in somatic cells. Why? Why not just fuse two somatic cells together to form a fertilized egg cell? Remember-exact chromosome numbers are critical to cell survival. If we fuse two 2n somatic cells, we will get one 4n cell or a cell in humans with 92 chromosomes. Too many chromosomes are as devastating to a cell as too few, and a 4n cell would quickly die. Therefore, gametes must always be *n* so that when two gametes fuse, the resulting fertilized cell is *2n*. In humans, when the 23 chromosomes of a sperm (22 autosomes and 1 sex chromosome) fuse (fertilize) an egg cell with 23 chromosomes (22 autosomes and 1 sex chromosome), a **zygote** is formed that contains 23 pairs of homologous chromosomes (44 autosomes and 2 sex chromosomes (sexosomes) (**Figure 9.2**). Note that from the male 2n male karyotype shown in Figure 9.1, two types of sperm (n) can be produced: 22 autosomes and an X sex chromosome or 22 autosomes and a Y sex chromosome. The 2n female karyotype depicted in Figure 9.1 shows that only one type of egg (n) can be produced: 22 autosomes and an X sex chromosome. Therefore, the sperm of the father determines the gender of the child in humans. If a 22 autosome X sexochrome sperm fertilizes a 22 autosome X sexochrome egg, you have a female zygote (22 pairs of autosomes and 2 XX sexochromes). However, if a 22 autosome Y sexochrome fertilizes a 22 autosome X sexochrome egg, you have a male zygote (22 pairs of autosomes and an X and a Y sexchromes (**Figure 9.3**).

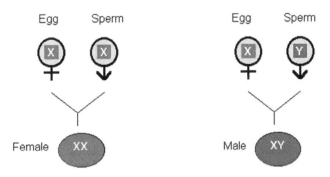

Figure 9.3

Thus, we realize that the *2n* number must be faithfully maintained in somatic cells as the *n* number must be faithfully maintained in gametes. As we learned earlier, the *2n* number of somatic cells is maintained through the process of mitosis. The *n* number of gametes, however, is maintained by a process known as *meiosis*. While the purpose of mitosis is to maintain *2n*, the purpose of meiosis is to reduce *2n* to *n* and whereas mitosis is cell replication, meiosis is reduction division.

In humans, meiosis occurs in the testes of males and the ovaries of females. Meiosis consists of two successive nuclear divisions: Meiosis I and Meiosis II.

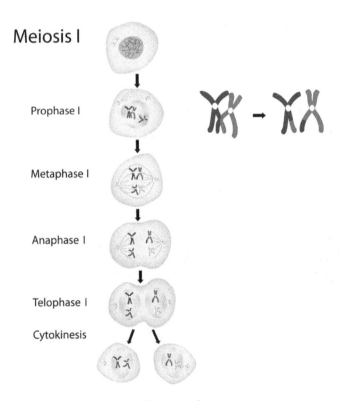

Figure 9.4

Meiosis I (**Figure 9.4**) Meiosis I separates homologous chromosomes, producing two haploid (n) cells. As a result, meiosis I is referred to as a reductional division. Even though in human cells during meiosis I each cell contains 46 **chromatids**, they are considered as being *n*, with 23 chromosomes

Meiosis II

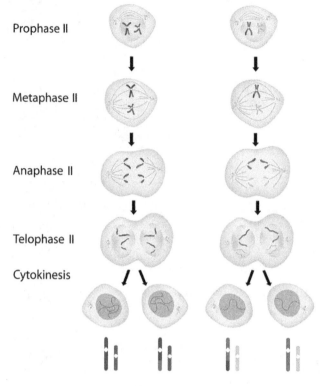

Prophase II

Metaphase II

Anaphase II

Telophase II

Cytokinesis

Figure 9.5

Meiosis II (**Figure 9.5**) Meiosis II is the second stage of the meiotic process, also known as equational division. Although the process is similar to mitosis, the genetic results are fundamentally different. The end result of Meiosis II is the production of four haploid cells (n) from the two haploid cells resulting from Meiosis I.

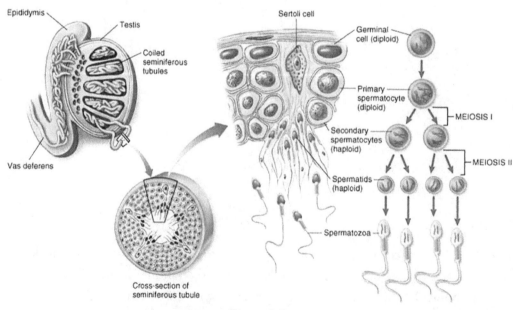

Figure 9.6

In the seminiferous tubules of the testes, a **germinal cell** becomes four haploid cells each known as a *spermatid*. The spermatids eventually mature into *spermatozoa* (sperm) in a process known as **spermatogenesis** (**Figure 9.6**). In the ovaries, a germinal cell matures into an *ovum* (egg) and three *polar bodies* that eventually die and disappear in a process called **oogenesis** (**Figure 9.7**).

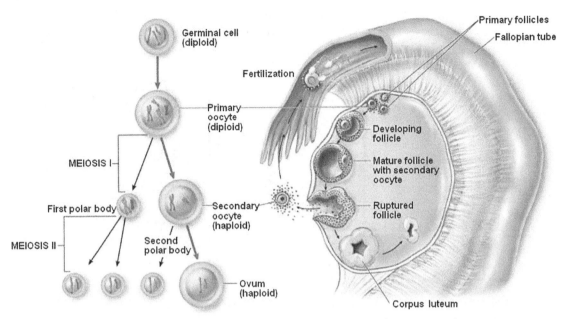

Figure 9.7 The process of oogenesis and the fertilization of an ovum in humans.

Table 9.1 *Meiosis vs. Mitosis*		
	Meiosis	**Mitosis**
End Result	Normally four cells, each with half the number of chromosomes as the parent cell	Two cells, having the same number of chromosomes as the parent cell
Function	Sexual reproduction, production of gametes	Cellular replication, growth, and repair
Happens in	Gonads	Most somatic cells
Genetically same as parent cell?	No	Nearly always
Crossing over?	Yes in Prophase I	Sometimes
Pairing of homologous chromosomes	Yes	No
Cytokinesis	Occurs twice, once in Telophase I and again in Telophase II	Occurs once in Telophase
Centromeres split	Occurs in Anaphase II	Occurs in Anaphase
Time required	Days or even weeks	12 to 24 hours

Life Cycles

Not all organisms reproduce sexually. Prokaryotes (bacteria and archaea) and some single-celled eukaryotic protists undergo **asexual reproduction** through a process called **binary fission** ("split in two"). Binary fission results in the reproduction of a living cell by dividing the cell into two cells, each with the potential to grow to the size of the original. An individual that reproduces asexually gives rise to a **clone**, a group of genetically identical individuals. Genetic differences can occasionally arise in asexually reproducing organisms as the result of mutations.

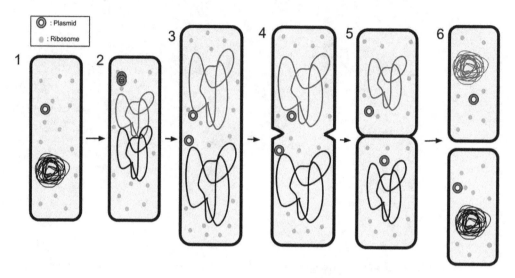

Figure 9.8 Binary Fission in Bacteria. (1) Before binary fission begins, the DNA is tightly coiled. (2) The DNA of the bacterium has replicated. (3) The DNA is pulled to separate poles of the bacterium as it increases in size to prepare for splitting. (4) The growth of a new cell wall begins the separation of the bacterium. (5) The new cell wall fully develops, resulting in the complete fission of the bacterium. (6) The new daughter cells have tightly coiled DNA, ribosomes, and plasmids.

The process of binary fission in bacteria involves a series of steps (**Figure 9.8**). Eukaryotic protists undergo multiple fission in which the nucleus of the parent cell divides several times by mitosis producing several nuclei. The cytoplasm then separates, creating multiple daughter cells.

In **sexual reproduction**, two parents give rise to offspring that have unique combinations of genes. In contrast to a clone, offspring of sexual reproduction vary genetically from their siblings and both parents. They are variations on a common theme of familial resemblance, not exact genetic copies. A *life cycle* is the generation-to-generation sequence of stages in the reproductive history of an organism, from conception to production of its own offspring.

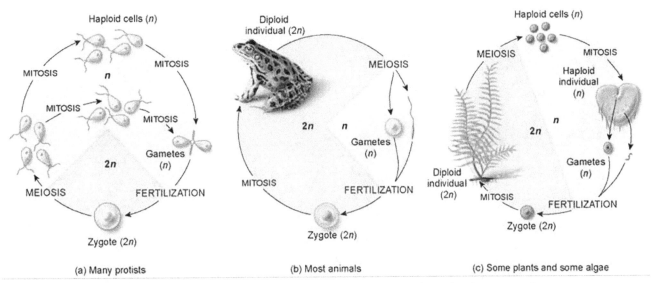

Figure 9.9 Sexual Life Styles. The common feature of all three cycles is the alternation of meiosis and fertilization, key events that contribute to genetic variation among offspring.

In sexually-reproducing organisms, the alternation of meiosis and fertilization is common. However, the timing of these two events in the life cycle varies, depending on the species. These variations can be grouped into three main types of life cycles (**Figure 9.9**). In the type that occurs in humans and most other animals, gametes are the only haploid cells (Figure 9.9b). Plants and some species of algae exhibit a second type of life cycle called alternation of generations (Figure 9.9c). This type has both diploid and haploid stages that are multi-cellular but that greatly differ from one another structurally.

A third type of life cycle occurs in most fungi and some protists, including some algae (Figure 9.9a). After gametes fuse and form a diploid zygote, meiosis occurs without a multicellular diploid offspring developing. Meiosis does not produce gametes but haploid cells that then divide by mitosis and give rise to either unicellular descendants or a haploid multicellular adult organism. Subsequently, the haploid organism carries out further mitosis, producing the cells that develop into gametes. The only diploid stage found in these species is the single-celled zygote.

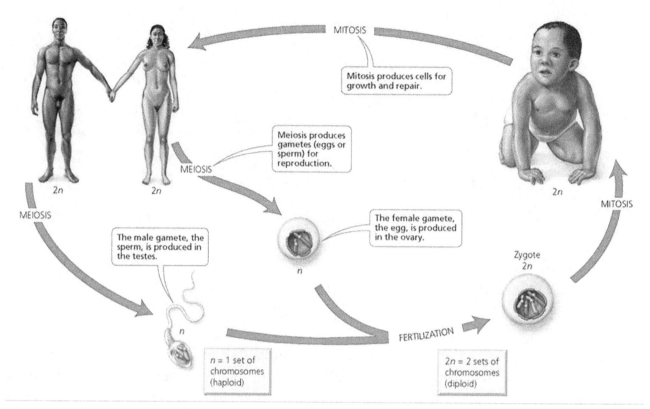

MITOSIS

Mitosis produces cells for growth and repair.

Meiosis produces gametes (eggs or sperm) for reproduction.

MEIOSIS

2n

2n

MEIOSIS

The male gamete, the sperm, is produced in the testes.

The female gamete, the egg, is produced in the ovary.

n

2n

MITOSIS

Zygote
2n

n

FERTILIZATION

n = 1 set of
chromosomes
(haploid)

2n = 2 sets of
chromosomes
(diploid)

Figure 9.10 The Human Life Cycle. In each generation, the number of chromosome sets doubles at fertilization but is halved during meiosis.

The human life cycle begins when a haploid sperm from the father fertilizes the haploid egg of the mother (**Figure 9.10**). The resulting fertilized egg (zygote) is diploid because it contains two haploid sets of chromosomes bearing genes representing the maternal and paternal family lines. Mitosis of the zygote and its descendant cells generates all the somatic cells of the body.

In general, the steps of the human life cycle are typical of many sexually reproducing animals. In fact, the processes of fertilization and meiosis are the hallmarks of sexual reproduction in animals, plants, fungi, and protists. Fertilization and meiosis alternate in sexual life cycles, maintaining a constant number of chromosomes from one generation to the next in each species and introducing genetic variation at each turn of the cycle.

Box 9.1
Variation Drives Evolution

Genetic variation is a measure of the genetic differences that exist within a population. The genetic variation of an entire species is often called genetic diversity. Genetic variations are the differences in DNA segments or genes between individuals, and each variation of a gene is called an allele. For example, a population with many different alleles at a single chromosome locus has a high amount of genetic variation.

Darwin recognized that population evolves through the differential reproductive success of its variant members. Those individuals best suited to their local environment leave the most offspring thereby transmitting their genes. Thus, natural selection results in the accumulation of genetic variations favored by the environment.

Genetic variation is caused by:

- **Mutation.** Mutation is the source of genetic variations. Mutations change and alter the structure and function of genes.

- **Random mating and fertilization** between members of the same species (sexual reproduction).

- **Crossing over** (recombination) between chromatids of homologous chromosomes. Crossing over is the exchange of genetic material between two non-sister chromatids during meiosis (**Figure 9.11**). Crossing over also accounts for genetic variation, because due to the swapping of genetic material during crossing over, the chromatids held together by the centromere are no longer identical. So, when the chromosomes go on to meiosis II and separate, some of the daughter cells receive daughter chromosomes with recombined alleles. Due to this genetic recombination, the offspring have a different set of alleles and genes than their parents do. In the diagram, genes B and b are crossed over with each other, making the resulting recombinants after meiosis Ab, AB, ab, and aB.

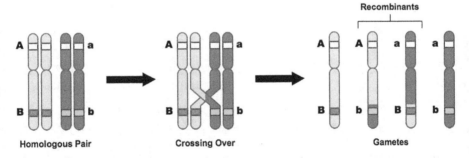

Figure 9.11 Crossing over occurs between prophase 1 and metaphase 1 and is the process where homologous chromosomes pair up with each other and exchange different segments of their genetic material to form recombinant chromosomes.

- **Independent assortment** during meiosis. In Meiosis I members of homologous pairs separate during anaphase. The separation of each pair is random with respect to all the other pairs. For each pair, there is a maternal and a paternal chromosome which came from the mother and father respectively. There is no mechanism to cause maternal or paternal chromosomes to either associate with or repel one another. Therefore the orientation of each chromosome pairs is random with respect to other chromosome pairs. This allows the offspring to be much more genetically variable than they would be if all the paternal and maternal chromosomes were inherited as a unit.

Figure 9.12 Genetic Variation in the Human Face. An enormous amount of phenotypic variation exists in the human face. This phenotypic variation is due at least partly to genetic variation within the human population.

Meiosis and sexual reproduction then shuffle the variations throughout the population (**Figure 9.12**). Variation allows some individuals within a population to adapt to the changing environment. Because natural selection acts directly only on phenotypes, more genetic variation within a population usually enables more phenotypic variation.

Some new alleles increase an organism's ability to survive and reproduce, which then ensures the survival of the allele in the population. Other new alleles may be immediately detrimental (such as a malformed oxygen-carrying protein), and organisms carrying these new mutations will die out. Neutral alleles are neither selected for nor against and usually remain in the population. Genetic variation is advantageous because it enables some individuals and, therefore, a population, to survive despite a changing environment.

Genetic variation is essential for natural selection because natural selection can only increase or decrease frequency of alleles that already exist in the population. In short, genetic variation drives evolution by means of natural selection.

Genetic Variation

In humans who have 23 pairs of chromosomes, the possible chromosomal combinations in the gametes is a staggering 2^{23} or 8,388,608. The chromosomes donated by the parents are combined, and in humans, this means that there are potentially $(2^{23})^2$, or 70,358,744,000,000 different chromosome combinations in the zygote. This number assumes that there was no crossing over between the nonsister chromatids before independent assortment. If a single crossing over event occurs, then $(4^{23})^2$, or 4,951,760,200,000,000,000,000,000,000 genetically different zygotes are possible for every couple. Keep in mind that crossing over can occur several times in each chromosome! As we said at the beginning, there will never be another you genetically.

In Summary

- Each of us is a unique genetic entity that has never been seen before on this planet and statistically will never be seen again.
- Genetics is the study of heredity, or how traits of living things are passed from one generation to the next.

- Parents endow their offspring with coded information in the form of hereditary units called genes. The entire set of genes you inherited from your parents is referred to as your genome that in totality encompasses some 30,000 genes.
- DNA and associated proteins called histones are found in the nucleus of a cell in interphase as thread-like structures known as chromatin.
- Genes are sections of DNA of the chromatin thread that code for specific proteins.
- The exact location of a gene within the DNA is called the gene's locus.
- The DNA inside the nucleus of each cell is the blueprint for not only that cell in particular but for the entire organism in general.
- Somatic cells are the cells that make up your entire physical body whereas gametes are specialized reproductive cells found in the testes of males (sperm) and the ovaries of females (eggs).
- In humans, somatic cells are diploid (2n or 46) containing 46 (23 pairs) chromosomes—22 pair of autosomes and 1 pair of sex chromosomes whereas gametes are haploid (n or 23) containing 23 chromosomes.
- To function properly, each somatic cell must contain 23 pairs (2n) of homologous chromosomes.
- Pairs of genes each on a different homologous chromosome are called alleles.
- The *n* number of gametes, however, is maintained by a process known as meiosis.
- In humans, meiosis occurs in the testes of males and the ovaries of females. Meiosis consists of two successive nuclear divisions: Meiosis I and Meiosis II.
- Prokaryotes (bacteria and archaea) and some single-celled eukaryotic protists undergo asexual reproduction through a process called binary fission.
- A life cycle is the generation-to-generation sequence of stages in the reproductive history of an organism, from conception to production of its own offspring.

Review and Reflect

1. ***Truly Unique*** Is the statement, *Genetically there will never be another you*, biologically accurate? Explain

2. ***Define It*** Imagine you have been contracted to write a biological dictionary. What would your entry be for the word *gene*?

3. ***Keep Defining*** Continue working on your biological dictionary by defining the words *diploid, haploid,* and *alleles*. Your definition should include an explanation as to why diploid cells are said to be *2n* whereas haploid cells are said to be *n*.

4. ***Keen on Karyotypes*** Study the karyotype shown in **Figure 9.13**. Is this the karyotype of a human? Defend your answer. Is this a karyotype of a male or female? Defend your answer.

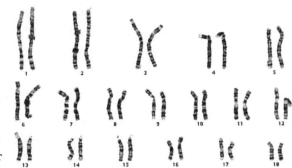

Figure 9.13

5. **Compare and Contrast** Compare and contrast mitosis and meiosis. How are they alike? How are they different?

6. **Determining Gender** Your friends are having a spirited discussion on human sexuality. One claims it is the father that determines the gender of a child while another claims it is the mother. How would you settle the disagreement?

7. **Compare and Contrast** Compare and contrast asexual reproduction with sexual reproduction.

8. **Sign of the Season** Some organisms such as *Daphnia* often undergo asexual reproduction at one time of year but sexual reproduction at another time. Explain how this is a useful adaptation for organisms to alternate between asexual and sexual reproduction.

9. **Drive On** Explain how genetic variation drives evolution by natural selection.

Create and Connect

1. **Asexual Diversity** The bdelloid rotifers (**Figure 9.14**) apparently have not reproduced sexually throughout the 40 million years of their evolutionary history. However, these rotifers do manage to acquire some genetic diversity. Investigate how an asexually reproducing organism such as these rotifers can acquire genetic diversity, and write a short essay explaining this phenomenon.

2. **Prove It** You have learned in this chapter that variation occurs among a population of living things but can you prove it? Design an experiment to prove variation occurs in any species of your choice except humans. Human experimentation is quite strictly regulated.

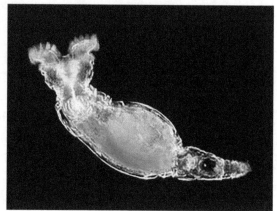

Figure 9.14

Guidelines

A. Following the tenets of a well-constructed experiment, your design should include the following components in order:
- The *Problem Question*. State exactly what problem you will be attempting to solve.
- Your *Hypothesis*. Although this is a fictitious experiment, word your hypothesis as realistically as possible.
- *Methods and Materials*. Explain exactly what you will do in your experiment including the materials necessary to accomplish the task. Be specific, take nothing for granted, and do not expect people to read your mind as they read your work.
- *Collecting and Analyzing Data*. Explain what type(s) of data will be collected, and what statistical tests might be performed on that data. It is not necessary to concoct fictitious data or imaginary observations.

B. Your instructor may provide additional details or further instructions.

CHAPTER 10

MENDELIAN GENETICS

When two plants, constantly different in one or several traits, are crossed, the traits they have in common are transmitted unchanged to the hybrids and their progeny, as numerous experiments have proven; a pair of differing traits, on the other hand, are united in the hybrid to form a new trait, which usually is subject to changes in the hybrids' progeny

—Gregor Mendel

Introduction

The end of the 20th century and the dawn of the 21st has seen the rise of advances in biotechnology that have allowed to us to better heal the world, feed the world, and fuel the world. Amazingly, all of this technology and biology originated from a humble garden on the grounds of St. Thomas's abbey in Brno, Czech Republic (formerly Moravia, Austria-Hungary) (**Figure 10.1**). In this quiet garden amid rows of garden peas an Augustinian friar named Gregor Johann Mendel would deduce the principles of heredity and establish the science of genetics.

Figure 10.1 St. Thomas's abbey in Brno. Mendel's experimental garden occupied about 5 acres of abbey land."

Rise of Genetics

The roots of present-day genetics can be traced to antiquity. The genetic power of selection and hybridization were undoubtedly employed by prehistoric plant growers and those who raised livestock, even though they did not understand the genetic principles involved. Babylonian stone tablets that are 6,000 years old have been interpreted as showing the pedigrees of several successive generations of horses, thus suggesting a conscious effort toward breeding improvement.

In later eras, Hippocrates, Aristotle, and other Greek philosopher scientists speculated on genetic precepts, but what little truth their speculation held was overshadowed by what we now know were many glaring errors. Stories of unusual hybrids were invented by the Greeks and repeated with additional imaginative flourishes by Pliny and other writers of their times. The giraffe was supposedly a hybrid between the camel and the leopard whereas an ostrich was imagined to result from the mating of the camel and the sparrow, and so on.

In the seventeenth and eighteenth centuries as the fledgling fields of *cytology* (study of cells) and *embryology* (study of the development of embryos) began to emerge, scientists struggled to explain and understand how traits were passed from parent to offspring and the biological mechanisms involved in this transfer.

Around the time of the Civil War in the United States, the peaceful gardens of a monastery in Brno in the Czech Republic a revolution of another sort—a revolution in genetics—was brewing as an unassuming Augustinian monk named Gregor Johann Mendel (**Figure 10.2**) quietly labored to understand the inheritance of traits in the common garden pea (*Pisum sativum*). Between 1856 and 1863 Mendel cultivated and genetically tested some 29,000 pea plants. His ground-breaking research was published in 1866. Unfortunately, his work was so revolutionary that biologists were not able to fully comprehend the significance of it and it was quickly forgotten.

Figure 10.2 Gregor Johann Mendel (1822-1884).

In 1900, Mendel's work was rediscovered and finally recognized for the monumental leap forward in genetic knowledge it represented. The first part of the 20th century saw a flurry of genetic revelations as discoveries piled one on the other. In the second half of the 20th century, researchers came to realize that the nucleus of the cell is the site of genetic information storage and that this information is stored on long strands of a double helix molecule known as deoxyribose nucleic acid or DNA. Short segments of this nuclear DNA, now known as genes, are the "factors" that Mendel worked so diligently to understand.

Mendel's work opened a door that led to a quantum leap forward in our understanding of heredity. Building on Mendel's cornerstone, pivotal advances over the last fifty years have revolutionized genetics, including the discovery of the molecular structure of DNA in 1953 by James Watson and Francis Crick, deciphering the genetic code, the development of recombinant DNA technology, and the mapping of the human genome. Armed with this knowledge, humankind currently finds itself immersed in the throes of the next great biological revolution—the biotechnology revolution.

Although Mendel's ground-breaking work dealt with plants, we now know that all life forms—plant, animal, and microbe—are shaped and regulated by DNA thus making genetics a powerful unifying force for understanding all aspects of biology. Modern biologists use the power of genetics to understand the form and function of individual organisms, to detail evolutionary relationships between species, to discern the intricacies of the growth and workings of populations, and to fathom the diversity-producing mechanisms of evolution.

Gregor Johann Mendel

In Mendel's time during the 1800s, the explanation for inheritance was the "blending" hypothesis. Most plant and animal breeders acknowledged that both parents contributed equally to a new individual. They mistakenly reasoned that parents of contrasting appearance always produced offspring of intermediate appearance. This concept, called the *blending concept of inheritance*, held that a cross between plants with red flowers and plants with white flowers would yield offspring that produced pink flowers. Were this hypothesis true, over many generations, a freely mating population would give rise to a uniform population of individuals with decreasing variation, something we don't see in nature. Also, the blending hypothesis fails to explain how traits often skip a generation only to reappear in later generations.

Johann Mendel was born to the farm, and as a child he worked as a gardener and studied beekeeping. As an adolescent, Mendel overcame financial hardship and illness to excel in high school and later, at the Olmutz Philosophical Institute.

In 1843 at age 21 he joined the Augustinian friars because it would allow him to obtain a free education and pursue a "life of the mind." He was given the name Gregor when he joined the friars. Mendel considered becoming a teacher but failed the necessary examination. In 1851, he left the abbey to pursue the study of physics and mathematics at the University of Vienna.

At the university, Mendel was strongly influenced by two professors. One was the physicist Christian Doppler, who encouraged experimentation and trained Mendel to use mathematics to explain natural phenomena. The other was botanist Franz Unger, who aroused Mendel's interest in the causes of variations in plants. After two years at the university, Mendel returned to the abbey and was assigned to teach at a local school. In this setting, Mendel was influenced not only teachers at the school who were enthusiastic about scientific research but also his fellow monks who shared a long-standing fascination with the breeding of plants.

Around 1857, Mendel began breeding garden peas in the abbey's experimental garden to study inheritance in plants. This was the serendipitous union of the right experimenter working with the right experimental subject being in the right place at the right time. Mendel proved to be a very careful and deliberate scientist who followed the scientific method very closely and kept very detailed and accurate notes and records. As important, his mathematical background allowed him to apply statistical methodology and the laws of probability to his findings.

Mendel's test organism, the garden pea, was a perfect choice for many reasons. The seeds were cheap and easy to obtain, came in a great number of varieties, produced a large number of offspring, and had a short generation time. Furthermore, pea plants will both self-fertilize, and cross-fertilize allowing their pollination process to be manually manipulated, but most importantly, the genome of the garden pea contains few enough genes that Mendel could track the inheritance pattern of the pea using only paper, pencil, and his analytical mind. This was critical as other researchers of his time attempting to understand the inheritance of sheep and insects failed because the genomes of those creatures are confusingly large and produce many traits that are subjective and not easily discerned.

Mendel chose 22 varieties of peas for his experiments because these peas clearly displayed only seven distinct traits (**Figure 10.3**). In genetic parlance, a heritable feature that varies among individuals, such as flower color or stem length is called a **character**. Each variant for a character, such as purple or white flower or tall or short stem is called a **trait**.

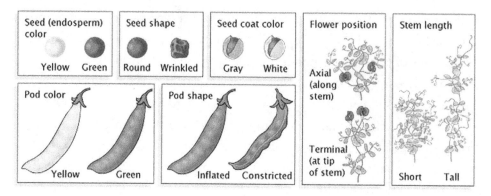

Figure 10.3

Mendel's Methods

Mendel chose to track only those characters that occurred in two distinct, alternative forms, such as purple and white flower color. He worked for two years to make sure that he started his experiments with varieties that, over many generations of self-pollination, produced only the same traits as the parent plant. Such plants are said to be *true-breeding*. For example, a plant with yellow seeds is true-breeding if the seeds produced by **self-pollination** in successive generations always give rise to plants that also produce yellow seeds.

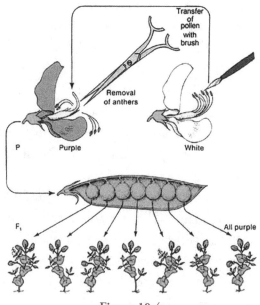

Figure 10.4

Once Mendel had established pure varieties, he began to cross breed them. This raises the question—how do you cross breed plants? Because pea plants self-fertilize, they have both male and female reproductive organs in the same flower. If Mendel wanted to cross a purple-flowered pea plant with a white-flowered pea plant, he would remove the male anthers containing the stamens (pollen sacs) from the purple flower so it could not self-fertilize by its own pollen. He would then transfer the pollen from the white flower (male) to the stigma of the purple flower (female) affecting **cross-pollination (Figure 10.4)**.

A mating, or crossing, of two true-breeding varieties is called **hybridization**. Thus, the 2n seeds that developed from the purple-white flower hybridization contain half a set of genes (n) from the male white flower and half a set of genes (n) from the female purple flower. To insure accuracy and verify results,

Mendel performed **reciprocal crosses**. That is, he reversed the gender of the flowers. The reciprocal cross of the purple-white hybridization detailed above would be to transfer the pollen from the purple flower to the female parts of the white flower. Mendel concluded it made no difference which way the cross is made.

In any genetic cross, the original parent types are referred to as the *P generation* (parental generation), and their hybrid offspring are the *F₁ generation* (first filial generation from the Latin word *filial* meaning "son."). In human terms, you and any siblings are the F_1 generation of your parents (P). Allowing the F_1 hybrids to self-pollinate or to cross-pollinate with other F_1 hybrids produces an *F₂ generation*. If you or your siblings have children, they will be the F_2 to your parents (P). Fortunately, Mendel carried out his crosses to the F_2 generation because the basic pattern of inheritance would have eluded him had he stopped with the F_1 generation. After eight years of time and over 10,000 crosses, Mendel's quantitative analysis of those F_2 plants allowed him to deduce two fundamental principles of heredity that have come to be called the law of segregation and the law of independent assortment.

Law of Segregation

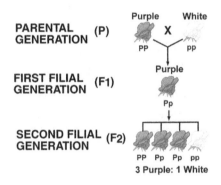

Figure 10.5

If the blending theory of inheritance were correct, the F_1 hybrids from a purple-white flower cross should have been pale purple, a trait intermediate between those of the P generation. However, those are not the results Mendel observed, and the results he did get puzzled him. When true breeding purple and white flowers are crossed, the F_1 generation is all purple. What had happened to the white trait? Had it somehow been lost? When Mendel allowed the F_1 plants to self-pollinate and planted their seeds, the white-flower trait reappeared in the F_2 generation (**Figure 10.5**), and the trait appeared with mathematical regularity.

Mendel used very large sample sizes and kept accurate records of his result: 705 of the F_2 plants had purple flowers, and 224 had white flowers, a ratio of approximately three purple to one white. Mendel reasoned that the "heritable factor" for white flower did not disappear in the F_1 plants but was somehow hidden or masked, when the purple flower factor was present. Mendel dubbed the purple flower color a **dominant trait**, and the white flower color a **recessive trait**.

Mendel established the same pattern of dominance in the six other traits of garden peas (**Figure 10.6**). From these experimental results Mendel deduced the law of segregation. Because Mendel knew nothing of DNA and genes, in the discussion that follows we will use modern terms instead of some of the

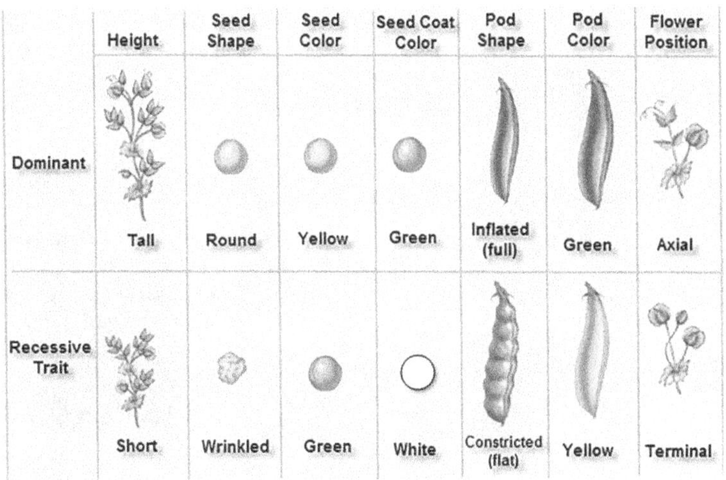

	Height	Seed Shape	Seed Color	Seed Coat Color	Pod Shape	Pod Color	Flower Position
Dominant	Tall	Round	Yellow	Green	Inflated (full)	Green	Axial
Recessive Trait	Short	Wrinkled	Green	White	Constricted (flat)	Yellow	Terminal

Figure 10.6 Results of Mendel's F1 crosses for seven characters in pea plants.

terms used by Mendel (For example, "gene" instead of Mendel's "inheritable factor."). Mendel's law of segregation is constructed of four related concepts:

Concept 1: *Alternative versions of genes for variations in inherited characters.* Geneticists know that the are alleles (gene pair) for flower color in pea plants, for example, exists in two versions, one for purple flowers and one for white flowers. We can relate this to the genes on DNA. The purple-flower allele and the white-flower allele are two DNA sequence variations possible at the flower-color locus on two homologous chromosomes. One of these alleles allows for the synthesis of purple pigment and one does not.

Concept 2: *Each organism inherits one gene of an allele pair from each parent.* In the case of flower color in peas, an F_1 individual could inherit the following possible combinations: PP, PW, WP, or WW.

Concept 3: *If the two alleles at a locus differ, then one, the dominant allele, determines the organism's appearance (phenotype)whereas the other, the recessive allele, has no noticeable effect on the organism's appearance.* Mendel's discovery of gene dominance clearly shows that any time you have the P gene, you get a purple flower—PP, PW or WP combinations will always yield a purple flower because the P (purple gene) is dominant. The only way a flower can be white in color is to have alleles that are WW.

It is important to understand that the physical appearance of an organism (called **phenotype**) is not always indicative of its genetic makeup (called **genotype**). For example, the presence of a single P will give you a purple flower phenotype. However, the purple flower phenotype can be achieved through one of three possible genotypes: PP, PW, or WP. Genotype determines phenotype whereas phenotype only hints at genotype.

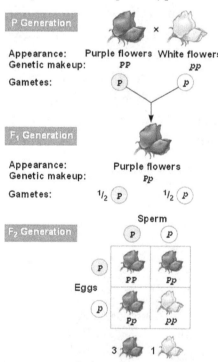

Figure 10.7 The Punnett square shown here is a diagrammatic device for predicting the allele composition of offspring from a cross between individuals of known genetic makeup. Capital letters symbolize the dominant allele and lower case letter symbolize the recessive allele. The nondominant allele is represented by the lower case of the letters used to represent the dominant allele.

Concept 4: This fourth and final concept is Mendel's law of segregation that states: *The two alleles for a heritable character segregate (separate from each other) during gamete formation and end up in different gametes.* The Law of Segregation states that every individual organism contains two alleles for each trait, and these alleles segregate (separate) during meiosis such that each gamete contains only one of the alleles. An offspring thus receives a pair of alleles for a trait by inheriting homologous chromosomes from the parent organisms: one allele for each trait from each parent (**Figure 10.7**).

If an organism has identical alleles for a particular character, that organism is true-breeding for that character and that allele is present in all gametes. But if different alleles are present as in the F_1 hybrids, then 50% of the gametes receive the dominant allele and 50% receive the recessive allele. Note that the phenotype ratio in Figure 10.7 is 3 purple:1 white whereas the genotype ratio is 1 pure purple:2 hybrid purple:1 pure white. A pair of identical alleles for a character is said to be **homozygous** for the gene controlling that character. However, an organism that has two different alleles for a gene is said to be **heterozygous**. In Figure 10.7, the PP and WW alleles are homozygous whereas the PW or WP alleles re heterozygous.

The Test Cross

Suppose a lost seed in Mendel's garden grew into a mature pea with purple flowers. One could not tell merely by looking at its purple phenotype whether it was homozygous (*PP*) or heterozygous (*Pp*) purple. To determine the genotype of the mystery pea's flower character, we can cross this plant with a white-flowered plant (*pp*). Knowing that purple (P) is dominant, if all the offspring of the cross have purple flowers then the purple-flowered mystery plant must be homozygous (PP) for the dominant allele, because a *PP* x *pp* cross produces all purple *Pp* offspring.

If both the purple and white phenotype appear among the offspring, then the purple-flowered plant must be heterozygous (hybrid). The offspring of a *Pp* x *pp* cross would be expected to have a genotype ratio of 1 *Pp*:1 *pp* (**Figure 10.8**). Breeding an organism of unknown genotype with a recessive homozygote is called a **test cross** because it can reveal the genotype of the unknown organism.

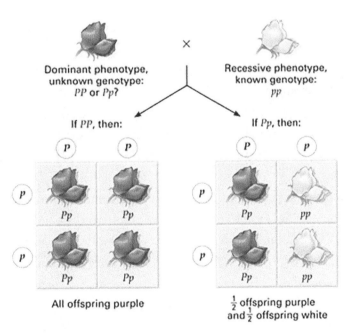

Figure 10.8 Mendel was the first to conceive of the idea of using a test cross to identify the genotype of an unknown plant.

Law of Independent Assortment

Mendel derived the law of segregation from experiments in which he tracked only a *single* character. All the F_1 progeny produced in his crosses of true-breeding parents were monohybrids, meaning that they were heterozygous for the one particular character being tracked in the cross. The cross between F_1 heterozygotes is referred to as a **monohybrid cross**.

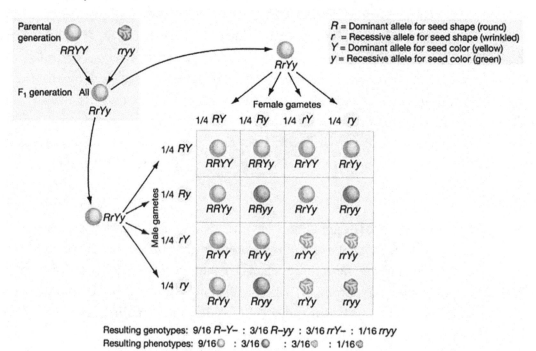

Resulting genotypes: 9/16 *R–Y–* : 3/16 *R–yy* : 3/16 *rrY–* : 1/16 *rryy*
Resulting phenotypes: 9/16 ○ : 3/16 ○ : 3/16 ○ : 1/16 ○

Figure 10.9 The phenotype ratio for a dihybrid cross is 9:3:3:1.

In his second round of experiments, Mendel took it to the next level and tracked *two* characters at one time. Imagine crossing two true-breeding pea varieties that differ in two characters such as a cross between a plant that produces round seeds (RR) that are yellow (YY) with a plant that produces wrinkled seeds (rr) that are green (yy) (**Figure 10.9**). The P cross of RRYY x rryy yields F_1 hybrids that are all RrYy. The F_1 hybrids can produce 4 kinds of female gametes (eggs)—RY, Ry, rY and ry—and 4 kinds of male gametes (pollen)—RY, Ry, rY, and ry. The Punnett square in Figure 10.8 shows the 16 possible fertilizations that could occur between the F_1 hybrids and the phenotype and genotype rations those 16 possible fertilizations would yield.

When Mendel conducted this experiment and classified the F_2 offspring, his results were close to the predicted 9:3:3:1 phenotype ratio, supporting his hypothesis that the alleles for one gene—controlling seed color or seed shape, in this example—are sorted into gametes independently of the alleles of other genes.

Mendel tested his seven pea characters in various dihybrid combinations and always observed a 9:3:3:1 phenotype ratio in the F_2 generation. In the dihybrid cross, the two different pairs of alleles segregate as if this were two monohybrid crosses. The results of Mendel's dihybrid experiments are the basis for what we now call the law of independent assortment that states: *Two or more genes assort independently. That is, each pair of alleles segregates independently of each other pair of alleles during gamete formation.* This law applies only to genes (allele pairs) located on different chromosomes (nonhomologous) or, alternatively, to genes that are very far apart on the same chromosome.

Box 10.1
The Probabilities of Peas

Probabilities are mathematical measures of possibilities. The empirical probability of an event is calculated by dividing the number of times the event occurs by the total number of opportunities for the event to occur. Empirical probabilities come from observations such as those of Mendel. An example of a genetic event is a round seed produced by a pea plant. Mendel demonstrated that the probability of the event "round seed" was guaranteed to occur in the F_1 offspring of true-breeding parents, one of which has round seeds and one of which has wrinkled seeds. When the F_1 plants were subsequently self-crossed, the probability of any given F_2 offspring having round seeds was now three out of four. In other words, in a large population of F_2 offspring chosen at random, 75 percent were expected to have round seeds, whereas 25 percent were expected to have wrinkled seeds (Figure 10.7). Using large numbers of crosses, Mendel was able to calculate probabilities and use these to predict the outcomes of other crosses.

Mendel demonstrated that the pea-plant characteristics he studied were transmitted as discrete units from parent to offspring. Mendel also determined that different characteristics were transmitted independently of one another (independent assortment) and could be considered in separate probability analyses. For instance, performing a cross between a plant with green, wrinkled seeds and a plant with yellow, round seeds produced offspring that had a 3:1 ratio of green:yellow seeds and a 3:1 ratio of round:wrinkled seeds. The characteristics of color and texture did not influence each other.

The *product rule* of probability can be applied to this phenomenon of the independent transmission of characteristics. It states that the probability of two independent events occurring together can be calculated by multiplying the individual probabilities of each event occurring alone. Imagine that you are rolling a six-sided die (D) and flipping a penny (P) at the same time. The die may roll any number from 1–6 ($D_\#$), whereas the penny may turn up heads (P_H) or tails (P_T). The outcome of rolling the die has no effect on the outcome of flipping the penny and vice versa. There are 12 possible outcomes, and each is expected to occur with equal probability: D_1P_H, D_1P_T, D_2P_H, D_2P_T, D_3P_H, D_3P_T, D_4P_H, D_4P_T, D_5P_H, D_5P_T, D_6P_H, D_6P_T.

Of the 12 possible outcomes, the die has a 2/12 (or 1/6) probability of rolling a two, and the penny has a 6/12 (or 1/2) probability of coming up heads. The probability that you will obtain the combined outcome 2 and heads is: $(D_2) \times (P_H) = (1/6) \times (1/2)$ or 1/12. The word "and" is a signal to apply the product rule. Consider how the product rule is applied to a dihybrid: the probability of having both dominant traits in the F_2 progeny is the product of the probabilities of having the dominant trait for each characteristic.

The *sum rule* is applied when considering two mutually-exclusive outcomes that can result from more than one pathway. It states that the probability of the occurrence of one event or the other, of two mutually-exclusive events, is the sum of their individual probabilities. The word "or" indicates that you should apply the sum rule. Let's imagine you are flipping a penny (P) and a quarter (Q). What is the probability of one coin coming up heads and one coming up tails? This can be achieved by two cases: the penny is heads (P_H) and the quarter is tails (Q_T), or the quarter is heads (Q_H), and the penny is tails (P_T). Either case fulfills the outcome. We calculate the probability of obtaining one head and one tail as $[(P_H) \times (Q_T)] + [(Q_H) \times (P_T)]$ $= [(1/2) \times (1/2)] + [(1/2) \times (1/2)] = 1/2$. You should also notice that we used the product rule to calculate the probability of P_H and Q_T and also the probability of P_T and Q_H, before we summed them. The sum rule can be applied to show the probability of having just one dominant trait in the F_2 generation of a dihybrid cross.

To use probability laws in practice, it is necessary to work with large sample sizes because small sample sizes are prone to deviations caused by chance. The large quantities of pea plants that Mendel examined allowed him to calculate the probabilities of the traits appearing in his F_2 generation. This discovery meant that when parental traits were known, the offspring's traits could be predicted accurately even before fertilization.

Non-Mendelian Inheritance

Any pattern of inheritance that follows Mendel's laws is said to be Mendelian. However, since Mendel's time geneticists have found examples of inheritance patterns that do not follow Mendel's laws—non-Mendelian inheritance. In the 20th century, geneticists built on Mendelian principles extending them to more diverse and complex organisms and patterns of inheritance more complex than those described by Mendel.

Mendel himself realized that he could not explain the more complicated patterns he observed in crosses involving other pea characters or other plant species. This does not diminish the usefulness of Mendelian genetics, however, because the basic principles of segregation and independent assortment apply even to more complex patterns of inheritance.

Inheritance of characters determined by a single gene deviate from simple Mendelian patterns when alleles are not completely dominant or recessive, when a particular gene has more than two alleles, or when a single gene produces multiple phenotypes.

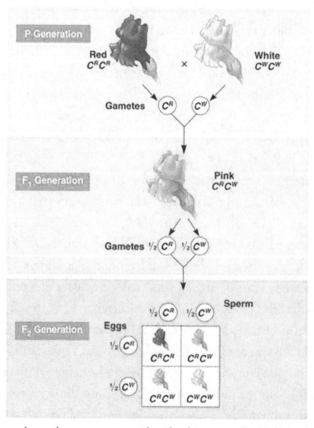

Figure 10.10 When red snapdragons are crossed with white ones, the F1 hybrids have pink flowers. Segregation of alleles into gametes of the F1 plants results in an F2 generation with a 1:2:1 ratio for both genotype and phenotype. Neither allele is dominant, so rather than using upper- and lowercase letters, we use the letter C with a superscript to indicate an allele for flower color: CR for red and CWfor white.

Degrees of Dominance For some genes, neither allele is completely dominant, and the F_1 hybrids have a phenotype somewhere between those of the two parental varieties. This phenomenon, called **incomplete dominance**, is seen when red snapdragons are crossed with white snapdragons (**Figure 10.10**). The intermediate phenotype results from flowers of the heterozygotes having less red pigment than the red homozygotes.

This might seem to provide evidence for the blending hypothesis discussed earlier. However, blending would predict that the red or white trait could never reappear among offspring from the pink hybrids, but that is not the case as shown in Figure 10.10.

A variation on dominance relationships between alleles is called **codominance**. In this variation, two alleles are fully expressed, affecting the phenotype in indistinguishable ways (**Figure 10.11**). It is important to understand that an allele is called *dominant* because it is seen in the phenotype, not because it somehow overpowers and subdues a recessive allele. When a dominant allele coexists with a recessive allele in a heterozygous manner, they do not interact at all. It is in the pathway from genotype to phenotype that dominance and recessive come in to play.

Although you might assume that the dominant allele for a particular character would be more common than the recessive allele, this is not always the case. Polydactyly, a condition in which a baby is born with extra fingers and toes, is considered a dominant allele yet only one baby out of 400 in the United States is born with this condition. The low frequency of polydactyly indicates that the recessive allele is far more prevalent in the population than the dominant allele.

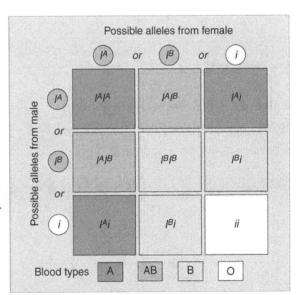

Figure 10.11 In the F1 hybrids (BW), both alleles are fully expressed resulting in mottled feathers

Multiple Alleles Mendel's peas had only two alleles for each character expressed, but most genes exist in more than two allelic forms. If there are only two alleles involved in determining the phenotype of a certain trait, but there are *three* possible phenotypes, then the inheritance of the trait illustrates either incomplete dominance or codominance. In these situations, a heterozygous (hybrid) genotype produces a third phenotype that is either a blend of the other two phenotypes (incomplete dominance) or a mixing of the other phenotypes with both equally expressed (codominance). If there are four or more possible phenotypes for a particular trait, then multiple alleles must exist in the population.

The ABO blood groups in humans are an example of multiple alleles. These groups are determined by three alleles of a single gene and are designated I^A, I^B, and i.

Figure 10.12 Multiple alleles for the ABO blood groups. The four blood groups result from different combinations of three alleles.

These alleles produce four blood types: A, B, AB, or O based on genetically determined proteins found on the surface of red blood cells (**Figure 10.12**).

Pleiotropy During his study of inheritance in pea plants, Mendel made several interesting observations regarding the color of various plant components. Specifically, Mendel noticed that plants with colored seed coats always had colored flowers and colored leaf axils. (Axils are the parts of the plant that attach leaves to stems.) Mendel also observed that pea plants with colorless seed coats always had white flowers and no pigmentation on their axils. In other words, in Mendel's pea plants, seed coat color was always associated with specific flower and axil colors.

Figure 10.13

Today, we know that Mendel's observations were the result of **pleiotropy** or the phenomenon in which a single gene contributes to multiple phenotypic traits. In this case, the seed coat color gene was not only responsible for seed coat color, but also for flower and axil pigmentation. The frizzle gene in chickens is an example of pleiotropy in animals (**Figure 10.13**). In 1936, researchers Walter Landauer and Elizabeth Upham observed that chickens that expressed the dominant frizzle gene produced feathers that curled outward rather than lying flat against their bodies. However, this was not the only phenotypic effect of this gene — along with producing defective feathers, the frizzle gene caused the fowl to have abnormal body temperatures, higher metabolic and blood flow rates, and greater digestive capacity. Furthermore, chickens who had this allele also laid fewer eggs than their wild-type counterparts, further highlighting the pleiotropic nature of the frizzle gene.

Dominance relationships, multiple alleles, and pleiotropy all have to do with the effects of a single gene. There are, however, two situations in which two or more genes are involved in determining phenotype: epistasis and polygenic inheritance.

Epistasis is the interaction between two or more genes to produce more than one phenotype. In 1908, British geneticists William Bateson and R. C. Punnett conducted research on the shape of chicken combs. They analyzed the three comb types of chickens know to exist at that time:

Chicken Varieties	Phenotype
Wyandotte	Rose comb
Brahmas	Pea comb
Leghorns	Single comb

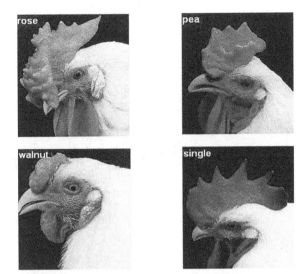

Figure 10.14 Genetic variation in chicken combs. (a) An RRpp genotype results in a rose comb; (b) An RRPP genotype results in a walnut comb; (c) A rrPP genotype results in a pea comb and (d) A rrpp genotype results in a single comb

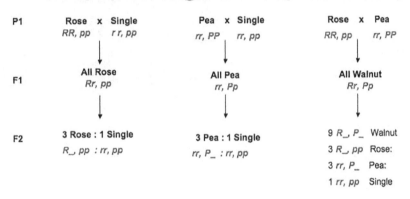

Figure 10.15

When Bateson and Punnett crossed a Wyandotte chicken with a Brahma chicken, all of the F_1 hybrids had a new type of comb that the duo termed a "walnut" comb (**Figure 10.14**). In this case, neither the rose comb of the Wyandotte nor the pea comb of the Brahma appeared to be dominant, because the F1 offspring had their own unique phenotype. Moreover, when two of those F1 progeny were crossed with each other, some of the members of the resulting F2 generation had walnut combs, some had rose combs, some had pea combs, and some had a single comb, like that seen in Leghorns (**Figure 10.15**). From these crosses, Bateson and Punnett concluded that the shape of the comb in chickens was the result of the interaction between two different genes—epistasis. Given the intricate and complicated molecular and cellular interactions responsible for an organism's development, it shouldn't be surprising that a single gene can affect and influence a number of characteristics.

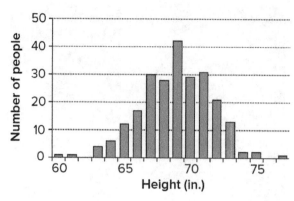

Figure 10.16 The variation in height within a group of high school seniors. A recent study using genomic methods identified at least 180 genes that affect height.

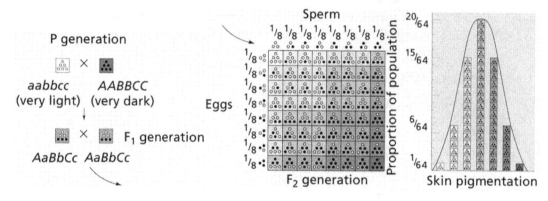

Figure 10.17 A simplified model for polygenic inheritance of skin color. In this model, three separately inherited genes affect skin color. The F1 heterozygotes (AaBbCc) each carry three dark skin alleles (black circles) and three light skin alleles (white circles). The Punnett square shows all the possible genetic combinations in gametes and offspring of many hypothetical matings between these heterozygotes. The gradation in phenotypes is shown on the graph to the right.

Polygenic Inheritance involves a group of nonepistatic genes working together to determine a range of phenotypes. Many characters in humans, for example, such as skin color, eye color, and height, are not one of two distinct characters, but instead a variation in the population in gradations along a continuum (**Figure 10.16**). Such characters are called *quantitative characters*, and their presence is indicative of polygenic inheritance. Human skin pigmentation is a polygenic trait that is determined by the additive effects of many different genes (**Figure 10.17**). The graph shows, as expected, most of the individuals having skin color in the intermediate range. Of course, environmental factors such as exposure to the sun, also affect the skin-color phenotype.

Figure 10.18 The acidity and free aluminum content of soil affect the color range of hydrangea flowers of the same genetic variety. The color ranges from deep blue (acidic soil) to deep pink (basic soil).

Nature vs. Nurture

Any organism is molded both by its nature (external environment) and its nurture (genetic makeup). A phenotype is not associated with a rigidly defined genotype, but rather with a range of possibilities due to environmental influences (**Figure 10.18**).

The phenotypic range is broadest for polygenic characters. Environment contributes to the quantitative nature of these characters, as seen in the continuous variation of human skin color. Geneticists refer to such characters as *multifactoral*, meaning that many factors, both genetic and environmental, collectively influence phenotype.

The nature vs. nurture question is often applied to heritability in humans. *Heritability* refers to the origins of differences between people. Individual development, even of highly heritable traits, such as eye color, depends on a range of environmental factors, from the other genes in the organism to physical variables such as temperature, oxygen levels etc., during its development.

The spectrum of hereditability ranges from situations such as Huntington's disease that is completely heritable to which language one speaks, which religion one practices, or which political party one supports that are not heritable at all. Some people may think of the degree of a trait being made up of two "buckets," genes and environment, each able to hold a certain capacity of the trait. But even for intermediate heritabilities, a trait is always shaped by both genetic dispositions and the environments in which people develop, merely with greater and lesser plasticities associated with these heritability measures.

Considering all that can occur in the pathways from genotypes to phenotypes, it is remarkable that Mendel could discover the fundamental principles governing the transmission of individual genes from parents to offspring. Mendel's laws of segregation and independent assortment explain heritable variations regarding alternative forms of alleles (Mendel's hereditary "particles") that are passed from generation to generation, according to simple rules of probability. This theory is equally valid for any organism with a sexual life cycle. From Mendel's abbey garden came a theory of particulate inheritance that serves as the cornerstone for modern genetics.

In Summary

- During Mendel's time the concept of inheritance by blending held sway.
- Around 1857, Mendel began breeding garden peas in the abbey's experimental garden to study inheritance in plants. His test subject was the garden pea (*Pisum sativum*).
- Mendel chose 22 varieties of peas for his experiments because these peas clearly displayed only seven distinct traits.
- A heritable feature that varies among individuals, such as flower color or stem length, is called a character. Each variant for a character, such as purple or white flower or tall or short stem, is called a trait.
- Mendel started his experiments with varieties that, over many generations of self-pollination, produced only the same traits as the parent plant. Such plants are said to be true-breeding. For example, a plant with yellow seeds is true-breeding if the seeds produced by self-pollination in successive generations always give rise to plants that also produce yellow seeds.

- To cross a purple-flowered pea plant with a white-flowered pea plant, Mendel would remove the male anthers containing the stamens (pollen sacs) from the purple flower so it could not self-fertilize by its own pollen. He would then transfer the pollen from the white flower (male) to the stigma of the purple flower (female) affecting cross-pollination.
- A mating, or crossing, of two true-breeding varieties is called hybridization.
- To insure accuracy and verify results, Mendel performed reciprocal crosses. That is, he reversed the gender of the flowers.
- The original parent types are referred to as the P generation (parental generation), and their hybrid offspring are the F_1 generation (first filial generation).
- Allowing the F_1 hybrids to self-pollinate or to cross-pollinate with other F_1 hybrids, produces an F_2 generation.
- Mendel reasoned that the "heritable factor" for white flower did not disappear in the F_1 plants but was somehow hidden or masked, when the purple flower factor was present. Mendel dubbed the purple flower color a dominant trait, and the white flower color a recessive trait.
- Mendel's law of segregation is constructed of four related concepts:
  - *Alternative versions of genes for variations in inherited characters.*
  - *Each organism inherits one gene of an allele pair from each parent.*
  - *If the two alleles at a locus differ, then one, the dominant allele, determines the organism's appearance (phenotype) whereas the other, the recessive allele, has no noticeable effect on the organism's appearance.*
  - *The two alleles for a heritable character segregate (separate from each other) during gamete formation and end up in different gametes.*
- The physical appearance of an organism (called phenotype) is not always indicative of its genetic makeup (called genotype).
- A pair of identical alleles for a character is said to be homozygous for the gene controlling that character. However, an organism that has two different alleles for a gene is said to be heterozygous.
- Breeding an organism of unknown genotype with a recessive homozygote is called a test cross because it can reveal the genotype of the unknown organism.
- The cross between F_1 heterozygotes is referred to as a monohybrid cross.
- The law of independent assortment that states: *Two or more genes assort independently. That is, each pair of alleles segregates independently of each other pair of alleles during gamete formation.*
- Inheritance of characters determined by a single gene deviate from simple Mendelian patterns when alleles are not completely dominant or recessive, when a particular gene has more than two alleles, or when a single gene produces multiple phenotypes.
- For some genes, neither allele is completely dominant, and the F_1 hybrids have a phenotype somewhere between those of the two parental varieties. This phenomenon is called incomplete dominance.
- In codominance, two alleles are fully expressed, affecting the phenotype in indistinguishable ways.
- Pleiotropy is a phenomenon in which a single gene contributes to multiple phenotypic traits.
- Epistasis is the interaction between two or more genes to produce more than one phenotype.

- Polygenic Inheritance involves a group of nonepistatic genes working together to determine a range of phenotypes.
- Multifactoral means that many factors, both genetic and environmental, collectively influence phenotype.
- Heritability refers to the origins of differences between people.

Review and Reflect

1. ***Pea and Man*** Around 1857, Mendel began breeding garden peas in the abbey's experimental garden to study inheritance in plants. In this chapter, we called this the "serendipitous union of the right experimenter working with the right experimental subject being in the right place at the right time." What is meant by that passage?

2. ***Pea and Pea*** Once Mendel had established pure varieties he began to cross breed them. How in the world do you cross breed pea plants?

3. ***Trading Gender*** Mendel was a careful and thoughtful scientist. To insure accuracy and verify results, he performed reciprocal crosses. What are reciprocal crosses?

4. ***Who is F_2?*** Pootsie Wootsie is on a picnic in the woods with her grandparents, her parents, her brother, and her sister. Who in this family group is the P generation? the F_1 generation? the F_2 generation?

5. ***Total Domination*** Mendel discovered that purple flower in pea plants is a dominant trait whereas white flower is a recessive trait. Explain the concept of dominance.

6. ***Define It*** Imagine you have been contracted to write a biological dictionary. What would your entries be for the words *phenotype* and *genotype*?

7. ***Go Our Separate Ways*** Explain Mendel's law of segregation.

8. ***Pure or Hybrid?*** In Mendel's garden, some peas that had purple flowers that had a genotype of *PP* whereas others with purple flowers had a genotype of *Pp*. Which of these genotypes is homozygous and which is heterozygous? How do you know?

9. ***Test It*** Mendel developed the concept of the test cross. What is a test cross and why are test crosses conducted?

10. ***You're so Independent*** Explain Mendel's law of independent assortment.

11. ***One or Two*** Mendel performed both monohybrid and dihybrid crosses. Explain the differences between these two different levels of genetic crosses.

Create and Connect

1. ***Talking With Mendel*** Imagine you are the time-traveling journalist, Buckaroo Scribere. You have traveled back to 1860 to interview Gregor Mendel. Detail what questions you would ask him and record what you presume would be his answers.

2. ***Nature vs. Nurture Defense*** Imagine you are the prosecuting attorney in a case where the defendant did some very inappropriate things. The defendant's attorney is going to go with the "Nature

vs. Nurture defense." That is, she will argue that her client is not guilty because her client's genes forced him to do it. As prosecutor, how would you counter that argument?

3. ***Walking With Mendel—Genetics Problems*** Solve the following problems using Mendelian genetic principles:

    A. In humans, normal skin pigmentation is due to a dominant gene (C) while albinism is due to a recessive gene (c). A normal man marries an albino woman. Their first child is an albino. What is the genotype of: the father? The mother? The child?

    B. An albino man marries a normal woman. They have 9 normal children. What is the genotype of: the father? the mother? the children?

    C. A normal man whose father was an albino marries an albino woman both of whose parents were normal. They have 3 children—2 normal and 1 albino. What are the genotypes of: the man's father? the man? the woman's father? the woman's mother?, the woman?

    D. In poultry rose comb is dependent upon a dominant gene (R) while single comb (r) is dependent upon a recessive gene. A rose comb rooster is mated with two rose comb hens. Hen A produces 14 chicks, all with rose comb. Hen B produces nine chicks, 7 with rose comb and 2 with single comb. What are the genotypes of: the rooster? Hen A? Hen B?

    E. When red snapdragons are crossed with white snapdragons, all the F1 generation have pink flowers. Explain this in terms of dominance (or the lack of dominance).

    F. In humans, aniridia (a type of blindness) is due to a dominant gene (A). Migraine headache is the result of a different dominant gene (M). A man with aniridia whose mother was not blind marries a woman who suffers from migraine but whose father did not. In what proportion of their children would both aniridia and migraine be expected to occur?

    G. Among Hereford cattle, there is a dominant gene called *polled*; the individuals having this gene lack horns. Suppose you purchase a herd consisting entirely of polled cattle, and you carefully determine that no cow in the herd has horns. However, some of the calves born that year do have horns. You remove those calves from the herd and make certain that no horned adult has gotten into your pasture. Despite these efforts, more horned calves are born the next year. What is the reason for the appearance of the horned calves? If your goal is to maintain a herd consisting entirely of polled cattle, what would you do?

    H. In hogs large, floppy ears (E) dominates small ears (e) and curly tail (T) dominates straight tail (t). A heterozygous large ear heterozygous curly tail boar is mated with a heterozygous large ear heterozygous curly tail sow. What are the possible phenotypes only of their offspring and in what ratio will these phenotypes occur?

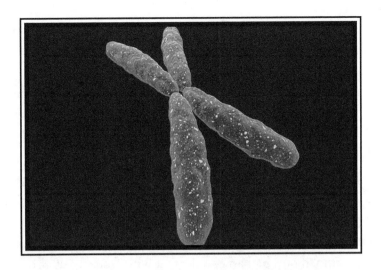

CHROMOSOMES AND THE GENETIC MATERIAL

I may finally call attention to the probability that the association of paternal and maternal chromosomes in pairs and their subsequent separation during reduction division as indicated above may constitute the physical basis of the Mendelian law of heredity.

—Walter Sutton

Introduction

After eight years of tedious work genetically testing over 29,000 crosses of garden peas, Mendel presented his work in 1865 as the short monograph, *Experiments With Plant Hybrids,* to the Brünn Society for Natural History and published it in the *Proceedings of the Brünn Society for Natural History* in the following year.

Hampered by his lack of standing in the scientific hierarchy of the time and by the fact that other scientists of the day could not fully comprehend and understand the significance of his work, the man who would later be righteously dubbed "The Father of Modern Genetics" and his brilliantly intuitive investigations into the workings of heredity were quietly and quickly relegated to the dustbin of scientific oddities.

The Mendel-Chromosome Connection

Gregor Mendel's "hereditary factors" were a purely abstract concept when he proposed their existence in 1860. At that time, no cellular structures had been identified that could be these imaginary units and most biologists of the time were skeptical about Mendel's proposed laws of inheritance.

Improvements in microscopy allowed cytologists to work out the process of mitosis in 1875 and meiosis in the 1890s. Following those discoveries, in 1900, Mendel's work resurfaced when independently of one another, Hugo de Vries, Erich von Tschermak, and Carl Correns rediscover Mendel's published, but long-neglected paper outlining the basic laws of inheritance.

The rediscovery of Mendel's laws in 1900 clarified inheritance, but Mendel had worked with traits of whole organisms. He did not investigate how characteristics are sorted and combined on a cellular level, where reproduction takes place. In 1902, the German scientist Theodor Boveri and the American Walter Sutton, working independently, suggested that chromosomes could be shown to bear the material of heredity. Mendelian concepts, as it turned out, had an excellent fit with facts about chromosomes.

Boveri (**Figure 11.1**) had previously shown that chromosomes remain organized units through the process of cell division, and he demonstrated that sperm and egg cells each contribute the same number of chromosomes. But does each chromosome have specific properties? Is a full complement of chromosomes necessary for reproduction and development?

In a series of experimental manipulations with sea urchin eggs, Boveri demonstrated that individual chromosomes uniquely impact development. Sea urchin eggs can be fertilized with two sperm. Boveri showed that daughter cells of such double unions possess variable numbers of chromosomes. Of the embryos that result, Boveri found that only the small percentage—about 11 percent—possessing the full set of 36 chromosomes would develop normally. A "specific assortment of chromosomes is responsible for normal development," wrote Boveri in 1902, "and this can mean only that the individual chromosomes possess different qualities."

Figure 11.1
Theodor Boveri (1862-1915)

In addition, Boveri recognized the Mendelian concepts of segregation and assortment could be interpreted to operate on a cellular level, with chromosomes containing the "factors"—as Mendel called the genes. The probability was "extraordinarily high," wrote Boveri in 1903, "that the characters dealt with in Mendelian experiments are truly connected to specific chromosomes."

An American graduate student, Walter Sutton (**Figure 11.2**), came to the same conclusion at about the same time. Sutton, working with marine life forms, had also become familiar with the process of "reduction division" (meiosis) that gives rise to reproductive germ cells, or gametes. In meiosis, the number of chromosomes is reduced by half in sperm and egg cells, with the original number restored in the zygote, or fertilized egg, during reproduction. This process was consonant with Mendel's idea of segregation. In 1902 Sutton suggested that "the association of paternal and

Figure 11.2
Walter Sutton (1877-1916)

maternal chromosomes in pairs and their subsequent separation during the reduction division...may constitute the physical basis of the Mendelian law of heredity." Sutton's "The Chromosomes in Heredity" was published in 1903.

The Boveri-Sutton Chromosome Theory, as it came to be known, was discussed and debated during the first years of the twentieth century. It was embraced by some but strongly rejected by others. By 1915 Thomas Hunt Morgan—initially a strong skeptic—laid the controversy to rest with studies of the fruit fly.

Thomas Hunt Morgan—Lord of the Flies

The first solid evidence associating a specific gene with a specific chromosome came early in the 20th century from the work of Thomas Hunt Morgan (**Figure 11.3**). Mendel had been insightful enough or lucky enough to choose an experimental test subject (garden pea) suitable for the research problem at hand. Morgan did the same. For his research, Morgan selected a species of fruit fly, *Drosophila melanogaster*, a common insect that feeds on the yeast growing on fruit.

Fruit flies are prolific breeders; a single breeding will produce hundreds of offspring, and a new generation can be bred every two weeks. Another advantage of the fruit fly is that it has only four pairs of chromosomes that are readily distinguishable with a light microscope. There are three pairs of autosomes and one pair of sex chromosomes. Female fruit flies have a pair of homologous X chromosomes, and males have one X chromosome and one Y chromosome (Figure 11.4).

Figure 11.3
Thomas Hunt Morgan (1866-1945)

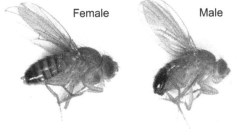

Figure 11.4 Female fruit flies tend to be larger in size and have a more pointed abdomen. Males are smaller with a rounded abdomen with a black tip.

Mendel could easily obtain different naturally-occurring pea varieties from seed suppliers, but Morgan faced stiff challenges in his attempts to gather different varieties of fruit flies. He faced the tedious task of carrying out many matings, and then microscopically inspecting large numbers of offspring in search of naturally occurring variant individuals. He complained, "Two years of work wasted. I have been breeding those flies for all that time and I've got nothing out of it." Morgan persisted.

After breeding millions of *Drosophila* in his laboratory at Columbia University, in 1910 Morgan noticed one fruit fly with a distinctive characteristic: white eyes instead of red (**Figure 11.5**). He isolated this specimen and mated it to an ordinary red-eyed fly. Although the F_1 generation of 1,237 offspring was all red-eyed save three, white-eyed flies appeared in larger numbers in the F_2 generation. Surprisingly, all white-eyed flies were male.

These results supported hypotheses of which Morgan himself was skeptical. He was at the time critical of the Mendelian theory of inheritance, mistrusted aspects of chromosomal theory,

Figure 11.5 The phenotype for a character most commonly observed in natural populations, such as red eyes in *Drosophila*, is called the *wild type*.

and did not believe that Darwin's concept of natural selection could account for the emergence of new species. But Morgan's discoveries with white- and red-eyed flies led him to reconsider each of these hypotheses.

In particular, Morgan began to entertain the possibility that association of eye color and sex in fruit flies had a physical and mechanistic basis in the chromosomes. The shape of one of *Drosophila's* four chromosome pairs was thought to be distinctive for sex determination. Males invariably possess the XY chromosome pair whereas flies with the XX chromosome are female. If the factor for eye color was located exclusively on the X chromosome, Morgan realized, Mendelian rules for inheritance of dominant and recessive traits could apply.

In brief, Morgan had discovered that eye color in *Drosophila* expressed a **sex-linked trait** (**Figure 11.6**). All first-generation offspring of a mutant white-eyed male and a normal red-eyed female would have red eyes because every chromosome pair would contain at least one copy of the X chromosome with the dominant trait. But half the females from this union would now possess a copy of the white-eyed male's recessive X chromosome. This chromosome would be transmitted, on average, to one-half of second-generation offspring—one-half of which would be male. Thus, second-generation offspring would include one-quarter with white eyes—and all of these would be male.

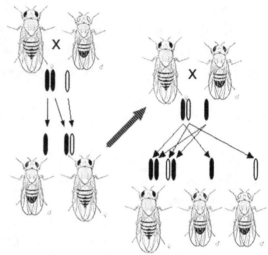

Figure 11.6 Morgan realized that red eyes were dominant to white eyes because all the F1 hybrids had red eyes. He then reasoned that the gene for white eyes must be carried on the X chromosome because half the males had white eyes and half had red eyes as Mendel's laws predicted. These diagrams are from Thomas Hunt Morgans's 1919 book *The Physical Basis of Heredity*

Intensive work led Morgan to discover more mutant traits—some two dozen between 1911 and 1914. With evidence drawn from cytology he was able to refine Mendelian laws and combine them with the theory—first suggested by Theodor Boveri and Walter Sutton—that the chromosomes carry hereditary information. In 1915, Morgan and his colleagues published *The Mechanism of Mendelian Heredity*. The major components of the chromosome theory of inheritance:

- Like beads on a string, discrete pairs of factors on chromosomes bear hereditary information. These factors—Morgan would later call them "genes"—segregate in gametes and combine during reproduction essentially as predicted by Mendelian laws.

- Certain characters are sex-linked because they arise on the same chromosome that determines gender.

- Other characteristics are also sometimes associated because, as paired chromosomes separate during germ cell development, genes proximate to one another tend to remain together. But sometimes, as a mechanistic consequence of reproduction, this linkage between genes is broken, allowing for new combinations of traits.

Morgan's experimental and theoretical work inaugurated research in genetics and promoted a revolution in biology. The evidence he gleaned from embryology and cell theory pointed the way toward a synthesis of genetics with evolutionary theory. Morgan himself explored aspects of these developments in later work, including *Evolution and Genetics* published in 1925, and *The Theory of the Gene* in 1926. He received the Nobel Prize in Physiology or Medicine in 1933.

Chromosomal Basis of Sex

Scientists have discovered a number of different chromosomal systems of sexual determination.

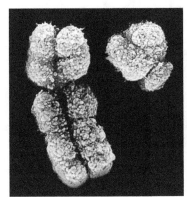

Figure 11.7 The human sex chromosomes— X on the left and Y on the right.

XX/XY System is the sex-determination system found in humans, most other mammals, some insects, and some plants. In this system, the sex of an individual is determined by a pair of sex chromosomes. Unlike the autosomes, the sex chromosomes are different in size and shape from the other (**Figure 11.7**).

Females have two of the same kind of sex chromosome (XX) and are called the **homogametic sex**. Males have two distinct sex chromosomes (XY) and are called the **heterogametic sex** (**Figure 11.8**). Researchers have sequenced the human Y chromosome and have identified 78 genes that code for about 25 proteins (some genes are duplicates). About half these genes are expressed only in the testis, and some are required for normal testicular functioning and the production of sperm.

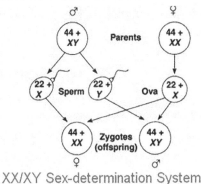

XX/XY Sex-determination System

Figure 11.8 In the XX/XY system, The sex of the offspring depends on whether the penetrating sperm cell contains an X chromosome or a Y.

In 1990, a British research team identified a gene on the Y chromosome required for the development of testes. They named the gene *SRY* for <u>sex</u> determining <u>region</u> of <u>Y</u>. In the absence of *SRY*, the gonads develop into ovaries. The biochemical, physiological, and anatomical features that distinguish male humans from female humans are complex, and many genes are involved in their development. In fact, *SRY* codes for a protein that regulates other genes.

XO System of Sex-determination

Figure 11.9 In the XX/XO system, there is only one type of sex chromosome, the X. Females are XX whereas males have only one sex chromosome (XO). Sex of the offspring is determined by whether the sperm cell contains an X chromosome or no sex chromosome."

XX/XO System is a sex-determination system found in a number of insects and a few mammals. In this variation of the XX/XY system, the females have two copies of the sex chromosome (XX), but males have only one (X0). The *0* denotes the absence of a second sex chromosome. Generally in this method, the sex is determined by amount of genes expressed across the two chromosomes (**Figure 11.9**).

ZW System is a sex-determination system found in birds, some fish, some reptiles, and some insects. The ZW sex-determination system is reversed compared to the XY system: females have two different kinds of chromosomes (ZW), and males have two of the same kind of chromosomes (ZZ) (**Figure 11.10**).

ZW sex-determination system

Figure 11.10 The sex chromosomes present in the egg (not sperm) determine the sex of the offspring. The sex chromosomes are designated Z and W. Females are ZW and males are ZZ.

Haplo-Diploid System is a sex-determination system found in ants and bees. Diploid individuals are female but may be sterile males. Males cannot have sons and do not have fathers. If a queen bee mates with one drone, her daughters share ¾ of their genes with each other, not ½ as in the XY and ZW systems (**Figure 11.11**).

Haplo-Diploid Sex Determination in Bees

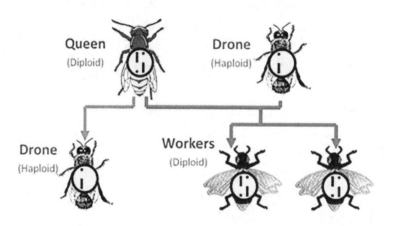

Sister-Workers are 75% - related to each other

Figure 11.11 In bees, females develop from fertilized eggs and are diploid. Males develop from unfertilized eggs and are haploid."

Box 11.1
Is the Y Chromosome Here to Stay?

Some view the uniquely male Y sex chromosome as a shadow of its former self and that because the genes contained on this relatively small strip of DNA are unimportant and disposable, it will continue to diminish and eventually disappear. However, a recent study concluded otherwise.

The lead author of the study states, "The Y chromosome has lost 90 percent of the genes it once shared with the X chromosome, and some scientists have speculated that the Y chromosome will disappear in less than 5 million years." She goes on to say, "Our study demonstrates that the genes that have been maintained, and those that migrated from the X to the Y, are important, and the human Y is going to stick around for a long while."

The Y chromosome is one of two sex chromosomes that humans and almost all mammals have. It is equipped with the genes that are responsible for the development of testes, and thus carried the blueprint for maleness. The idea that this chromosome is practically superfluous — or is at least approaching a state of redundancy — partly stems from the fact that it doesn't recombine its genetic information with another chromosome to maintain an optimal genetic tool kit like the other 22 chromosomes do. It is also suspected that the Y chromosome's modest 27 unique genes compared with thousands contained on other chromosomes have to do with the mating record of males not being as successful as females' throughout history, which means fewer types of Y chromosomes have been passed on to subsequent generations. This has led to the view that the Y chromosome has become a genetic "wasteland" that's waiting to be put out of its misery in a matter of geological seconds.

The current study compared the Y chromosome of eight African and eight European men and found patterns of variation indicating that rather than diminishing into obscurity, natural selection is maintaining its genetic content because it plays such an important role in male fertility. This lack of genetic variation of the Y chromosome is observed across the entire world and is actually what makes this sex chromosome so interesting and informative; by hardly changing over millions of years, it becomes a unique genetic time capsule that can be used to chart the course of human history, the authors explain.

Furthermore, the Berkley researchers showed that the Y chromosome's limited variation and small size doesn't entirely have to do with fewer males successfully passing on their genes to the next generation. They calculated that such a scenario would require less than a fourth of males to have passed on their Y chromosome to the next generation throughout history, which they view as highly unlikely. Instead, they figured that evolutionary pressure was also involved with cutting the chromosome down to size. Researchers conclude that a model of purifying selection acting on the Y chromosome to remove harmful mutations, in combination with a moderate reduction in the number of males that are passing on their Y chromosomes, can explain low Y diversity."

Of the 27 genes that are on the Y chromosome, 17 have been around for over 200 million years while the rest were acquired more recently. Male infertility has been linked to one or more copies of these newer genes getting lost along the reproductive way.

A gene located on either sex chromosome is called a *sex-linked gene*. Those located on the Y chromosome are called *Y-linked genes*. The Y chromosome containing around 200 genes is passed along virtually intact from a father to all his sons. On the other hand, the human X chromosome contains approximately 1,100 genes known as *X-linked genes*. The fact that males and females inherit a different number of X chromosomes leads to a pattern of inheritance different from that produced by genes on autosomes.

Genetic Recombination and Linkage

Genetic linkage is the tendency of alleles that are close together on a chromosome to be inherited together during the meiosis phase of sexual reproduction. Genes whose loci are nearer to each other are less likely to be separated onto different chromatids during chromosomal crossover and are therefore said to be genetically *linked*. In other words, the nearer two genes are on a chromosome, the lower is the chance of a crossing over occurring between them, and the more likely they are to be inherited together.

Genetic linkage was first discovered by the British geneticists Edith Rebecca Saunders, William Bateson and Reginald Punnett shortly after Mendel's laws were rediscovered. Mendel's law of independent assortment applies to the genes that are situated in separate chromosomes. When genes for different characters are located in the same chromosome, they are tied to one another and are said to be linked. They are inherited together by the offspring and will not be assorted independently. Thus, the tendency of two or more genes of the same chromosome to remain together in the process of inheritance is called linkage. Bateson and Punnet (1906), while working with sweet pea (*Lathyrus odoratus*) observed that flower color and pollen shape tend to remain together and do not assort independently as per Mendel's law of independent assortment.

When two different varieties of sweet pea—one having red flowers and round pollen grain and other having purple flower and long pollen grain were crossed, the F₁ plants were purple flowered with long pollen (purple long characters were respectively dominant over red and round characters) (**Figure 11.12**).

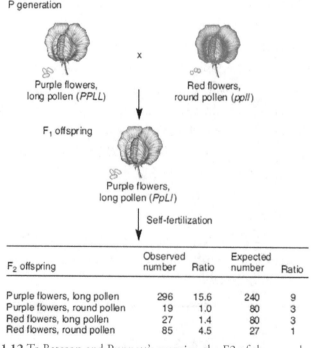

P generation

Purple flowers,
long pollen (*PPLL*)   x   Red flowers,
round pollen (*ppll*)

F₁ offspring

Purple flowers,
long pollen (*PpLl*)

Self-fertilization

F₂ offspring	Observed number	Ratio	Expected number	Ratio
Purple flowers, long pollen	296	15.6	240	9
Purple flowers, round pollen	19	1.0	80	3
Red flowers, long pollen	27	1.4	80	3
Red flowers, round pollen	85	4.5	27	1

Figure 11.12 To Bateson and Punnett's surprise, the F2 of the cross between F1 hybrids deviated strikingly from the expected 9:3:3:1 results.

Punnett and Bateson hypothesized that the F_1 had produced more PL and pl gametes than the recombinant gametes Pl and pL. Because these gametic types were the parental types, the researchers thought that physical coupling between the two dominants P and L and the two recessive p and l might have prevented the two genes from independent assortment. However they did not know the nature of the coupling.

The confirmation of Bateson and Punnett's hypothesis had to await Thomas Hunt Morgan and his work with *Drosophila*. Morgan had established the concept of sex chromosome linkage, and he went looking for it in autosomes. Morgan wanted to know whether the genes for body color and wing size are genetically linked, and if so, how this affects their inheritance. The alleles for body color are b^+ (gray wild-type) and b (black), and those for wing size are vg^+ (normal wild-type) and vg (vestigial or miniature) (**Figure 11.13**).

Figure 11.13 Morgan's P generation consisted of (a) wild type and (b) double mutant black body and vestigial wings.

Morgan mated true-breeding P (parental) flies—wild-type $b^+ b^+ vg^+ vg^+$ with black vestigial-winged flies $b b vg vg$ to produce heterozygous F_1 hybrids ($b^+ b vg^+ vg$), all of which were wild-type in appearance as predicted. He then mated wild-type F_1 dihybrid females with homozygous recessive males ($bb vgvg$). That test cross revealed the genotype of the eggs made by the dihybrid female.

The test cross male sperm contributes only recessive alleles, so the phenotype of the offspring reflects the genotype of the female's eggs:

Sperm: *bvg*

Eggs: *b⁺vg⁺   bbvg   b⁺vg   bvg⁺*

The test cross revealed four phenotypes in the F_2 generation:
- *Wild type* (Gray-normal) b⁺ b vg⁺ vg
- *Black-vestigial* bb vgvg
- *Gray-vestigial* b⁺ b vgvg
- *Black-normal* bb vg⁺ vg

Table 11.1				
Morgan's Predicted Ratios and Actual Data				
Predicted ratio if genes are located on different chromosomes.	1	1	1	1
Predicted ratio if genes are located on the same chromosome and parental alleles are always inherited together.	1	1	0	0
Data from Morgan's experiment.	965	944	206	185

Morgan's data revealed a striking difference between predicted results and actual data. To explain these results he concluded that since most offspring had a parental (P generation) phenotype, body color and wing size are genetically linked on the same chromosome. However, the production of a relatively small number of offspring with nonparental phenotypes indicated that some mechanism occasionally breaks the linkage between specific alleles of genes on the same chromosome. In other words, the genes were linked but there was something else stirring the genetic pot.

Crossing Over and Linkage Mapping

At first, Thomas Hunt Morgan wasn't sure why his data didn't match his predicted results when he crossed F_1 dihybrid females with homozygous recessive males for body color and wing length. He proposed that some process must occasionally break the physical connection between specific alleles of genes on the same chromosome. Further experimentation showed that a process now called **crossing over**, accounts for the recombination of linked genes. In crossing over, which occurs while replicated homologous chromosomes are paired during prophase of meiosis 1, a set of proteins orchestrates an exchange of corresponding segments of one maternal and one parental chromatid. In effect, when crossing over occurs, end portions of two nonsister chromatids trade places (**Figure 11.14**).

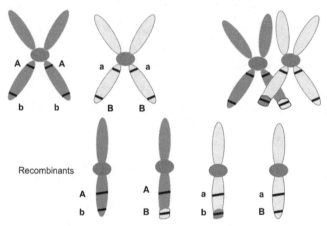

Figure 11.14 Crossing over occurs between prophase 1 and metaphase 1 and is the process whereby homologous chromosomes pair up with each other and exchange different segments of their genetic material to form recombinant chromosomes.

Crossing over accounts for genetic variation, because due to the swapping of genetic material during crossing over, the chromatids held together by the centromere are no longer identical. So, when the chromosomes go on to meiosis II and separate, some of the daughter cells receive daughter chromosomes with recombined alleles. Due to this genetic recombination, the offspring have a different set of alleles and genes than their parents do. In the diagram, genes B and b are crossed over with each other, making the resulting recombinants after meiosis Ab, AB, ab, and aB.

Now, putting all these ideas together, we see that the different recombinant chromosomes resulting from crossing over may bring alleles together in new combinations, and the subsequent events of meiosis distribute the recombinant chromosomes to gametes in a multitude of combinations. Random fertilization then increases even further the number of variant allele combinations that can be created. This abundance of genetic variation provides the raw material on which natural selection works. If new traits conferred by a particular combination of alleles are better suited for a given environment, organisms possessing those genotypes will be expected to thrive and leave more offspring. Ultimately, the interplay between environment and genotype will determine which genetic combinations persist over time.

As Morgan studied more linked genes, he saw that the proportion of recombinant progeny varied considerably, depending on which linked genes were being studied, and he thought that these variations in crossover frequency might somehow indicate the actual distances separating genes on the chromosomes. Morgan assigned the study of this problem to a student, Alfred Sturtevant. Morgan asked Sturtevant still an undergraduate at the time, to make some sense of the data on crossing-over between different linked genes. In one night, Sturtevant developed a method for describing relations between genes that is still used today. In Sturtevant's own words, "In the latter part of 1911, in conversation with Morgan, I suddenly realized that the variations in strength of linkage, already attributed by Morgan to differences in the spatial separation of genes, offered the possibility of determining sequences in the linear dimension of a chromosome. I went home and spent most of the night (to the neglect of my undergraduate homework) in producing the first chromosome map."

Sturtevant reasoned that the percentage of recombinant offspring, the *recombinant frequency*, depends on the distance between genes on a chromosome. Assuming crossing over to be a random event, Sturtevant predicted that "the farther apart two genes are, the higher the probability that a crossing over will occur between them and therefore the higher the recombination frequency." Sturtevant then proceeded to assign relative positions to genes on the same chromosomes—that is, to map gene locations.

A genetic map based on recombination frequencies is called a *linkage map*. Some genes on a chromosome are so far from each other that a crossover between them is virtually certain. The observed frequency of recombination in crosses involving two such genes can have a maximum value of 50%, a result indistinguishable from that for genes on different chromosomes. Despite being physically connected by being on the same chromosome, the genes are genetically unlinked and thus assort independently as if they were on different chromosomes.

Using recombination data, Sturtevant was able to map numerous *Drosophila* genes in linear arrays. He found that the genes clustered into four groups of linked genes (linkage groups) fitting the fact that light microscopy had revealed four pairs of chromosomes in *Drosophila*. Each chromosome has a liner array of specific genes, each with its own locus (**Figure 11.15**).

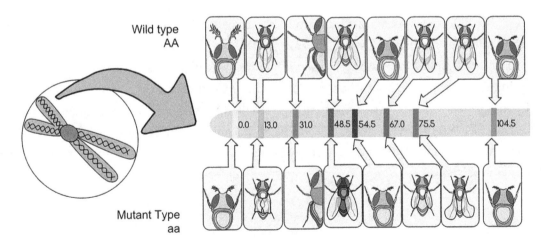

Figure 11.15 This gene linkage map shows the relative positions of allelic characteristics on the second Drosophila chromosome. The alleles on the chromosome form a linkage group due to their tendency to form together into gametes. The distance between the genes (map units) are equal to the percentage of crossing-over events that occurs between different alleles. 1% recombination frequency = 1 map unit.

The frequency of crossing over is not uniform over the length of a chromosome, as Sturtevant assumed, and therefore map units do not correspond to actual physical distances. A linkage map does portray the order of genes along a chromosome, but it does not accurately portray the precise location of those genes. Other methods enable geneticist to construct *cytogenetic maps* of

Figure 11.16 Fragment of a stained chromosome showing a barcode-like banding pattern.

chromosomes that locate genes with respect to chromosomal features, such as stained bands, that can be seen through a microscope (**Figure 11.16**). Today, most researchers sequence whole genomes to map the location of genes of a given species. The entire nucleotide sequence is the ultimate physical map of a chromosome, revealing the physical distances between gene loci in DNA nucleotides.

The Genetic Material

DNA is like a computer program but far, far more advanced than any software ever created.

—Bill Gates

Early in the 20th century geneticists came to understand that Mendel's "factors" were genes and that those genes were carried on chromosomes. The question then became...What is the nature of the genes themselves? During the late 1920s, bacteriologist Fredrick Griffith (**Figure 11.17**) was attempting to develop a vaccine against *Streptococcus pneumoniae*, which causes pneumonia in mammals. During his experiments, he noted that when these bacteria are grown on culture plates, some called S strain bacteria, produce shiny, smooth colonies whereas others, called R strain bacteria produce colonies that have a rough appearance. Microscopic examination revealed that the S strain bacteria have a capsule (mucous covering) that makes them smooth, but the R strain does not.

Figure 11.17
Frederick Griffith (1879-1941)

When Griffith injected mice with the S strain, the mice died, but they did not die if he injected them with R strain (**Figure 11.18**). Griffith thought the capsule might be responsible for the *virulence* (ability to kill) of the S strain, so he injected mice with heat-killed S strain. The mice did not die. Finally, Griffith injected the mice with a mixture of heat-killed S strain and live R strain, the mice unexpectedly died. In fact, Griffith recovered living S strain bacteria from the bodies of the dead mice. These results clearly demonstrated that something had passed from the dead S strain to the living R strain transforming them into virulent bacteria.

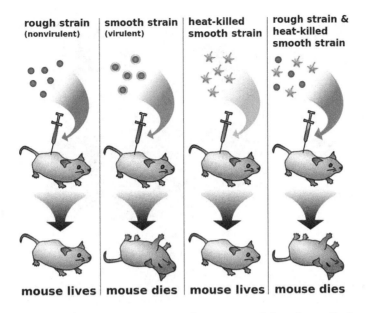

Figure 11.18 Griffith's Transformation Experiment. S strain caused the mice to die, but R strain did not. Heat-killed S strain did not kill the mice but heat-killed S strain plus live R strain did kill the mice.

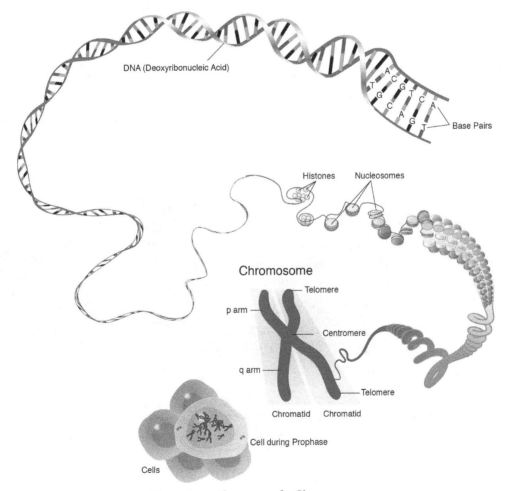

Figure 11.19 Structure of a Chromosome.

Figure 11.20 The research team that proved DNA is the genetic material: (a) Oswald Avery (1877-1955); (b) Colin MacLeod (1909-1972) and (c) Maclyn McCarty

By the 1940's, it was known that genes were on chromosomes and that chromosomes contain both DNA and proteins called histones (**Figure 11.19**). The question of the time was whether protein or DNA was the genetic material. In 1944, after 16 years of research, Oswald Avery, Maclyn McCarty, and Colin MacLeod (**Figure 11.20**), published a paper demonstrating that the transforming substance that allows *Streptococcus* to produce a capsule and be virulent is DNA. This conclusion was based on a number of facts:

1. DNA from S strain bacteria caused R strain bacteria to be transformed to the point where they produce a capsule and become virulent.
2. The addition of DNAase, an enzyme that digests DNA, prevents transformation from occurring.
3. The addition of enzymes that degrade protein and RNase, an enzyme that digests RNA had no effect on transformation. This shows that neither protein nor RNA is the genetic material.

In the 1950s, Alfred Hershey and Martha Chase (**Figure 11.21**) Firmly established DNA as the genetic material. Hershey and Chase used a virus called a T phage virus, composed of radioactively labeled DNA and capsid coat proteins, to infect *E. coli* bacteria. They discovered that the radioactive DNA tracers, but not protein ended up inside the bacterial cells, causing them to be transformed (**Figure 11.22**). Since only the genetic material could have caused this transformation, Hershey and Chase concluded that DNA must be the genetic material.

By the 1050s, DNA was known to be the genetic material of all living organisms. That left two final question to be answered: What is the structure of DNA and how does it function as the genetic material?

Figure 11.21 The research team that firmly established DNA as the genetic material: Alfred Hershey (1908-1997) and Martha Chase (1927-2003).

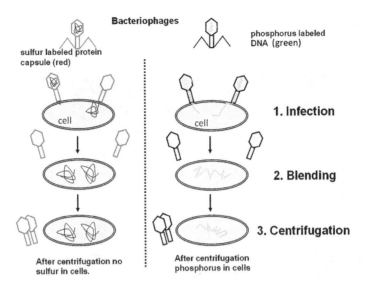

Figure 11.22 Hershey and Chase found that the radioactive phosphorous tagged DNA entered the cell and triggered transformation, but the sulfur labeled protein coat did not.

Structure of DNA

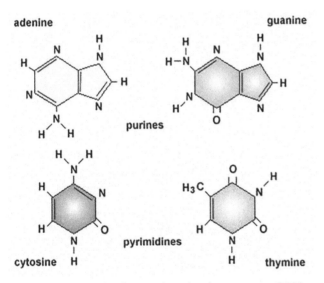

Figure 11.23 The four nucleotides that comprise DNA.

Figure 11.24
Edward Chargaff (1905-2002)

By the early 1950s, researchers knew that DNA contains four different nucleotides: two with *purine* bases, *adenine* (A) and guanine (G) that have a double ring; and two with *pyrimidine* bases, *thymine* (T) and *cytosine* (C) that have a single ring (**Figure 11.23**). In the early 1950s, biochemist Edward Chargaff (**Figure 11.24**) discovered two rules that helped lead to the discovery of the double helix structure of DNA:

Rule One: In DNA, the number of guanine units equals the number of cytosine units, and the number of adenine units equals the number of thymine units. This hinted at the base pair makeup of DNA.

Rule Two: The relative amounts of guanine, cytosine, adenine and thymine bases vary from one species to another.

Table 11.2 *Chargaff's Species Data*				
DNA Composition in Various Species (%)				
Species	**A**	**T**	**G**	**C**
Homo sapiens (humans)	31.0	31.5	19.1	18.4
Drosophila melanogaster (fruit fly)	27.3	27.6	22.5	22.5
Zea Mays (corn)	25.6	25.3	24.5	24.6
Neurospora crassa (fungus)	23.0	23.3	27.1	26.6
Escherichia coli (bacteria)	24.6	24.3	25.5	25.6
Bacillus subtilis (bacteria)	28.4	29.0	21.0	21.6

The nucleotide content of DNA is not fixed across species, and DNA does have the *variability* between species required for it to be the genetic material. Within each species, however, DNA was found to have the *constancy* required of the genetic material in that all members of a species have the same base composition.

Although only one of four bases is possible at each nucleotide position in DNA, the sheer number of bases and the length of most DNA molecules are more than sufficient to provide for variability. It has been calculated that each human chromosome typically contains about 140 million base pairs. This provides for a staggering number of possible sequences of nucleotides. Because any of the four possible nucleotides can be present at each nucleotide position, the total number of possible nucleotide sequences is $4^{140,000,000}$. No wonder each species has its own unique base percentages and each individual their own unique genome.

Rosalind Franklin (**Figure 11.25**) studied the structure of DNA using X-rays. Maurice Wilkins, a colleague of Franklin's, showed one of her crystallographic patterns to James Watson, who immediately grasped its significance.

James Watson, an American, was on a postdoctoral fellowship at Cavendish Laboratories in Cambridge, England, when he began to work with the biophysicist Francis Crick (**Figure 11.26**). Watson and Crick knew that DNA is a polymer of nucleotides, but they did not know how the nucleotides were arranged within the molecule. From Rosalind Franklin's X-ray diffraction studies of DNA, they deduced that DNA is a double helix with sugar-phosphate backbones on the outside and paired bases on the inside.

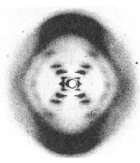

Figure 11.25 Rosalind Franklin (1920-1958) and the X-ray diffraction pattern her research produced. The crossed (X) pattern in the center told investigators that DNA is a helix, and the dark portions at the top and bottom told them that some feature is repeated over and over.

Figure 11.26 James Watson (1928-present) on the left and Francis Crick (1916-2004) on the right. These men worked out a model of the structure of DNA for which they were awarded the Nobel Prize in 1962

According to Watson and Crick's model, the two DNA strands of the double helix are *antiparallel*, meaning that the sugar-phosphate groups that are chained together by hydrogen bonds are oriented in the opposite direction (**Figure 11.27**). Each nucleotide possesses a phosphate group located at the 5' position of the sugar. Nucleotides are joined by linking the 5' phosphate of one nucleotide to a free hydroxyl (—OH) located at the 3' position of the nucleotide of the preceding nucleotide giving the molecule directionality. Antiparallel simple means that while one strand of DNA runs 5' to 3', the other strands runs in a parallel but opposite direction.

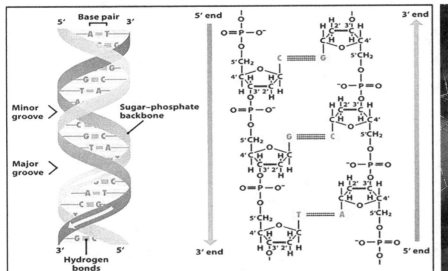
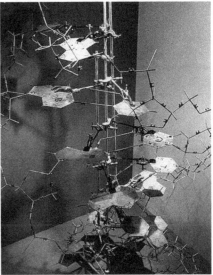

Figure 11.27 Watson-Crick Model of DNA Structure. (a) Diagrammatic view showing the antiparallel nature of the DNA molecule; (b) The crude metal model Watson and Crick used to work out their hypotheses about the structure of DNA.

The Watson-Crick model agreed with Chargaff's rules that state A=T and G=C. Figure 11.27 shows that A is hydrogen-bonded to T, and G is hydrogen-bonded to C. This complimentary base pairing means that a purine (large, two-ring base) is always bonded to a pyrimidine (smaller, one-ring base). This antiparallel pairing arrangement of the two strands ensures that the bases are oriented properly space-wise to allow their interaction.

Watson and Crick published their findings in a one-page paper, with the understated title "A Structure for Deoxyribose Nucleic Acid," in the British scientific weekly *Nature* on April 25, 1953, illustrated with a schematic drawing of the double helix by Crick's wife, Odile. A coin toss decided the order in which they were named as authors.

Replication of DNA

Once the structure of DNA had been worked out, questions, as usual, arose. During mitosis and meiosis, the DNA must be replicate. How does this happen? There were several hypotheses as to how DNA might replicate, but in 1958 Matthew Messelson and Franklin Stahl conducted a series of experiments that supported the hypothesis that DNA undergoes *semiconservative replication*. DNA replication is considered semiconservative because each daughter DNA double helix contains one strand from the parental DNA helix and a new strand. DNA replication requires three main steps: unwinding, complementary base pairing, and joining (**Figure 11.28**).

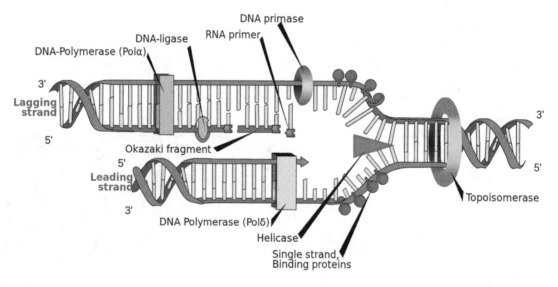

Figure 11.28 Topoisomerases are enzymes that regulate the overwinding or underwinding of DNA. DNA becomes overwound ahead of a replication fork. If left unabated, this torsion would eventually stop the ability of DNA or RNA polymerases involved in these processes to continue down the DNA strand.

Unwinding An enzyme, *DNA helicase* unwinds DNA and separates the parental strands. This creates two replication forks that move away from each other. These separated strands now become the template to create two new DNA molecules. DNA is stable as a helix, but not as single strands. Single-stranded binding proteins (SSB) attach to the newly separated DNA and prevent it from re-forming the helix so replication can continue.

Complementary Base Pairing An enzyme, *DNA primase*, places short RNA primers on the strands to be replicated. *DNA polymerase* recognizes this RNA target and begins DNA synthesis, allowing new nucleotides to form complementary base pairs with the old strand and connecting the new nucleotides together in a chain. DNA polymerase also proofreads the strands and can correct any errors.

The parental strands are antiparallel to each other, and each of the new daughter strands must also be antiparallel to its matching parental strand, but this creates a problem. DNA can only be synthesized in a 5' to 3' direction. One strand, the *leading strand*, is exposed so that the synthesis in a 5' to 3' direction is easier and replication is continuous. The other new strand, the *lagging strand*, in the fork must be synthesized in the opposite direction, requiring DNA polymerase to synthesize the new strand in short 5' to 3' segments with periodic starts and stops. These short strands are called Okazaki fragments after Japanese scientist Reji Okazaki, who discovered them.

Joining The enzyme *DNA ligase* is the "glue" that melds all the Okazaki fragments together, resulting in the two double helix molecules that are identical to each other and the original molecule. In eukaryotes, DNA replication begins at numerous sites along the length of the linear chromosome and replication bubbles spread bidirectionally until they meet. The chromosomes of eukaryotes are long, making replication a time-consuming process—500 to 5,000 base pairs per minute—but there are many individual sites that bubble up and speed up the process.

A DNA polymerase is very accurate and makes a mistake approximately once per 100,000 base pairs at most. However, this error rate would result in many errors accumulating over the course of repeated cell divisions. DNA polymerase is also capable of checking for accuracy, or proofreading the daughter strand it is making. It can recognize a mismatched nucleotide and remove it from a daughter strand by reversing direction and removing several nucleotides. Once it has removed the mismatched nucleotide, it changes direction again and resumes making DNA.

The Genetic Code—Function of DNA and RNA

That brings us to the final question...How can DNA found only in the nucleus regulate and control the rest of the cell? In other words, how does DNA function? Although scientists knew that DNA somehow directed protein production, they did not initially know specifically how the code was translated. Logically, the code would have to be at least a *triplet code;* that is, each coding unit, or **codon**, would need to be made up of three nucleotides to provide sufficient variety to encode 20 different amino acids.

In 1961, Marshall Nirenberg and J. Heinrich Matthei (**Figure 11.29**) performed an experiment that laid the groundwork for cracking the genetic code. In the experiment, an extract from bacterial cells that could make protein even when no intact living cells were present was prepared. Adding a synthetic RNA, poly-U, to this extract caused it to make a protein composed entirely of the amino acid phenylalanine. Therefore, the codon for phenylalanine was known to be UUU. Later, they were able to translate just three nucleotides at a time; in that way it was possible to assign an amino acid to each codon (**Figure 11.30**). This experiment cracked the first codon of the genetic code and showed that RNA controlled the production of specific types of protein.

Figure 11.29 Marshall Nirenberg (1927-2010) to the right and J. Heinrich Matthe (1929-present) to the left.

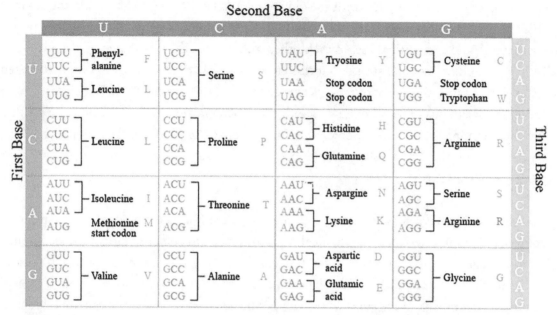

Figure 11.30 Messenger RNA codons. Notice that each of the codons (in boxes) is composed of three letters representing the first base, second base, and third base.

The genetic code is a masterpiece of scientific discovery because it is a key that unlocks the very basis of biological life. The genetic code has a number of different features:

1. The code is *universal*. The genetic code is universal to all living things. The universal nature of the code provides strong evidence that all living organisms share a common evolutionary heritage.
2. The code is *degenerate*. This term means that most amino acids have more than one codon. This redundancy (degeneracy) of the code helps protect against harmful mutations.
3. The code is *unambiguous*. Each triplet codon has only one meaning.
4. The code has *start* and *stop signals*. There is only one start codon, but there are three stop codons.

In the code, genetic information flows from the DNA in the nucleus to messenger RNA (mRNA) out into the cytoplasm where ribosomal RNA (rRNA) and transfer RNA (tRNA) translate it into protein. Together, the flow of information from DNA→ RNA→ protein→ phenotype is known as the *central dogma of molecular biology*. The code is translated in two steps: transcription and translation.

Transcription

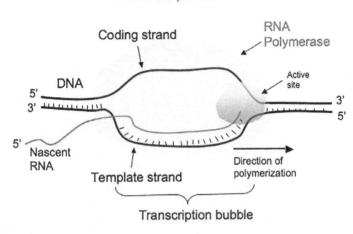

Figure 11.31

Transcription (writing) In ancient times before printing was invented, people called scribes copied documents. The "scribe" in the genetic code is messenger RNA (mRNA). Transcription proceeds in three stages: (**Figure 11.31**).

1. *Initiation* is the beginning of transcription. It commences when the enzyme *RNA polymerase* binds to a region of a gene known as the *active site* or *promoter*. This signals the DNA to unwind so the enzyme can "read" the bases in one of the DNA strands. This creates a *transcription bubble* that separates the two strands of DNA.

2. *Elongation* is the addition of nucleotides to the mRNA strand. RNA polymerase reads the unwound DNA strand and builds the mRNA molecule, using complementary base pairs. There is a brief time during this process when the newly formed RNA is bound to the unwound DNA. During this process, an adenine (A) in the DNA binds to a uracil (U) in the RNA. RNA polymerase reads down the DNA template strand in a 5' to 3' direction and continues until the RNA polymerase comes to a DNA stop signal, where termination occurs.

3. *Termination* is the ending of transcription and occurs when RNA polymerase crosses a stop (termination) sequence in the gene. The *nascent strand* or *primary transcript RNA* is complete, and it detaches from DNA. Primary transcript RNA is processed by adding a cap to the 5' end and 3' poly A (AAA) tail and removing *introns* (non-protein coding regions). The single strand is now a mature and functional mRNA and moves out of the nucleus through the nuclear pores and into the cytoplasm.

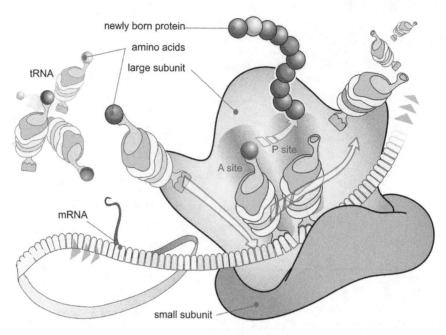

Figure 11.32

Translation is the second and final step needed to express a gene into a protein. During translation, the sequence of codons (nucleotide triplets) in the mRNA is held by a ribosomal RNA (rRNA) as transfer RNA (tRNA) connects a series of amino acids into a polypeptide. The translation process is divided into three steps: (**Figure 11.32**)

1. *Initiation*: When a small subunit of a ribosome charged with a tRNA + the amino acid methionine encounters an mRNA, it attaches and starts to scan for a start signal. When it finds the start sequence AUG, the codon (triplet) for the amino acid methionine, the large subunit joins the small one to form a complete ribosome, and protein synthesis is initiated.

2. *Elongation*: A new tRNA+amino acid enters the ribosome, at the next codon downstream of the AUG codon. If its **anticodon** matches the mRNA codon, it base pairs and the ribosome can link the two amino acids together (**Figure 11.33**).(If a tRNA with the wrong anticodon and therefore the wrong amino acid enters the ribosome, it cannot base pair with the mRNA and is rejected.) The ribosome then moves one triplet forward, and a new tRNA+amino acid can enter the ribosome and the procedure is repeated.

3. *Termination*: When the ribosome reaches one of three stop codons, for example, UGA, there are no corre-

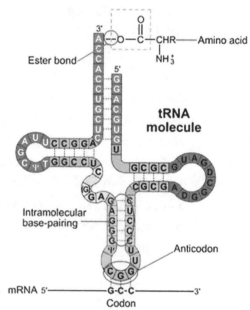

Figure 11.33 Structure of tRNA. The tRNA carrying an amino acid attaches to the mRNA on the ribosome. The triplet anticodon on the tRNA matches up to the triplet codon on the mRNA.

sponding tRNAs to that sequence. Instead termination proteins bind to the ribosome and stimulate the release of the polypeptide chain (the protein), and the ribosome dissociates from the mRNA. When the ribosome is released from the mRNA, the large and small subunit dissociate. The small subunit can now be loaded with a new tRNA+methionine and start translation once again. Some cells need large quantities of a particular protein. To meet this requirement they make many mRNA copies of the corresponding gene and have many ribosomes working on each mRNA. After translation, the protein will usually undergo some further modifications before it becomes fully active.

Translation occurs at ribosomes. Some ribosomes (polyribosomes) remain free in the cytoplasm, and some become attached to rough endoplasmic reticulum (ER). The first few amino acids of a polypeptide chain act as a signal peptide that indicates where the polypeptide belongs in the cell or if it is to be secreted from the cell.

Human Heredity

The problem with the human gene pool is that there is no lifeguard.

—David Gerrold

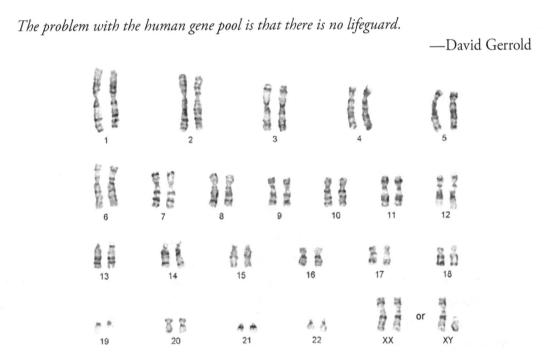

Figure 11.34 In a karyotype, 22 pairs of autosomes plus XX sexosomes is a female genome whereas 22 pairs of autosomes plus XY sexosomes is a male genome. The XY chromosome pair are nonhomologous.

A **karyotype** of the human chromosomes reveals a genome of 22 pairs (44) of homologous autosomes and 1 pair (2) of sex chromosomes (sexosomes) (**Figure 11.34**). A karyotype is a picture of all the chromosomes from a person's cells. A karyotype gives information about the number of chromosomes a person has, the structure of their chromosomes and the sex of the individual. Spriralling through those 46 human chromosomes are some 25,000 to 30,000 genes. As discussed previously, you are a unique genetic singularity in the flow of humankind from our biological origins to future generations yet unborn.

Mendel worked out the inheritance of garden peas with paper and pencil, but the hereditary patterns of complex animals, especially humans has proven far more difficult to discern. The study of human heredity has advanced slowly and proven difficult to study for a variety of reasons:

1. The large number of genes in human cells.

2. The long generation time in humans. Generation time in bacteria is measured in hours, fruit flies in days, mice in months, but decades in humans. That means decades between the P generation and the F_1 and then F_2 generations. Humans breed slowly.

3. The small number of offspring that each P generation of humans produces. A female fruit fly with a life span of 40-50 days can lay 2000 eggs whereas a female mouse with a lifespan of one to two years can produce 40 offspring a year. A human female produces only four offspring at best over an 80-year lifespan.

4. Legal, social, and moral concerns prevent conducting controlled breeding experiments with humans.

We can't conduct controlled breeding experiments with humans so we must study human inheritance patterns using indirect methods:

- **Population sampling** involves sampling a small portion of the population for a certain trait(s) and then extrapolating (expanding) those results to the whole population. Sampling is usually done because it is impossible to test every single individual in the population. It is also done to save time, money and effort while conducting the research.

- **Twin studies** allow comparisons between monozygotic (MZ or identical) twins and dizygotic (DZ or fraternal) twins, and are conducted to evaluate the degree of genetic and environmental influence on a specific trait. MZ twins are the same sex and share 100% of their genes. DZ twins can be the same- or opposite-sex and share, on average, 50% of their genes. Identical twins form during the first mitotic division of a fertilized egg cell. Each daughter cell of the fertilized egg cell goes on to form a complete human individual. Non-identical twins are the result of two separate fertilized egg cells each going on to form a complete human individual.

- **Family studies** have been used in genetics since its beginnings. Mendelian ratios were based on research about the relationship between garden peas. Mendel demonstrated that offspring get their genetic information from their parents. That is the foundational principle of family studies. A **pedigree** (family tree) is a graphic construct of the relationship between family members for a certain genetic trait (**Figure 11.35**). Pedigrees are commonly used in families to find out the probability of a child having a disorder in a particular family. The goals of pedigree analysis are to discover where the genes in question are located (X chromosome, Y chromosome, or autosomes), and to determine whether a trait is dominant or recessive. If a trait/phenotype shows a 50/50 ratio between men and women, it is autosomal, but it is considered X-linked if most of the males in a pedigree are afflicted with the disorder. Another use of pedigrees is to establish whether a trait is dominant or recessive.

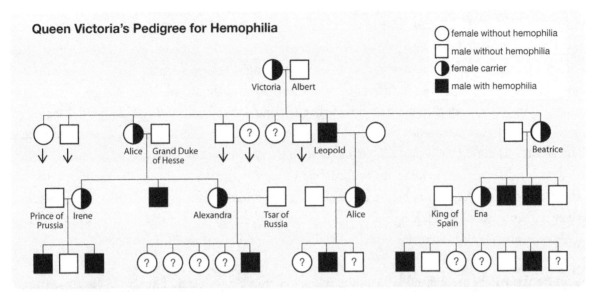

Figure 11.35 Hemophilia acquired the name the "royal disease" due to the high number of descendants of Queen Victoria afflicted by it. The first instance of hemophilia in the British Royal family occured on the birth of Prince Leopold on 7th April, 1853, Leopold was the fourth son and eighth child of Queen Victoria and Prince Albert of Saxe-Coburg-Gotha. No earlier occurrence of the disease in the Royal family had been known, it is assumed that a mutation occurred in the sperm of the Queen's father, Edward Augustus, Duke of Kent.

Human Genetic Disorders

Given the complexity of the human genome, it is little wonder that things sometimes go wrong and every person carries some dysfunctional genes. In the right (or wrong) combinations, defective genes can cause *genetic disorders*. Geneticists have identified over 3,000 genetic disorders. A genetic disorder is a condition caused in whole or in part by a change in the DNA sequence away from the normal sequence. Genetic disorders can be caused by a mutation in one gene (monogenic disorder), by mutations in multiple genes (multifactorial inheritance disorder), by a combination of gene mutations and environmental factors, or by damage to chromosomes (changes in the number or structure of entire chromosomes, the structures that carry genes).

As we unlock the secrets of the human genome (the complete set of human genes), we are learning that nearly all diseases have a genetic component. Some diseases are caused by mutations that are inherited from the parents and are present in an individual at birth, like sickle cell disease. Other diseases are caused by acquired mutations in a gene or group of genes that occur during a person's life. Such mutations are not inherited from a parent but occur either randomly or due to some environmental exposure (such as cigarette smoke). These include many cancers, as well as some forms of neurofibromatosis.

Genetic disorders typically involve the inheritance of a particular mutated disease-causing gene, such as sickle cell disease, cystic fibrosis, and Tay-Sachs disease. The mutated gene is passed down through a family, and each generation of children can inherit the gene that causes the disease. Rarely, one of these monogenic diseases can occur spontaneously in a child when his/her parents do not have the disease gene, or there is no history of the disease in the family. This can result from a new mutation occurring in the egg or sperm that gave rise to that child.

Most genetic disorders, however, are *multifactorial inheritance disorders*, meaning they are caused by a combination of inherited mutations in multiple genes, often acting together with environmental factors. Examples of such diseases include many commonly-occurring diseases, such as heart disease and diabetes, which are present in many people in different populations around the world.

Research on the human genome has shown that although many commonly occurring diseases are usually caused by inheritance of mutations in multiple genes at once, such common diseases can also be caused by rare hereditary mutations in a single gene. In these cases, gene mutations that cause or strongly predispose a person to these diseases run in a family, and can significantly increase each family member's risk of developing the disease. One example is breast cancer, where inheritance of a mutated BRCA1 or BRCA2 gene confers a significant risk of developing the disease. Geneticists categorize genetic disorders into three groups: monogenetic disorders, multifactorial inheritance disorders, and chromosome disorders.

Monogenetic disorders are caused by a point mutation in a single gene. The mutation may be present on one or both chromosomes (one chromosome inherited from each parent). Examples of monogenic disorders are: sickle cell disease, cystic fibrosis, polycystic kidney disease, and Tay-Sachs disease. Monogenic disorders are relatively rare in comparison with more commonly-occurring diseases, such as diabetes and heart disease. A major distinction among monogenic disorders is between "dominant" and "recessive" diseases. Dominant diseases are caused by the presence of the disease gene on just one of the two inherited parental chromosomes. In dominant diseases, the chance of a child inheriting the disease is 50 percent. In a family situation, for example, if the parents have four children, it may be possible that two of those children inherit the disease gene. Examples of dominant diseases are Huntington's disease, Marfan syndrome, and polydactyly (**Figure 11.36**).

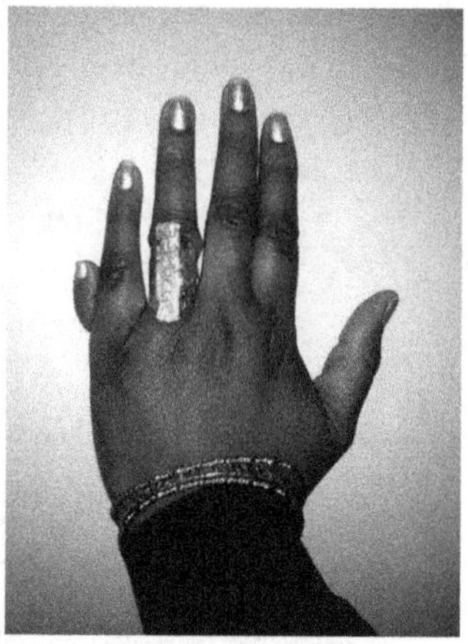

Figure 11.36 Preaxial polydactyly is a condition in which the extra digit has normal sensation but no joint and hence cannot not move independently

Recessive diseases require the presence of the disease gene on both of the inherited parental chromosomes. In this case, the chance of a child inheriting a recessive disease is 25 percent. In the family example, if the parents have four children, it may be more likely that only one child will develop the disease. Examples of recessive diseases include cystic fibrosis, Tay-Sachs disease, albinism, sickle cell disorder, and PKU (phenylketonuria). Some monogenetic disorders such as hemophilia, Duchene muscular dystrophy, color blindness, and androgenic alopecia (male pattern baldness).

Multifactorial inheritance disorders are caused by a combination of small inherited variations in genes, often acting together with environmental factors. Heart disease, diabetes, and most cancers are examples of such disorders. Behaviors are also multifactorial, involving multiple genes that are affected by a variety of other factors. Researchers are learning more about the genetic contribution to behavioral disorders such as alcoholism, obesity, mental illness and Alzheimer's disease.

Chromosome disorders are caused by an excess or deficiency of the genes that are located on chromosomes, or by structural changes within chromosomes. Down syndrome, for example, is caused by an extra copy of chromosome 21 (called trisomy 21), although no individual gene on the chromosome is abnormal (**Figure 11.37**). Turner's syndrome is the complete absence of one X in females (XO instead of XX), and Kleinfelter's syndrome is caused by one extra X chromosome in males (XXY instead of XY). In Chapter 13 we will examine the role of biotechnology in relieving the pain, suffering and even death caused by some genetic disorders.

Figure 11.37 Down syndrome is a chromosomal condition associated with intellectual disability, a characteristic facial appearance, and weak muscle tone (hypotonia) in infancy. People with Down's syndrome have a variety of birth defects and intellectual disabilities ranging from mild to moderate."

In Summary

- The German scientist Theodor Boveri and the American Walter Sutton, working independently, suggested that chromosomes could be shown to bear the material of heredity.
- Sutton's "The Chromosomes in Heredity" was published in 1903.
- Thomas Hunt Morgan discovered that eye color in *Drosophila* is expressed a sex-linked trait.
- Scientists have discovered a number of different chromosomal systems of sexual determination:
  - XX/XY System
  - XX/XO System
  - ZW System
  - Haplo-Diploid System
- A gene located on either sex chromosome is called a sex-linked gene whereas those located on the Y chromosome are called Y-linked genes and those on the X chromosome are X-linked genes.
- Genetic linkage is the tendency of alleles that are close together on a chromosome to be inherited together during the meiosis phase of sexual reproduction.
- A process called crossing over accounts for the recombination of linked genes.
- Alfred Sturtevant constructed the first genetic map using recombination frequencies.
- In the 1950s, Alfred Hershey and Martha Chase firmly established DNA as the genetic material.
- In the early 1950s, biochemist Edward Chargaff discovered two rules that helped lead to the discovery of the double helix structure of DNA:
- Rosalind Franklin studied the structure of DNA using X-rays.
- James Watson and Francis Crick were the first to work out the structure of DNA.
- DNA undergoes semiconservative replication. DNA replication is considered semiconservative because each daughter DNA double helix contains one strand from the parental DNA helix and a new strand.
- DNA replication requires three main steps: unwinding, complementary base pairing, and joining.
- In the genetic code, genetic information flows from the DNA in the nucleus to messenger RNA (mRNA) out into the cytoplasm where ribosomal RNA (rRNA) and transfer RNA (tRNA) translate it into protein.
- The code is translated in two steps: transcription and translation.

- The study of human heredity has advanced slowly and proven difficult to study for a variety of reasons.
- We can't conduct controlled breeding experiments with humans so we must study human inheritance patterns using indirect methods.
- A genetic disorder is a condition caused in whole or in part by a change in the DNA sequence away from the normal sequence.
- Geneticists categorize genetic disorders into three groups: monogenetic disorders, multifactorial inheritance disorders, and chromosome disorders.

Review and Reflect

1. *Rise of Genetics* Starting with Mendel and ending with Watson and Crick, create a genetics timeline. Your timeline should include all the major figures we discussed in this chapter and any minor players you feel should be recognized. Give a brief synopsis of the work of each individual or research team.

2. *Mendel Revisited* Which of Mendel's laws relates to the inheritance of alleles for a single character? Which law relates to the inheritance of alleles for two characters in a dihybrid cross?

3. *A Fortuitous Mutation* Why was it fortuitous that a mutant white-eyed fruit fly showed up in Thomas Hunt Morgan's famous fly room? Propose a possible reason that the first naturally occurring mutant fly Morgan saw involved a gene on a sex chromosome.

4. *Compare and Contrast* Compare and contrast the sex determination system for humans, birds, grasshoppers, and ants.

5. *Puzzling Data* Table 11.1 shows the data Thomas Hunt Morgan got when he crossed gray body normal wing wild type flies with black body vestigial wing flies. Why was he puzzled by the results and what hypothesis did he propose to explain the discrepancy between predicted results and actual ratios?

6. *Genetic Cartography* Genes *A*, *B*, and *C* are located on the same chromosome. Testcrosses show that the recombination frequency between *A* and *B* is 28% and between *A* and *C* is 12%. Draw a line to represent a chromosome and place these genes in their proper linear order on the chromosome.

7. *Follow the Rules* What are Chargaff's rules and how did these rules help determine the structure of DNA?

8. *Hidden Disorders* Neither Barbie or Ken has Duchene muscular dystrophy, but their firstborn son does. What is the probability that a second child will have the disorder? What is the probability if the second child is a boy? A girl?

9. *Hidden Connections* As strange as it sounds, there is a connection between the incidence of sickle cell anemia and the incidence of malaria. Explain this connection.

10. *Color-blind Mystery* Why are there more colorblind males than there are colorblind females? Explain what has to happen for a male to be colorblind and for a female to be colorblind.

11. **Too Many Genes** **Figure 11.38** is the karyotype of a female with Down's syndrome. How can you tell the person is a female and how can you tell they have Down's syndrome? Use Figure 11.34 for comparison.

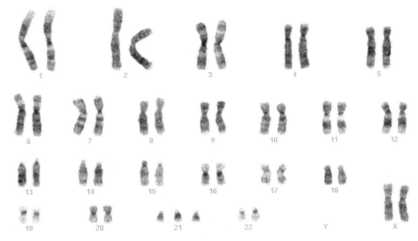

Figure 11.38

Create and Connect

1. **In the Fly Room** Thomas Hunt Morgan wanted to analyze the behavior of two alleles of a fruit fly eye-color gene. In crosses similar to those done by Mendel with pea plants, Morgan and his colleagues mated a wild-type (red-eyed) female with a mutant white-eyed male. The F_1 hybrids all had red eyes. If we perform a reciprocal cross of a white-eyed female with a red-eyed (wild-type) male what phenotypes and genotypes do you expect in the F_1 hybrids and in what percentages?

2. **Roll On** (**Figure 11.39**) The ability to roll your tongue (left) or not (right) is genetically determined. Is this phenotype dominant or recessive? Use population sampling techniques to answer this question.

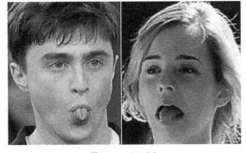

Figure 11.39

3. **Pedigree Analysis** Study the pedigree of three generations of Queen Victoria's offspring for the genetic disorder hemophilia as shown in Figure 11.35. Analyze this pedigree and answer the questions.

Questions

A. Did Queen Victoria have hemophilia? Explain

B. Hemophilia is a point mutation but is it dominant or recessive? Explain

C. What pattern do you see across the three generations shown in the pedigree? Explain

D. Is the defective gene carried on an autosome or a sex chromosome? Is hemophilia a sex-linked disorder? Explain

CHAPTER 12

GENE REGULATION AND EXPRESSION

The capacity to blunder slightly is the real marvel of DNA. Without this special attribute, we would still be anaerobic bacteria and there would be no music.

—Lewis Thomas

Introduction

In 1902, British physician Archibald Garrod (**Figure 12.1**) was the first to suggest that genes dictate phenotypes through enzymes that catalyze specific chemical reactions in the cell. Garrod postulated that the symptoms of an inherited disorder reflect a person's inability to make a particular enzyme. He later referred to such disorders as "inborn error of metabolism." As an example, Garrod pointed to the hereditary disorder alkaptonuria. In this disorder, the urine is black because it contains the chemical alkapton that darkens on exposure to air.

Alkapton build-up is the result of the afflicted body's inability to metabolically process the amino acids phenylalinine and tyrosine. Garrod reasoned that most people inherit the proper enzymes to metabolize alkapton whereas people with alkaptonuria have not inherited the proper metabolic enzymes.

Figure 12.1
Archibald Garrod (1857-1936)

Garrod pioneered the field of inborn errors of metabolism with his study of alkaptonuria and expanded his studies to include cystinuria, pentosuria, and albinism so strikingly illustrated by the snake in the chapter-opening photo.

Gene Expression

Garrod's work was ahead of its time, but research into gene expression decades later supported his hypothesis that a gene is expressed with the production of a specific enzyme. Biochemists accumulated much evidence that cells synthesize and degrade most organic molecules via metabolic pathways in which each chemical reaction in a sequence is catalyzed by a specific enzyme. Gene expression then is the process by which coded information from a gene is used in the synthesis of a *functional* gene product, mainly proteins. A gene is not expressed if the product it produces is not functional. The process of gene expression is used by every life form known to generate the macromolecular machinery of life.

In the 1930s, George Beadle and Boris Ephrussi speculated that in *Drosophila*, each of the various mutations affecting eye color pigments prevents pigment synthesis at a specific step by preventing production of the enzyme that catalyzes that step. However, neither the chemical reactions nor the enzymes that catalyze them were known at the time, and the complexity of *Drosophila* proved a drawback to developing experiments that would demonstrate a link between specific genes and their chemical products.

In 1941, George Beadle and Edward Tatum (**Figure 12.2**) turned to a simpler creature, in which specific products of metabolism could be directly studied. A bread mold, *Neurospora crassa*, proved ideal. *Neurospora* can be cultured together with sugar, inorganic salts, and the vitamin biotin. This fungus has a short life cycle, and reproduces sexually and replicates asexually—that is, sexual reproduction gives rise to spores. In addition, *Neurospora* possesses only one set of unpaired chromosomes, so that any mutation is immediately expressed.

In what became a celebrated experiment, Beadle and Tatum first irradiated a large number of *Neurospora* and thereby produced some organisms with mutant genes. They then crossed these potential mutants with non-irradiated *Neurospora*. Normal products of

George Wells
Beadle
(1903 - 1989)

Edward Lawrie
Tatum
(1909 - 1975)

Figure 12.2 The research team that develop the foundation concept of one gene-one enzyme.

this sexual recombination could multiply in a simple growth medium. However, Beadle and Tatum showed that some of the mutant spores would not replicate without the addition of a specific amino acid—arginine. They developed four strains of arginine-dependent *Neurospora*—each of which, they showed, had lost the use of a specific gene that ordinarily facilitates one particular enzyme necessary to the production of arginine.

Beadle and Tatum's fairly simple experiment was a keystone in the development of molecular biology. In its basic form, the concept that genes produce enzymes had been first put forth as early as 1901 by Archibald Garrod—as Beadle acknowledged when he and Tatum were awarded the Nobel Prize in Physiology or Medicine in 1958. Although Garrod's work had been largely ignored, Beadle and Tatum's research, more than three decades later, was immediately recognized because of their experimental evidence.

From Beadle and Tatum's work arose a basic hypothesis: One gene specifies the production of one enzyme. This idea has proven exceptionally fruitful, but also much debated and eventually modified. Today, it is usually said, more accurately, that each gene specifies the production of a single polypeptide—that is, a protein or protein component. Thus, two or more genes may contribute to the synthesis of a particular enzyme. In addition, some products of genes are not enzymes per se, but structural proteins.

Today we know of countless examples in which a mutation in a gene causes a faulty enzyme that eventually leads to an identifiable condition. The albino snake that ushered in this chapter lacks a key enzyme called tyrosinase in the metabolic pathway that produces melanin, a dark pigment. The absence of melanin causes white scales, as well as other effects throughout the snake's body. Its eyes are pink because there is no melanin present to mask the reddish color of the blood vessels there.

As researchers learned more about proteins and how they are coded for, they made revisions to Beadle and Tatum' s one gene-one enzyme hypothesis. For one thing, not all proteins are enzymes. Some proteins are structural, such as keratin that forms animal hair, hooves, and nails whereas others, such as insulin, are hormonal. Thus, biologists modified Beadle and Tatum to one gene-one protein. However, they then discover that many proteins are constructed from more than one gene. For example, hemoglobin, the oxygen-transporting protein found on the red blood cells of vertebrates, contains two kinds of polypeptides indicating that each polypeptide is coded for by two different genes. Again Beadle and Tatum was revised and restated as *one gene-one polypeptide.*

Coding for Proteins

At this point in our discussion of gene expression, some of the things we learned in the previous chapter need to be summarized:

- Genes provide the instruction for making specific proteins, but a gene does not build a protein directly.
- The bridge between DNA and proteins synthesis is the nucleic acid RNA.
- RNA is chemically similar to DNA except that it contains ribose instead of deoxyribose as its sugar and have the nitrogenous base uracil (U) rather than thymine (T).
- DNA is a double helix whereas RNA is a single strand.
- Each nucleotide along a DNA strand has A, G, C, or T as its base and each nucleotide along an RNA strand has A, G, C, or U as its base.
- The nucleotide code from the DNA is translated into an amino acid code on ribosomes in the cytoplasm.
- Transcription is the process of forming mRNA from the nucleic DNA.
- Translation is the process of tRNA reading the triplet code on the mRNA and bringing in amino acids to be linked into the polypeptide coded for from the nuclei DNA.
- The genetic code is nearly universal, shared by organisms from the tiniest bacteria to complex plants and animals.
- The universality of the genetic code is evolutionarily significant because it suggests that the code was present in the common ancestor(s) of all present-day organisms.

The basic mechanics of transcription and translation are similar for bacteria and eukaryotes with one exception. Because bacteria do not have nuclei, their DNA is not separated from the cytoplasm by nuclear membranes from ribosomes and other protein-synthesizing components. This lack of compartmentalization allows translation of a mRNA to begin while its transcription is still in progress.

Mutations—Altering the Code

Genetic diversity, the driving force of natural selection, is increased by changes to the genetic code known as mutations. A **mutation** is the permanent altering of a nucleotide sequence in the genome of an organism. There are two categories of mutations:

1. **Hereditary mutations** are inherited from a parent and are present throughout a person's life in virtually every cell in the body. These mutations are also called germline mutations because they are present in the parent's gametes (egg or sperm) that are also known as germ cells. When an egg and a sperm cell unite, the resulting fertilized egg cell receives DNA from both parents. If this DNA has a mutation, the offspring that grows from the fertilized egg will have the mutation in each of his or her cells.

2. **Acquired mutations** or **somatic mutations** occur at some time in an organism's life and are present only in certain cells, not in every cell in the body. These changes can be caused by environmental factors such as ultraviolet radiation from the sun, or can occur if a mistake is made as DNA copies itself during cell division. Acquired mutations in somatic cells (body cells) cannot be passed on to the next generation.

Mutations may be classified as small-scale or large-scale. Small-scale mutations may be either point mutations, substitutions, insertions, deletions or frameshift mutations.

- **Point mutations** are small-scale mutations that result in changes to a single nucleotide pair of a gene (**Figure 12.3**). If a point mutation occurs in a gamete or in a cell that gives rise to gametes, it may be transmitted to offspring and future generations. A point mutation that does not change the encoded amino acid is said to be a *silent mutation* whereas a *missense mutation* would be a mutation in which a change in one DNA base pair results in one DNA base pair results in

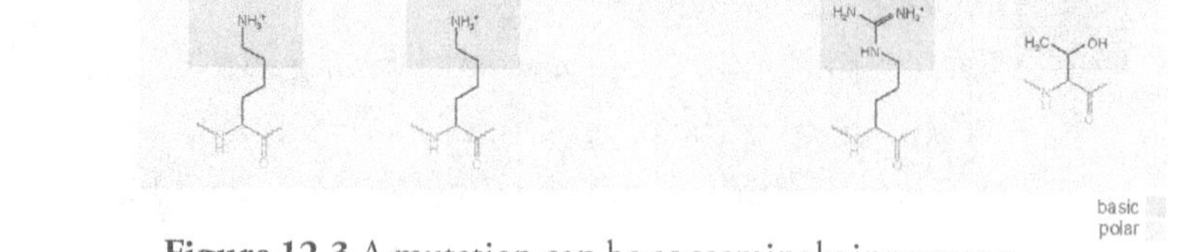

Figure 12.3 A mutation can be as seemingly innocuous as the replacement of a single nucleotide.

the substitution of one amino acid for another in the protein made by a gene, and a *nonsense mutation* which is also a change in just one base causes a complete stoppage of polypeptide formation.

If a mutation has an adverse effect on the functionality of the protein it codes for, the mutant condition is referred to as a genetic disorder. As an example, we can trace the genetic basis for the disorder sickle-cell anemia to the mutation of a single nucleotide pair in the gene that encodes for the ß-globin polypeptide of hemoglobin. This point mutation leads to the production of an abnormal protein (**Figure 12.4**). Sickle cell anemia is an inherited form of anemia — a condition in which there aren't enough healthy red blood cells to carry adequate oxygen throughout your body.

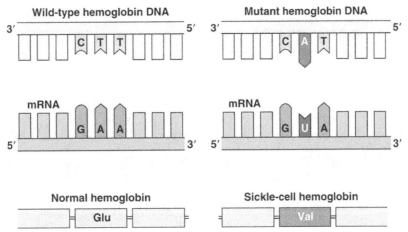

Figure 12.4 The replacement of a single nucleotide (A for T) leads to a mistake in the mRNA that results in a nonfucntional protein.

Normally, your red blood cells are flexible and round, moving easily through your blood vessels. In sickle cell anemia, the red blood cells become rigid and sticky and are shaped like sickles or crescent moons. These irregularly shaped cells can get stuck in small blood vessels, which can slow or block blood flow and oxygen to parts of the body (**Figure 12.5**).

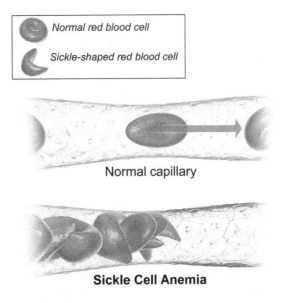

Figure 12.5 The shape of the mutant red blood cells causes them to not move smoothly and easily through the capillaries. They tend to accumulate at junctions between capillaries where they form clots.

- **Substitutions** occur when one nucleotide and its partner are replaced with another pair of nucleotides (**Figure 12.6**). Substitution mutations are usually missense mutations in that the altered codon still codes for an amino acid, just not the correct amino acid.
- **Insertions** (**Figure 12.7**) are mutations where extra base pairs are inserted into a new place on the DNA whereas deletions (**Figure 12.8**) are mutations in which a section of DNA is lost or deleted. These mutations have a disastrous effect on the resulting protein more often than substitutions do.
- **Frameshift mutations** occur when insertions or deletions alter the reading frame of the triplet grouping of nucleotides on the mRNA that is read during translation (**Figure 12.9**). In Figure 12.9 a deletion has altered a codon causing the codon to be parsed incorrectly.

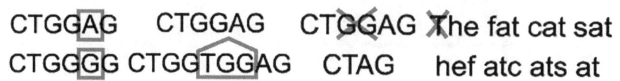

| Figure 12.6 | Figure 12.7 | Figure 12.8 | Figure 12.9 |

Large-scale mutations are known as *chromosome anomalies*. A chromosome anomaly is a missing, extra, or irregular portion of chromosomal DNA. An abnormal number of chromosomes is called **aneuploidy**. Aneuploidy occurs when an individual either is missing a chromosome from a pair (monosomy) or has more than two chromosomes of a pair (trisomy, tetrasomy). Aneuploidy disorders in humans will be discussed in detail in Chapter 13.

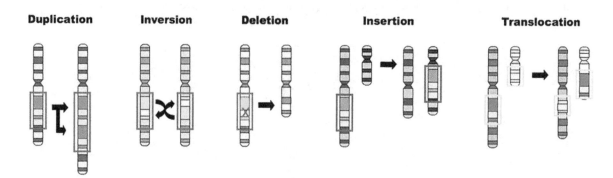

Figure 12.10

Chromosome structure may be altered in several ways: (**Figure 12.10**)
- *Deletions*: A portion of the chromosome is missing or deleted.
- *Duplications*: A portion of the chromosome is duplicated, resulting in extra genetic material.
- *Translocation*: A portion of one chromosome is transferred to another chromosome.
- *Inversions*: A portion of one chromosome has broken off, turned upside down, and reattached, resulting in the genetic material being inverted.
- *Insertions*: A portion of one chromosome has been deleted from its normal place and inserted into another chromosome.

Causes of Mutations

DNA replication is a truly amazing biological phenomenon. Consider the countless number of times that your cells divide to make you who you are—not just during development, but even now, as a fully mature adult. Then consider that every time a human cell divides and its DNA replicates, it has to copy and transmit the exact same sequence of 3 billion nucleotides to its daughter cells. Finally, consider the fact that in life (literally), nothing is perfect. While most DNA replicates with fairly high fidelity, mistakes do happen, with polymerase enzymes sometimes inserting the wrong nucleotide or too many or too few nucleotides into a sequence. Fortunately, most of these mistakes are fixed through various processes. Repair enzymes recognize structural imperfections between improperly paired nucleotides, cutting out the wrong ones and putting the right ones in their place. But some replication errors make it past these mechanisms, thus becoming permanent mutations. These altered nucleotide sequences can then be passed down from one cellular generation to the next, and if they occur in cells that give rise to gametes, they can even be transmitted to subsequent organismal generations. Moreover, when the genes for the DNA repair enzymes themselves become mutated, mistakes begin accumulating at a much higher rate. In eukaryotes, such mutations can lead to cancer.

Mutations arise in a number of ways:

1. **Spontaneous mutations** are point mutations that occur during DNA replication or recombination. These errors can lead to nucleotide pair substitution, insertions, or deletions, as well as to mutations affecting longer stretches of DNA. In many cases, the error will be corrected by DNA proofreading and repair systems. Otherwise, the incorrect base will be used as a template in the next round of replication. The spontaneous mutation rate varies. Large genes provide a large target and tend to mutate more frequently. A study of the five coat color loci in mice showed that the rate of mutation ranged from 2×10^{-6} to 40×10^{-6} mutations per gamete per gene. Data from several studies on eukaryotic organisms shows that in general, the spontaneous mutation rate is $2\text{-}12 \times 10^{-6}$ mutations per gamete per gene. Given that the human genome contains 100,000 genes, we can conclude that we would predict that 1-5 human gametes would contain a mutation in some gene.

2. **Induced mutations** are those caused by some sort of mutagen. **Mutagens** are physical and chemical agents that act on DNA from the outside. Mutations may be induced by exposure to ultraviolet rays and alpha, beta, gamma, and X-ray radiation, by extreme changes in temperature, and by certain mutagenic chemicals. Nucleotide analogs are chemicals similar to normal DNA nucleotides but that pair incorrectly during DNA replication. Other chemical mutagens interfere with correct DNA replication by inserting themselves into the DNA and distorting the double helix. Still other mutagens cause chemical changes in bases that change their pairing properties.

Gene Regulation

A cell only express a subset of its genome. This is a critical feature in multicellular organisms. Imagine if your skin starting expressing heart muscle protein that is normally only expressed in heart muscle cells. Obviously, gene expression is precisely regulated.

Cruising through freshwater lakes and ponds in Central and South America is *Anableps anableps*, a fish known locally as "cuatro ojos" ("four eyes") (**Figure 12.11**). The eye's upper half is well-suited for aerial vision and the lower half for aquatic vision. The molecular explanation for this is that the cells of the two parts of the eye express a slightly different set of genes involved in vision, even though these two groups of cells are quite similar and contain identical genomes. Precise regulation of the genes involved makes this remarkable biological feat possible.

Figure 12.11 The four-eyed fish does not have four eyes, but it does have four pupils. Each eye's two pupils are separated by a band of iris.

Genes can't control an organism on their own; rather, they must interact with and respond to the organism's environment. Some genes are *constitutive genes*, or always "on," regardless of environmental conditions. Such genes are among the most important elements of a cell's genome, and they control the ability of DNA to replicate, express itself and repair itself. These genes also control protein synthesis and much of an organism's central metabolism. In contrast, regulated genes are needed only occasionally — but how do these genes get turned "on" and "off"? What specific molecules control when they are expressed?

It turns out that the regulation of such genes differs between prokaryotes and eukaryotes. In prokaryotes, most regulatory proteins are negative and therefore turn genes off. In 1961, French microbiologists Francois Jacob and Jacques Monod demonstrated that *Escherichia coli* is capable of regulating the expression of its genes. They observed that the genes in a metabolic pathway, called *structural genes*, are grouped on a chromosome and transcribed at the same time. Armed with these facts, they proposed the *operon model* to explain gene regulation in prokaryotes.

A key advantage of grouping genes of related function into one transcription unit (operon) is that a single "off-on switch" can control the whole cluster of functionally related genes; in other words, these genes are coordinately controlled. A classic example is the *lac* operon. The operon consists of a *regulatory gene* that encodes a repressor protein that binds to the *operator site*, and a promoter that initiates transcription of a particular gene. It would be wasteful to produce enzymes when there is no lactose available or if there is a preferable energy source available, such as glucose. The *lac* operon uses a two-part control mechanism to ensure that the cell expends energy producing the enzymes encoded by the *lac* operon only when necessary.

The disaccharide lactose (milk sugar) is available to *E. coli* that live in the human gut if the host drinks milk. Lactose metabolism begins with hydrolysis of the disaccharide into its component monosaccharides (glucose and galactose), a reaction catalyzed by the enzyme ß-galactosidase. The gene for ß-galactosidase (*lacZ*) is part of the *lac* operon (**Figure 12.12**). If lactose is not present, the regulatory gene produces a repressor protein that represses the operator so the enzyme can be produced. However, in the presence of lactose, an inducer inactivates the repressor and the operon produces enzyme.

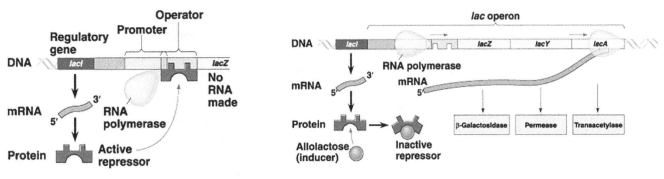

(a) Lactose absent, repressor active, operon off

(b) Lactose present, repressor inactive, operon on

Figure 12.12 The lac operon in E. coli. (a) In the absence of lactose the lac repressor switches off the operon by binding to the operator; (b) In the presence of lactose, the lac repressor is inactivated and the operon is activated to produce the enzyme.

Regulation of the *lac* operon involves *negative* control of genes because the operons are switched off by the active form of the repressor protein. Gene regulation is said to be *positive* only when a regulatory protein interacts directly with the genome to switch transcription on. Interestingly, in eukaryotes, the default state of gene expression is "off" rather than "on," as it is in prokaryotes.

Both prokaryotes and eukaryotes continually turn genes on and off in response to signals from their internal and external environments. Regulation of gene expression is also essential for cell specialization in multicellular organisms that are made up of many different types of cells. To perform its own distinct role, each cell type must maintain a specific program of gene expression in which genes certain genes are expressed, and others are not.

A typical human cell might express about 20% of its polypeptide-coding genes at any given time, and highly differentiated cells, such as nerve or muscle cells, express even a smaller fraction of their genes. The genome in all the cells of an organism is identical, so differences between cell types is not due to different genes, but to differential gene expression, the expression of different genes by cells with the same genome.

In all organisms, gene expression is commonly controlled at transcription, and regulation at this stage often occurs in response to signals coming from outside the cell, such as hormones or signaling molecules. The greater complexity of eukaryotic cell structure and function provides greater opportunities for regulating gene expression. In eukaryotes, there are five primary levels of gene regulation: three pertain to the nucleus (chromatin structure, transcriptional control, and posttranscriptional control) and two pertain to the cytoplasm (translational control and posttranslational control) (**Figure 12.13**).

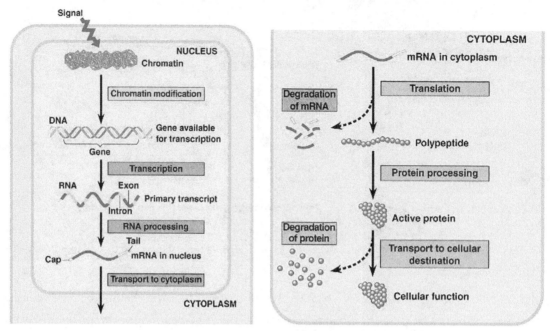

Figure 12.13 Stages in eukaryotic cells at which gene expression can be regulated. The nuclear envelope separating transcription from translation in eukaryotic cells offers an opportunity for post-transcriptional control that is absent in prokaryotes. In addition, eukaryotes have a greater variety of control mechanisms operating before transcription and after translation.

Chromatin structure If the DNA from all 46 chromosomes in a human cell nucleus was laid out end to end, it would measure approximately two meters. However, the diameter would be only 2 nm. Considering that the size of a typical human cell is about 10 μm (100,000 cells lined up to equal one meter), DNA must be tightly packaged to fit in the cell's nucleus. At the same time, it must also be readily accessible for the genes to be expressed. During some stages of the cell cycle, the long strands of DNA are condensed into compact chromosomes. There are a number of ways that chromosomes are compacted to fit in the cell's nucleus and be accessible for gene expression. In the first level of compaction, short stretches of the DNA double helix wrap around a core of eight histone proteins at regular intervals along the entire length of the chromosome. Each beadlike histone-DNA complex is called a **nucleosome**. A DNA molecule in this form is about seven times shorter than the double helix without the histones (**Figure 12.14**).

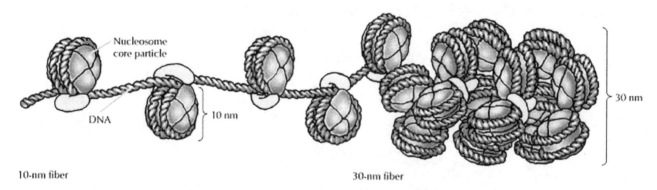

Figure 12.14 Molecular Assemblage of Nucleosomes. The DNA (red) is wrapped around the histone (blue) and both form the nucleosome core particle. This structure is locked in mammals by the linker histone H1 (yellow).

The next level of compaction occurs as the nucleosomes are coiled into a 30-nm chromatin fiber. This coiling further shortens the chromosome so that it is now about 50 times shorter than the extended form. In the third level of packing, the nucleosomes fold into a 30 nm in diameter structure called the solenoid.

The degree to which chromatin is compacted greatly affects the accessibility of the chromatin to the transcriptional process. Active genes in eukaryotic cells are associated with more loosely packed chromatin called *euchromatin* whereas more tightly packed DNA, called *heterochromatin*, contains mostly inactive genes (**Figure 12.15**). When DNA needs to be transcribed, a group of proteins called the *chromatin remodeling complex* pushes aside (unpacks) the histone portion of a nucleosome thereby exposing a length of DNA (gene) allowing transcription to begin.

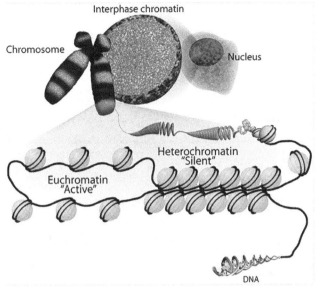

Figure 12.15 Histones regulate accessibility to DNA; chromatin becomes genetically active when histones no longer block access to DNA. Heterochromatin is said to be "silent" because histones are blocking the DNA. Euchromatin is said to be "active" because the DNA has been unpacked and is accessible to the transcription machinery."

Transcriptional control Chromatin modifying enzymes provide initial control of gene expression by making a region of DNA either more or less able to bind the transcription machinery. Once the chromatin is optionally modified, the initiation of transcription is the next major step at which gene expression is regulated.

Recall that RNA polymerase attaches on the promoter sequence at the "upstream" end of the gene, and then proceeds to transcribe the gene, synthesizing a mature mRNA. Associated with most eukaryotic genes are multiple *control elements*, segments of noncoding DNA that serve as binding sites for proteins called *transcription factors* that, in turn, regulate transcription. Control elements and the transcription factors they bind are critical to the precise regulation of gene expression seen in different cell types. A cell has many different kinds of transcription factors, and a variety of transcription factors may be active at a single promoter. Transcription factors may promote transcription or if key transcription factors are absent repress transcription.

Even if all the transcription factors are present, transcription may not begin without the assistance of a DNA-binding protein called a *transcription activator*. These activators bind to regions of DNA called *enhancers* that may be located some distance from the promoter. A hairpin loop in the DNA brings the transcription activators attached to the enhancer into contact with the transcription factor complex (**Figure 12.16**). Likewise,

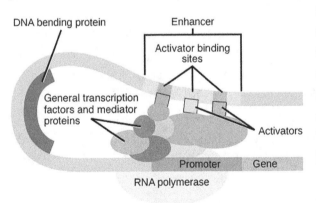

Figure 12.16 Eukaryotic Transcription Factors. Transcription requires that transcription factors bind to the promoter and transcription activators bind to an enhancer. The DNA loops and mediator proteins act as a bridge linking activators to factors. Only then does transcription begin.

the binding of repressors within the promoter may prohibit the transcription of certain genes. Most genes are subject to regulation by both activators and repressors.

Posttranscriptional control of gene expression occurs in the nucleus and includes mRNA splicing and controlling the speed with which mRNA leaves the nucleus. RNA processing in the nucleus and the export of mature mRNA to the cytoplasm provide several opportunities for regulating gene expression that are not available in prokaryotes. One example is alternative RNA splicing depends on which RNA segments are treated as *exons* (expressed sequences that code for polypeptides) and which as *introns* (intervening sequences that do not code for polypeptides) (**Figure 12.17**).

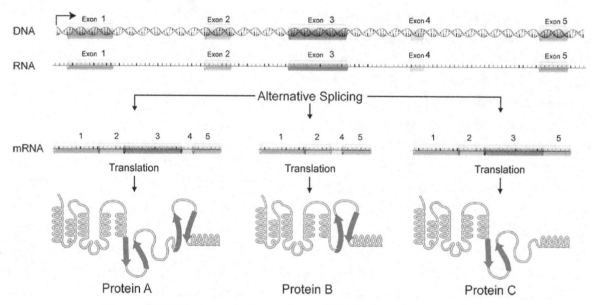

Figure 12.17 The removal of exons from mRNA leads to alternative splicing resulting in different translations and ultimately different proteins.

Translational control of gene expression occurs in the cytoplasm where translation presents another opportunity for regulating gene expression. For some mRNAs, the initiation of translation can be blocked by regulatory proteins that bind to specific sequences or structures within the untranslated region (UTR) at the 5' end or the 3' end, preventing the attachment of the ribosome.

Alternatively, translation of *all* of the mRNAs in a cell can be regulated simultaneously. Such "global" control usually involves the activation or inactivation of one or more of the protein factors required to initiate translation.

Posttransitional control, which also takes place in the cytoplasm, occurs after protein synthesis. Often, eukaryotic polypeptides must be processed to yield functional protein molecules. In addition, many proteins undergo chemical modifications that make them functional. For example, regulatory proteins are commonly activated or inactivated by the reversible addition of phosphate groups and sugars. Many proteins must be transported to target destinations in the cell in order to function. Regulation might occur at any of the steps involved in modifying or transporting a protein.

Lastly, the length of time each proteins functions in the cell is strictly regulated by means of **selective degradation**. To time stamp a particular protein for destruction, the cell commonly attaches molecules of a small protein called *ubiquitin* to the protein. Giant protein complexes called *proteasomes* then recognize the ubiquitin-tagged proteins and degrade them.

Box 12.1
Epigenetics—Same but Different

Figure 12.18

Bob and Bill are identical twins in their early-30s. They both have thin, brown hair, are slightly balding wear about the same size clothes and shoes (**Figure 12.18**). As children, their parents emphasized their similarities by often dressing them alike and giving them the same sports and academic opportunities. Once they became teenagers, however, things began to change. Bill preferred jeans and raggedy T-shirts while Bob preferred slacks and polos. Their personalities began to change as well. Bill was outgoing and spontaneous whereas Bob was more reserved and thoughtful, and now Bill is showing elevated PSA levels in his blood indicating the possibility of prostate cancers while Bob's numbers are perfect.

How is it possible that two people with basically the same genome and raised alike can grow to be so different? Any person is the product of two forces: nature and nurture. Nature, your genome, provides all the genetic traits you possess, your phenotype. Nature is internal whereas nurture is external. Nurture is all aspects of your personal and social life from the way you were reared to your educational experiences to your preferences for everything from food to mates; everything about your life other than what is determined by your genes.

Researchers tell us that there may be a third force at work that can affect a person's overall health and well-being. Researchers working with identical twins believe there is a bridge between nature and nurture in the form of epigenetics. Identical twins present a unique opportunity to study epigenetics because they are clones resulting from the initial division of a single fertilized egg. Assuming a similar upbringing, their gradual differences over time can be attributed to their disparate control of genes (**Figure 12.19**).

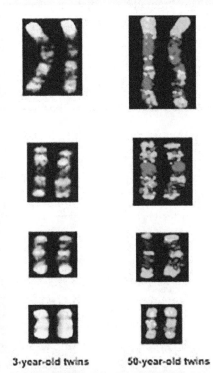

3-year-old twins   50-year-old twins

Figure 12.19 One twin's epigenetic tags are dyed green, and the other twin's tags are dyed red. An overlap in green and red shows up as yellow. The 50-year twins have more epigenetic tags in different places than do the 3-year old twins.

Epigenetics is the study of heritable changes in gene expression (active versus inactive genes) that does not involve changes to the underlying DNA sequence — a change in phenotype without a change in genotype — that affects how cells read the genes. Epigenetic change is a regular and natural occurrence but can also be influenced by several factors including age, and the environment/lifestyle.

Epigenetic modifications can manifest as commonly as the manner in which cells terminally differentiate to end up as skin cells, liver cells, brain cells, etc. Or, epigenetic change can have more damaging effects that can result in diseases like cancer. At least three systems including DNA methylation (a methyl group attaches to the cytosine base of DNA), histone modification and chromatin remodeling, and non-coding RNA (ncRNA)-associated gene silencing are currently considered to initiate and sustain epigenetic change.

New and ongoing research is continuously uncovering the role of epigenetics in a variety of human disorders and fatal diseases. Interest in epigenetics has led to new findings about the relationship between epigenetic changes and a host of disorders including various cancers, mental retardation associated disorders, immune disorders, neuropsychiatric disorders and pediatric disorders. Identical twin discordance for autism, psychiatric disorders, and cancer have been shown to have different DNA methylation on certain genes.

Epigenetic changes may be reversible. A study using rats showed that rat pups that are licked and nurtured by their mothers become calm adults. Rat pups that are not nurtured are anxious. Injecting a calm rat with a drug that adds methyl groups creates an anxious rat. Conversely, injecting an anxious rat with a drug that removes methyl groups creates a calm rat. Thus we see that the differences between Bill and Bob are most likely the cumulative effect of their nature (epigenetics) rather than changes in their nurture.

Genome sequencing has revealed that protein-coding DNA accounts for only 1.5% of the human genome and a similarly small percentage of the genomes of many other multicellular eukaryotes. A very small fraction of the non-coding DNA consists of genes for RNAs such as ribosomal RNA and transfer RNA the remainder was initially termed "junk DNA." However, recent data contradicts this notion. A massive study of the entire human genome completed in 2010 showed that roughly 75% of the genome is transcribed at some point in any given cell.

These and other results suggest that a significant amount of the genome may be transcribed into non-protein-coding RNAs (ncRNAs) including a variety of small RNA (sRNA) molecules. Scientists now know that sRNA molecules represent an important form of gene regulation that functions at multiple levels of gene expression. How do sRNA molecules regulate gene expression?

1. The transcribed RNA can form loops as hydrogen bonding occurs between its bases.
2. The double-stranded RNA molecules (dsRNA) is diced up by enzymes in a cell to form sRNA molecules.
3. Some of these sRNA molecules regulate transcription whereas others are involved in the regulation of translation and some alter the compaction of DNA so that some genes are inaccessible to the transcription machinery of the cell.
4. Small RNAs are the source of microRNAs (miRNAs), small snippets of RNA that can bind to and disable the translation of mRNA in the cytoplasm.

5. Small RNAs are also the source of small-interfering RNA (siRNAs) that join with an enzyme to form an active silencing complex. This activated complex targets specific mRNAs in the cell for breakdown, preventing them from being expressed.

By using a combination of miRNA and siRNA molecules, a cell can fine-tune the amount of product being expressed from a gene, much as a dimmer switch can regulate the brightness of the lights in a room. Because both miRNA and siRNA molecules interfere with the normal gene expression pathways, the process is often referred to as *RNA interference.*

In Summary

- Gene expression then is the process by which coded information from a gene is used in the synthesis of a *functional* gene product, mainly proteins.
- Beadle and Tatum proposed that one gene specifies the production of one enzyme.
- Beadle and Tatum's theorem was revised and is now restated as *one gene-one polypeptide.*
- The basic mechanics of transcription and translation are similar for bacteria and eukaryotes with one exception. The lack of compartmentalization in bacteria allows translation of a mRNA to begin while its transcription is still in progress.
- A mutation is the permanent altering of a nucleotide sequence of the genome of an organism.
- There are two kinds of mutations: hereditary mutations and acquired mutations.
- Mutations may be classified as small-scale or large-scale. Small-scale mutations may be either point mutations, substitutions, insertions, deletions, or frameshift mutations.
- Large-scale mutations are known as chromosome anomalies.
- An abnormal number of chromosomes is called aneuploidy.
- Chromosome structure may be altered by deletions, duplications, translocations, inversions, and insertions.
- Mutations may be categorized as spontaneous or induced.
- Francois Jacob and Jacques Monod proposed the operon model to explain gene regulation in prokaryotes.
- The genome in all the cells of an organism is identical so differences between cell types is not due to different genes, but to differential gene expression, the expression of different genes by cells with the same genome.
- In eukaryotes, there are five primary levels of gene regulation: three pertain to the nucleus (chromatin structure, transcriptional control, and posttranscriptional control) and two pertain to the cytoplasm (translational control and posttranslational control).

Review and Reflect

1. ***Walking with Beadle and Tatum*** Why did Beadle and Tatum propose the one gene-one enzyme theory? Why was their research considered so important and why has it been revised into one gene-one polypeptide?

2. ***Define It*** Imagine you have been contracted to write entries for a biological encyclopedia. Write a short paragraph explaining *genetic expression*.

3. ***Compare and Contrast*** Compare and contrast hereditary mutations with acquired mutations.

4. ***Flip the Script*** (**Figure 12.20**) This figure diagrammatically depicts which of the following chromosome anomalies: inversion, insertion, translocation, or duplication. Defend your answer.

5. ***Out of the Blue*** Some mutations are categorized as spontaneous mutations. Does this mean they just magically happen out of the blue? Explain

6. ***Turned On or Turned Off?*** You are taking a test over the material covered in this chapter. How would you answer the following question?

    A certain mutation in E. coli changes the lac operator so that the active repressor cannot bind. How would this affect the cell's production of ß-galactosidase?

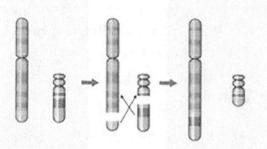

Figure 12.20

7. ***Speak or Remain Silent*** Heterochromatin is said to be "silent" whereas euchromatin "speaks." What does this mean?

8. ***On the Clock*** One way the cell has of regulating proteins is by selective degradation. Explain how selective degradation allows cells to regulate proteins.

9. ***Nature or Nurture?*** Your older rescue dog Roscoe is a very nervous and easily excitable dog that may have been abused as a puppy. From what you have learned of epigenetics, do you think your dog's flighty personality is due to his nature or nurture? Explain.

10. ***Nothing but Junk*** For some time scientists regarded the large amount of non-coding DNA as "junk DNA." What do scientists now understand non-coding DNA to be and why is it important in gene regulation?

Create and Connect

1. ***Go Red! Go White!*** Depending on environmental conditions, the bacterium *Serratia marcescens* will grow as either red colonies or white colonies on culture plates. In other words, different environmental conditions cause a difference in gene expression in *Serratia*. Design an experiment to determine the environmental factor that causes this difference in gene expression in *Serratia*. HINT: The environmental factor is either temperature or light.

   ### Guidelines

   A. Following the tenets of a well-constructed experiment, your design should include the following components in order:

   - The *Problem Question*. State exactly what problem you will be attempting to solve.
   - Your *Hypothesis*. Although this is a fictitious experiment, word your hypothesis as realistically as possible.

- *Methods and Materials.* Explain exactly what you will do in your experiment including the materials necessary to accomplish the task. Be specific, take nothing for granted, and do not expect people to read your mind as they read your work.
- *Collecting and Analyzing Data.* Explain what type(s) of data will be collected, and what statistical tests might be performed on that data. It is not necessary to concoct fictitious data or imaginary observations.

B. Your instructor may provide additional details or further instructions.

2. ***Dioxin Disaster*** Trace amounts of dioxin were present in Agent Orange, a defoliant sprayed on forests to reveal troop movements during the Vietnam War. Animal testing revealed that dioxin can cause birth defects, cancer, liver and thymus damage, and immune system suppression, sometimes leading to death. Dioxin acts like a steroid hormone, entering a cell and binding to a cytoplasmic receptor that then binds to a cell's DNA. Discuss:

A. How might you determine whether a specific illness is related to dioxin exposure?

B. How does the mechanism of damage explain the variety of dioxin's effects on different body systems and in different animals?

C. How might you determine if a particular individual human became ill as a result of exposure to dioxin?

3. ***Shining a Light on Gene Regulation*** The flashlight fish (Anomalopidae) (**Figure 12.21**) has an organ under its eye that emits light (bioluminescence) that serves to attract prey and startle predators. The light does not emit from the fish but from bacteria that live in the organ in a mutualistic relationship with the fish. The bacteria must multiply to a certain density (a "quorum") in the organ at which point they begin to emit light. A group of six or so genes called *lux* genes produce products that are necessary for light emission. Given that these bacteria are regulated together, propose a hypothesis for how the genes are organized and regulated.

Figure 12.21

BIOTECHNOLOGY

Humans have entered a new stage of self-designed evolution in which we will be able to change and improve our DNA.

—Stephen Hawking

Introduction

Throughout the course of human history, numerous notable biological or scientific advances have occurred. The first was the beginning of agricultural during which humankind learned how to grow crops and domesticate animals. The Renaissance was marked by the beginning of true science with the understanding of scientific method and the development of experimental design. Next, we saw the invention of machines during the Industrial Revolution (that continues to this day) followed closely by the discovery of the cause of infectious disease and the perfection of antibiotics. Today we live in the next great leap forward—applied genetics and biotechnology.

Applied Genetics

Humans have been applying genetic principles for millennia, long before Mendel perfected his laws and Watson and Crick understood the structure of DNA. *Applied genetics* is the manipulation of the hereditary characteristics of an organism usually for human benefit or purpose.

Classical Applied Genetics

Some of the methods of applying genetics—selective breeding, inbreeding, and hybridization—have been around for a centuries and should thus be catagorized as *classical applied genetics*:

Selective breeding, also known as artificial selection, is a process used by humans to develop new organisms with desirable characteristics. Breeders select two parents that have beneficial phenotypic traits to reproduce, yielding offspring with those desired traits. Selective breeding can be used to produce tastier fruits and vegetables, crops with greater resistance to pests, and larger animals that can be used for meat.

Selective breeding of both plants and animals has been practiced since early prehistory; key species such as wheat, rice, and dogs have been significantly different from their wild ancestors for millennia, and maize, which required especially large changes from teosinte, its wild form, was selectively bred in Mesoamerica.

Selective breeding in animals involves selecting animals with a homogeneous appearance, behavior, and other characteristics to develop particular *breeds* through culling animals with particular traits and selecting for further breeding those with other traits. Purebred animals have a single, recognizable breed, and purebreds with recorded lineage are called pedigreed. Crossbreeds are a mix of two purebreds, whereas mixed breeds are a mix of several breeds, often unknown.

Animal breeding begins with breeding stock, a group of animals used for the purpose of planned breeding. When individuals are looking to breed animals, they look for certain valuable traits in purebred stock for a certain purpose or may intend to use some type of crossbreeding to produce a new type of stock with different, and, it is presumed, superior abilities in a given area of endeavor. Purebred breeding aims to establish and maintain stable traits that animals will pass to the next generation. By "breeding the best of the best," employing a certain degree of inbreeding, considerable culling, and selection for "superior" qualities, one could develop a bloodline superior in certain respects to the original base stock.

For example, to breed chickens, a typical breeder must consider egg production, meat quantity, and new, young birds for further reproduction. Thus, the breeder has to study different breeds and types of chickens and analyze what can be expected from a certain set of characteristics before he or she starts breeding them. Therefore, when purchasing initial breeding stock, the breeder seeks a group of birds that will most closely fit the purpose Romans. Treatises as much as 2,000 years old give advice on selecting animals for different purposes, and these ancient works cite still older authorities.

The term "artificial selection" was coined by Charles Darwin in his famous work on evolution, *On the Origin of Species*, but the practice itself clearly predates Darwin by thousands of years. Darwin was interested in the process as an illustration of his proposed wider process of natural selection. Darwin noted that many domesticated animals and plants had special properties that were developed by intentional animal and plant breeding from individuals that showed desirable characteristics and discouraging the breeding of individuals with less desirable characteristics. As some of the earliest forms of biotechnology, both plant and animal breeding intended.

Perhaps the earliest example of selective breeding in animals is the domestic dog (*Canis familiaris*). It is unknown exactly when and where dogs were first domesticated, but humans have been breeding dogs for at least 14,000 years. Scientists believe that the domestic dog evolved from the wild gray wolf (*Canis lupus*), and through artificial selection, humans were able to create hundreds of different dog breeds. As people domesticated and bred dogs, they favored specific traits, like size or intelligence, for certain tasks, such as hunting, shepherding, or companionship. Selective breeding has led to the 155 existing breeds of dogs we have today. As a result, many dog breeds vastly differ in appearance, a unique phenomenon in the animal world, as different breeds of a single species generally resemble each other. The Chihuahua and the Dalmatian, for instance, are both dogs, yet they share few physical attributes (**Figure 13.1**). Selective breeding has also been practiced in agriculture for thousands of years. Almost every fruit and vegetable eaten today is a product of artificial selection. Cabbage, broccoli, cauliflower, Brussels sprouts, and kale are all vegetables derived from the same plant, *Brassica oleracea*, also known as wild cabbage (**Figure 13.2**).

Figure 13.1

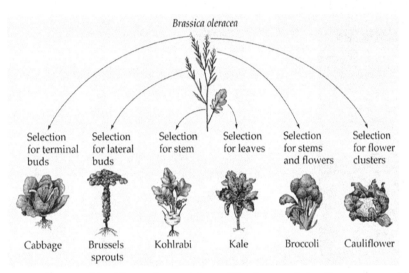

Figure 13.2 By isolating wild cabbage plants with specific characteristics, farmers were able to create a variety of vegetables from a single source, each with differing flavors and textures.

Figure 13.3 Over time, selective breeding modified teosinte's few fruitcases (left) into modern corn's rows of exposed kernels (right).

Corn, or maize, is an unusual product of selective breeding. Unlike rice, wheat, and cabbage, which have clear ancestors, there is no wild plant that looks like corn. The earliest records of maize indicate that the plant was developed in southern Mexico 6,000-10,000 years ago from a grass called teosinte. Scientists believe that early Mexican farmers selected only the largest and tastiest kernels of teosinte for planting, rejecting punier kernels. This process allowed Mexicans to develop corn very quickly, as small changes in the plant's genetic makeup had dramatic effects on the grain's taste and size (**Figure 13.3**). Despite their physical dissimilarities, teosinte and corn only differ by about 5 genes. Today, corn is a staple in diets across the world. In 2015-2016, an estimated 38 billion bushels of maize was produced around the world, primarily in the United States, China, and Brazil.

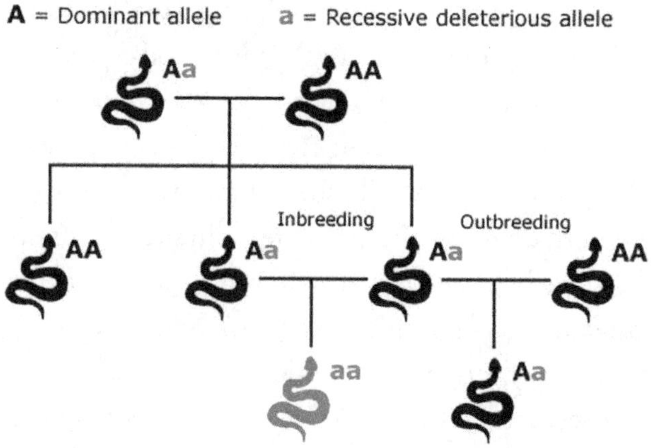

Inbreeding involves mating animals or crossing plants that are closely related genetically to maintain a particular trait such as coat color in dogs. Inbreeding is a technique used in selective breeding. In livestock breeding, breeders may use inbreeding when, for example, trying to establish a new and desirable trait in the stock, but will need to watch for undesirable characteristics in offspring, which can then be eliminated through further selective breeding or culling.

Inbreeding is used to reveal deleterious recessive alleles, which can then be eliminated through *assortative breeding* or culling. In plant breeding, inbred lines are used as stocks for the creation of hybrid lines. Inbreeding in plants also occurs naturally in the form of self-pollination. Inbreeding results in **homozygosity** thus increasing the chances of offspring being affected by recessive or deleterious traits. Homozygosity leads to a decreased biological fitness of a population (called inbreeding depression), or the ability of the population to survive and reproduce. The avoidance of expression of such deleterious recessive alleles caused by inbreeding via is the main selective reason for outbreeding (**Figure 13.4**).

Figure 13.4 Interbreeding allows the deleterious gene to be expressed whereas outbreeding depresses the expression of that gene.

A = Dominant allele   a = Recessive deleterious allele

Figure 13.5 Hercules the liger is the result of the intraspecific cross between a lion and a tiger

Hybridization has two meanings genetically. First, hybridization can mean the result of interbreeding between two animals or plants of different taxonomic groups (genus, species). Hybrids between different species within the same genus are sometimes known as *interspecific hybrids* or crosses. Hybrids between different sub-species within a species are known as *intra-specific hybrids* (**Figure 13.5**). Hybrids between different genera are sometimes known as *intergeneric hybrids*.

Interspecific hybrids are bred by mating two species, normally from within the same genus. The offspring display traits and characteristics of both parents (**Figure 13.6**). The offspring of an interspecific cross are very often sterile; thus, preventing the movement of genes from one species to the other, keeping both species distinct. Sterility is often attributed to the different number of chromosomes the two species have, for example, donkeys have 62 chromosomes, while horses have 64 chromosomes, and mules have 63 chromosomes.

Figure 13.6 A mule is the result of an interspecific cross between a male donkey (jack) and a female horse (mare). A hinny is the offspring of a male horse (stallion) and a female donkey (jenny). A mule and hinny are examples of reciprocal hybrids.

Mules and other normally sterile interspecific hybrids cannot produce viable gametes because differences in chromosome structure prevent appropriate pairing and segregation during meiosis, meiosis is disrupted, and viable sperm and eggs are not formed. However, fertility in female mules has been reported with a donkey as the father.

Depending on the parents, there are several different types of hybrids:

- *Single cross hybrids* result from the cross between two pure bred lines and produces an F_1 generation called an F_1 hybrid. The cross between two different homozygous lines produces an F_1 hybrid that is heterozygous; having two alleles, one contributed by each parent and typically one is dominant and the other recessive. The F_1 generation is also homogeneous, producing offspring that are all similar to each other.

- *Double cross hybrids* result from the cross between two different F_1 hybrids.

- *Three-way cross hybrids* result from the cross between one parent that is an F_1 hybrid, and the other is from an inbred line.

- *Triple cross hybrids* result from the crossing of two different three-way cross hybrids.

The second type of hybrid consists of crosses between populations, breeds or cultivars within a single species. This second meaning is often used in plant and animal breeding. In plant and animal breeding, hybrids are commonly produced and selected because they have desirable characteristics not found or inconsistently present in the parent individuals or populations. Hybrids are sometimes stronger than either parent variety, a phenomenon most common with plant hybrids, which when present is known as *hybrid vigor* (**heterosis**) or heterozygote advantage. This rearranging of the genetic material between populations or races is often called **hybridization**.

Modern Applied Genetics

Other methods of applying genetics have been developed more recently and are known as *modern applied genetics*. Modern applied genetics involves several different approaches:

Figure 13.7 Tangos, thin-skinned seedless mandarins with a tangy-sweet flavor, are the result of a radiation induced mutation.

Mutation induction involves using X-rays or chemical compounds to purposefully cause mutations. By inducing mutations, scientists have been able to increase genetic variation, which breeders depend on to produce crops with desirable traits, such as resistance to diseases and insects. Unlike recombinant DNA techniques, induced mutation does not add any foreign genetic material into the plant.

Essentially, induced mutations produce results that could have occurred through naturally occurring mutations and selection of desirable progeny. However, as mutations occur naturally at low frequencies and randomly in the DNA, it would require much more time to achieve such results. Mutations can be induced in a variety of ways, such as by exposure to ultraviolet or ionizing radiation or chemical mutagens.

Since the 1950s, over 2,000 crop varieties have been developed by inducing mutations to randomly alter genetic traits and then selecting for improved types among the progeny. Many common foods, such as popular red grapefruit varieties, are the result of induced mutations (**Figure 13.7**). Recently, methods have been developed to allow scientists to efficiently identify and select for mutations in specific genes. This is enhancing the utilization of induced mutation in plant breeding, as changes in genes known to produce specific results can be identified rather than having to screen very large populations of plant to find the rare individuals with the desired trait.

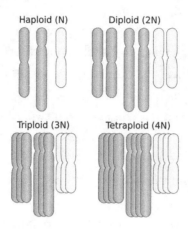

Figure 13.8 In multicellular eukaryotes, somatic (body) cells are diploid (2n) and gametes are haploid (n). Some natural or induced situations exist where somatic cells are triploid (3n) or even tetraploid (4n).

Polyploidy is a situation in which the cells of an organism contain more than two paired (homologous) sets of chromosomes. Eukaryotes are diploid (2n), meaning they have two sets of chromosomes—one set inherited from each parent. However, polyploidy is found in some organisms and is especially common in plants (**Figure 13.8**). Polyploids can arise naturally when errors during cell division create cells with more than one set of chromosomes. Alternatively, polyploidy can be induced using chemicals that increase the chance of errors in cell division (**Figure 13.9**). Polyploid plants exhibit hybrid vigor as evidenced by:

- higher yield of fruit
- greater resistance to disease
- greater resistance to drought

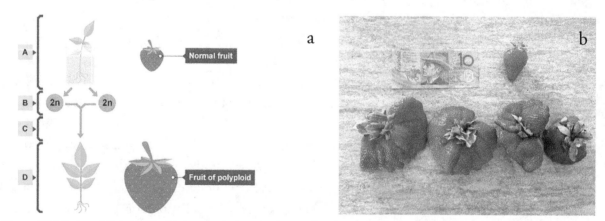

Figure 13.9 Induced Polyploidy. (a) At step A the strawberry plant is exposed to a chemical. At step B, due to the effect of the chemical, errors in cell division produce diploid gametes (2n) instead of the normal haploid gametes (n). As a result, at step C two diploid gametes fuse at fertilization. Finally at step D a new plant develops which has four sets of chromosomes (4n); (b) Normal strawberry top in relation to Australian currency and induced tetraploid strawberries along the bottom.

Cloning is the production of genetically identical copies of DNA, cells, or entire organisms through some asexual means. Clones are exact genetic copies. Every single bit of their DNA is identical. Clones can happen naturally—identical twins are just one of many examples. Or they can be made in the lab. Many people first heard of cloning when Dolly the Sheep showed up on the scene in 1997. Artificial cloning technologies have been around for much longer than Dolly, though.

There are two ways to make an exact genetic copy of an organism in a lab: *artificial embryo twinning* and *somatic cell nuclear transfer*.

Artificial embryo twinning is a relatively low-tech way to make clones. As the name suggests, this technique mimics the natural process that creates identical twins. In nature, twins form very early in development when the embryo splits in two. Twinning happens in the first days after egg and sperm join, while the embryo is made of just a small number of unspecialized cells. Each half of the embryo continues dividing on its own, ultimately developing into separate, complete individuals. Since they developed from the same fertilized egg, the resulting individuals are genetically identical.

Artificial embryo twinning uses the same approach, but it is carried out in a Petri dish instead of inside the mother. A very early embryo is separated into individual cells, which are allowed to divide and develop for a short time in the Petri dish. The embryos are then placed into a surrogate mother, where they finish developing. Again, since all the embryos came from the same fertilized egg, they are genetically identical.

Somatic cell nuclear transfer (SCNT), also called nuclear transfer, uses a different approach than artificial embryo twinning, but it produces the same result: an exact genetic copy, or clone, of an individual. This was the method used to create Dolly the Sheep.

To make Dolly, researchers isolated a somatic cell from an adult female sheep. Next, they removed the nucleus and its DNA content from an egg cell. Then they transferred the nucleus from the somatic cell to the egg cell. After a couple of chemical tweaks, the egg cell, with its new nucleus, was behaving just like a freshly fertilized egg. It developed into an embryo, which was implanted into a surrogate mother and carried to term. The transfer step is most often done using an electrical current to fuse the membranes of the egg and the somatic cell (**Figure 13.10**).

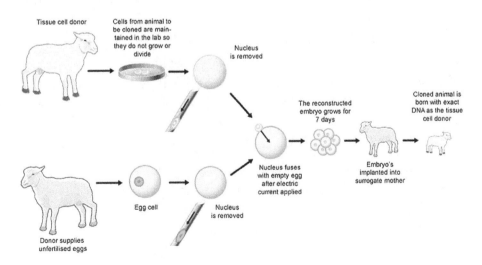

Figure 13.10 Dolly the sheep was an exact genetic replica (clone) of the adult female sheep that donated the somatic cell. She was the first-ever mammal to be cloned from an adult somatic cell.

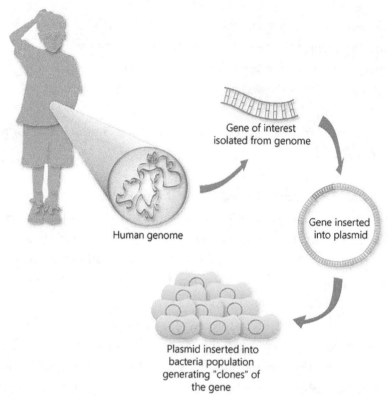

Gene of interest
isolated from genome

Human genome

Gene inserted
into plasmid

Plasmid inserted into
bacteria population
generating "clones" of
the gene

Figure 13.11 Gene cloning involves using specific enzymes to isolate a gene of interest from the genome. The gene is then inserted into a bacterial plasmid that serves as a vector. The vector plasmid carries the gene into bacterial cells where it begins to produce the protein it is encoded to make.

When scientists clone an organism, they are making an exact genetic copy of the whole organism, as described above. When scientists clone a gene, they isolate and make exact copies of just one of an organism's genes. Cloning a gene usually involves copying the DNA sequence of that gene into a smaller, more easily manipulated piece of DNA, such as a plasmid (**Figure 13.11**). Gene cloning provides scientists with an essentially unlimited quantity of any individual DNA segments derived from any genome. This material can be used for a wide range of purposes, including those in both basic and applied biological science. A few of the most important applications are:

- *Genome organization and gene expression* Gene cloning has led directly to the deciphering of the complete DNA sequence of the genomes of a very large number of species and an exploration of genetic diversity within individual species. This work has been done mostly by determining the DNA sequence of large numbers of randomly cloned fragments of the genome, and assembling the overlapping sequences.

- *Production of recombinant products* **Recombinant DNA** is the general name for a piece of DNA that has been created by the combination of at least two strands. The DNA sequences used in the construction of recombinant DNA molecules can originate from any species. For example, plant DNA may be joined to bacterial DNA, or human DNA may be joined with fungal DNA. In addition, DNA sequences that do not occur anywhere in nature may be created by the chemical synthesis of DNA, and incorporated into recombinant molecules. Using recombinant DNA technology and synthetic DNA, literally any DNA sequence may be created and introduced into any of a very wide range of living organisms (**Figure 13.12**).Many useful human proteins are currently available as recombinant products from transgenic organisms. These include (1) medically useful proteins whose administration can correct a defective or poorly expressed gene. For example, recombinant factor VIII, a blood-clotting factor deficient in some forms of hemophilia, and recombinant insulin, used to treat some forms of diabetes.

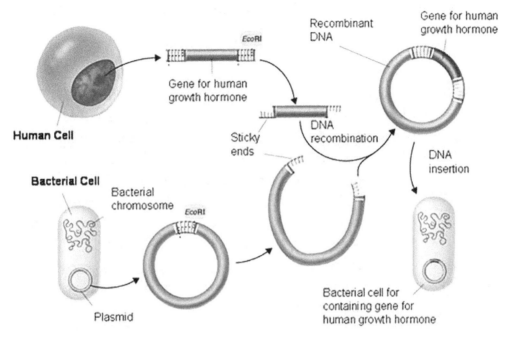

Figure 13.12 Recombinant DNA techniques (often called genetic engineering) allow scientists to create new species by making genomes that do not exist in nature.

Gene pharming, the use of transgenic farm animals to produce pharmaceutical products, is being pursued by a number of drug companies. Genes that code for therapeutic and diagnostic proteins are incorporated into an animal's DNA, and the proteins desired appear in the animal's milk, (2) proteins that can be administered to assist in a life-threatening emergency such as tissue plasminogen activator, used to treat strokes, and (3) recombinant subunit vaccines, in which a purified protein can be used to immunize patients against infectious diseases, without exposing them to the infectious agent itself, such as hepatitis B vaccine, and (4) recombinant proteins as standard material for diagnostic laboratory tests.

• *Transgenic organisms* cloned genes may be inserted into organisms, generating transgenic organisms, also termed genetically modified organisms (GMOs). Although most GMOs are generated for purposes of basic biological research, such as the transgenic mouse, a number of GMOs have been developed for commercial use, ranging from animals and plants that produce pharmaceuticals or other compounds (pharming), herbicide-resistant crop plants, and fluorescent tropical fish for home entertainment such as those in the chapter-opening photo that contain the gene that codes for green fluorescent protein (GFP), originally extracted from jellyfish. In agriculture, commercial varieties of important crops (including soy, maize/corn, sorghum, canola, alfalfa, and cotton) have been developed that incorporate a recombinant gene that results in resistance to the herbicide glyphosate (trade name *Roundup*) and simplifies weed control. These crops are in common commercial use in several countries.

Recombinant DNA techniques have also resulted in insect-resistant crops. *Bacillus thuringeiensis* is a bacterium that naturally produces a protein (Bt toxin) with insecticidal properties. The bacterium has been applied externally to crops as an insect-control strategy for many years, and this practice has been widely adopted in agriculture and gardening. Recently, plants have been developed that express a recombinant form of the bacterial protein that may effectively control some insect predators, resulting in transgenic plants that produce their own insecticides internally.

• *Gene Therapy* is the therapeutic delivery of genes into a patient's cells to treat various disorders. The first attempt at modifying human DNA was performed in 1980, but the first successful and approved nuclear gene transfer in humans was performed in May 1989. The first therapeutic use of gene transfer as well as the first direct insertion of human DNA into the nuclear genome was performed in a trial starting in September 1990. Between 1989 and February 2016, over 2,300 clinical trials had been conducted, more than half of them in phase I.

Gene therapy may be conducted *ex vivo* in which the gene(s) are inserted into cells that have been removed from the body and then returned, or *in vivo* in which the gene is delivered into the cells by recombinant viruses or DNA complex vectors (**Figure 13.13**). It should be noted that not all medical procedures that introduce alterations to a patient's genetic makeup can be considered gene therapy. Bone marrow transplantation and organ transplants, in general, have been found to introduce foreign DNA into patients.

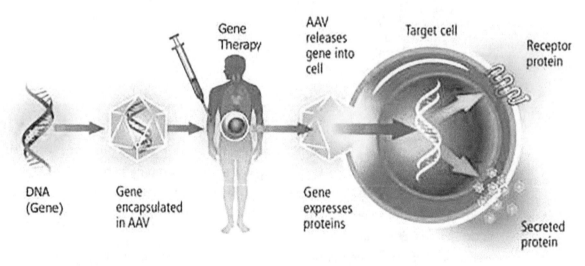

Figure 13.13 Gene therapy using adeno-associated viral vectors (AAV). The human gene to be transferred is inserted into the virus genome. The virus vector inserts the gene into the cell where it becomes part of the genome and begins producing the protein for which it codes.

Gene therapy may be classified into two types: *somatic cell gene therapy (SCGT)* in which genes are inserted into somatic cells only or *germline gene therapy (GGT)* in which genes are inserted into gametes (egg or sperm) only.

Box 13.1
The GMO Controversy

A genetically modified organism (GMO) is any organism whose genetic material has been altered through recombinant DNA techniques. GMOs are used in the production of pharmaceuticals, commercial products, and food. The use of GMOs in food production, however, has proven to be very controversial.

The dispute involves buyers, biotechnology companies, governmental regulators, nongovernmental organizations, and scientists. The key areas of controversy related to GMO food are whether GM food should be labeled, the role of government regulators, the effect of GM crops on health and the environment, the effect on pesticide resistance, the impact of GM crops for farmers, and the role of GM crops in feeding the world population.

There is a scientific consensus that currently available food derived from GM crops poses no greater risk to human health than conventional food, but that each GM food needs to be tested on a case-by-case basis before introduction. Nonetheless, members of the public are much less likely than scientists to perceive GM foods as safe.

No reports of ill effects have been proven in the human population from ingesting GM food. Although labeling of GMO products in the marketplace is required in many countries, it is not required in the United States, and no distinction between marketed GMO and non-GMO foods is recognized by the US Food and Drug Administration (FDA). It has been argued that while GM foods could potentially help feed 842 million malnourished people globally, laws to require labeling of foods containing genetically modified ingredients, could have the unintended consequence of interrupting the process of spreading GM technologies to impoverished countries that suffer from food security problems.

The first genetically modified seeds were planted in the United States in 1996, and despite the controversy surrounding them, genetically modified plants have taken root in our world. Currently, over 200 million hectacres worldwide are planted to GM crops. GM soybeans account for 79%, GM cotton for 70%, and GM maize (corn) for 32% of worldwide production.

As with any new technology, members of society have the responsibility to become informed about genetically modified plants, in order to make decisions about their responsible use and regulation.

In Summary

- Applied genetics is the manipulation of the hereditary characteristics of an organism usually for human benefit or purpose.
- Some methods of applying genetics—selective breeding, inbreeding, and hybridization—have been around for a long time and are thought of as *classical applied genetics.*
- Selective breeding of both plants and animals has been practiced since early prehistory.
- Selective breeding in animals involves selecting animals with a homogeneous appearance, behavior, and other characteristics to develop particular breeds through culling animals with particular traits and selecting for further breeding those with other traits.

- Purebred breeding aims to establish and maintain stable traits that animals will pass to the next generation.
- Through artificial selection, humans were able to create hundreds of different dog breeds.
- Almost every fruit and vegetable eaten today is a product of artificial selection.
- Inbreeding involves mating animals or crossing plants that are closely related genetically to maintain a particular trait.
- Inbreeding results in homozygosity thus increasing the chances of offspring being affected by recessive or deleterious traits.
- Hybrids between different species within the same genus are sometimes known as interspecific hybrids or crosses. Hybrids between different sub-species within a species are known as intra-specific hybrids.
- There are several types of hybrids: single cross, double cross, three-way cross, and triple cross.
- Other methods of applying genetics have been developed more recently and are known as *modern applied genetics.*
- Modern applied genetics involves several different approaches: mutation induction, and cloning.
- Mutation induction involves using X-rays or chemical compounds to purposefully cause mutations.
- Polyploidy is a situation in which the cells of an organism contain more than two paired (homologous) sets of chromosomes.
- Cloning is the production of genetically identical copies of DNA, cells, or entire organisms through some asexual means.
- There are two ways to make an exact genetic copy of an organism in a lab: artificial embryo twinning and somatic cell nuclear transfer.
- Gene cloning has led directly to the deciphering of the complete DNA sequence of the genomes of a very large number of species and an exploration of genetic diversity within individual species.
- Recombinant DNA is the general name for a piece of DNA that has been created by the combination of at least two strands. The DNA sequences used in the construction of recombinant DNA molecules can originate from any species.
- Many useful human proteins are currently available as recombinant products from transgenic organisms.
- Cloned genes may be inserted into organisms, generating transgenic organisms, also termed genetically modified organisms (GMOs).
- Gene therapy is the therapeutic delivery of genes into a patient's cells to treat various disorders.
- Gene therapy may be conducted *ex vivo* in which the gene(s) are inserted into cells that have been removed from the body and then returned, or *in vivo* in which the gene is delivered into the cells by recombinant viruses or DNA complex vectors.
- Gene therapy may be classified into two types: somatic cell gene therapy (SCGT) in which genes are inserted into somatic cells only or germline gene therapy (GGT) in which genes are inserted into gametes (egg or sperm) only.

Review and Reflect

1. ***Pig Breeder's Bureau*** The Pig Breeders Bureau has given you the task of breeding pigs that have the specific traits of large, floppy ears and spotted skin. Create a sequential diagram showing how selective breeding, inbreeding, and hybridization might be used to accomplish this task.

2. ***Pollen Prevention*** Hybrid seed corn fields are planted in a repeating 4 female rows:1 male row:4 female rows as shown in **Figure 13.14**. Note that the male rows in the field have tassels but that the tassels have been removed from the female rows. Why? Explain how this planting pattern results in hybrid seeds being formed.

Figure 13.14.

3. ***Corn from Grass?*** Explain how farmers are thought to have developed maize from the grass teosinte thousands of years ago in what is now southern Mexico. Reference Figure 13.3.

4. ***Too Close for Comfort*** Inbreeding involves mating animals or crossing plants that are closely related genetically to maintain a particular trait. Discuss the pros and cons of inbreeding.

5. ***Breeding Strange Animals*** The unusual animal pictured in **Figure 13.15** is a "zonkey." Research this animal and determine if it is an interspecific hybrid or an intra-specific hybrid.

6. ***Practical Polyploidy*** The huge strawberries depicted in Figure 13.9b are the result of induced polyploidy. Explain how these strawberries were developed and detail the benefits of induced polyploidy in horticulture and agriculture.

7. ***Cloning Roscoe*** Your beloved dog, Roscoe, has died. Shortly before his death, a tissue sample was taken. Imagine you have the expertise, experience, and equipment to clone your dead dog. Explain how you would go about it.

Figure 13.15

8. ***Clone Wars*** In the Great Hall of Genetics a debate is raging. Some wish to outlaw the practice of cloning whereas others such as you want to maintain the practice. Defend gene cloning by explaining the benefits humankind derives from the process.

9. ***Gene Repair*** Imagine you are a genetic counselor, and you have just told a young couple that their infant daughter is a candidate for gene therapy to correct a genetic disorder she was born with. Explain to them in general terms what gene therapy is, how gene therapy works, and hopefully what will be the outcome.

Create and Connect

1. ***The GMO Controversy—A Position Paper*** The use of GM crops has proven a boon to increasing food production around the world, but the use of GMOs to produce food is a very controversial topic. Research the situation and write a position paper on this controversy.

   Guidelines

   A. Compose a position paper, not an <u>opinion</u> paper. Defend your position with as many facts, figures, quotes, and pertinent information as possible.

   B. Your work will be evaluated not on the "correctness" of your position but the quality of the defense of your position.

   C. Your instructor may provide additional details or further instructions.

2. ***Designer Genes*** The use of recombinant DNA techniques allows scientists to genetically engineer the combination of different types of cells and create whole new species of transgenic plants, animals, and microbes. If you could engineer the cross between the cells of any plant or animal, what transgenic creature would you create? Diagram and explain your creature creation.

   Guidelines

   A. Your creature must be the combination of at least two plants and/or animals but no more than four plants and/or animals.

   B. Give your creature a unique and original name.

   C. Explain what "parent" creatures you started with and why you started with them.

   D. Explain what unique features your new creature has.

3. ***Amazing Maize*** So much corn is grown in the United States that it is known as "King of the Crops." Nearly all the corn grown commercially in this country is hybrid corn. Research this power crop and write a report on hybrid corn (maize). Your report should include how seed companies produce the F_1 seeds they sell to farmers.

   Guidelines

   A. Format your report in the following manner:
      - *Title page* (including your name and lab section)
      - *Body of the Report* (include pictures, charts, tables, etc. here as appropriate). The body of the report should be a minimum of 2 pages long—double-spaced, 1 inch margins all around with 12 pt font.
      - *Literature Cited* A minimum of 2 references required. Only <u>one</u> reference may be from an online site. The *Literature Cited* page should be a separate page from the body of the report and it should be the last page of the report. Do NOT use your textbook as a reference.

   B. The instructor may provide additional details and further instructions.

HISTORY OF EVOLUTIONARY THOUGHT

I was a young man with uninformed ideas. I threw out queries and suggestions, wondering all the time over everything, and to my astonishment the ideas took like wildfire. People made a religion of them.

—Charles Darwin

Introduction

As biologists contemplate the workings of the natural world, they have come to realize that the whole of life rests on four fundamental cornerstones, and that to comprehend the whole of life one must first understand each of these cornerstones and how they integrate with each other:

1. ***Life is unified*** All organisms—plant, animal, or microbe—on the planet are controlled by essentially the same molecular system:
   - a universal genetic code
   - genetic information in the form of DNA and this information is transmitted across generations
   - protein synthesis via ribosomes
   - proteins serving as enzymes and catalysts
   - DNA triplets coding for the same amino acid
   - chemical energy provided by the ATP-ADP cycle

2. **_Life is diverse_** Earth is populated with tens of millions of different species with ever more species slowly coming into existence as other species fade into extinction.

3. **_Living things possess survival adaptations_** Through evolution, every species is equipped with unique adaptations that allow it to survive within its particular environment.

4. **_Life has a rich history_** The first life forms appeared on this planet at least 3.4 billion years ago. Since then, new forms have continued to appear (originate), diversify, and eventually go extinct. In the process, this never-ending parade of species leaves behind traces of their former existence both in the fossil record and in the genes of their living relatives.

Only one set of ideas integrates these four fundamental concepts of biology into a complete explanation of the nature of life, and that is modern evolutionary theory. It should be noted that there has been much misapplication and misunderstanding of the word "theory" relative to the concept of biological evolution. The word "theory" in the phrase the "modern theory of evolution" is _not_ used in the experimental sense to indicate an untested/unproven hypothesis, but rather to indicate an integrating and unifying set of ideas as in _Cell Theory_. Furthermore, in this book we define the word **evolution** in its broadest sense to indicate change over time (L. _evolutus_, unroll like a scroll). Given this definition, it is hard to deny biological evolution (change) has occurred on this planet. To wander through any large natural history museum or even to hold a single fossil in one's hands is to possess overwhelming evidence that life on this planet has evolved over the course of time.

As with any notable scientific advance, the concept of biological evolution did not suddenly spring forth from the thoughts or actions of a single person. No scientist changes the worldview in a vacuum by themselves. To grasp the magnitude of modern evolutionary thought, one needs to understand the piece-by-piece assemblage of ideas, evidence, and investigation on which that thought is based. That is the very purpose of this chapter.

Rise of Evolutionary Thought

The evidence for biological change over long periods of time is indisputable. Even ancient people understood this and attempted to explain the mechanisms by which such changes occur. The development of evolutionary theory is a long and convoluted tale encompassing thousands of years of time and the contribution of many individuals along the way. Rooted in antiquity, the possibility and processes of evolution were conceived as philosophical ideas during the times of the ancient Greeks and Romans, but have been developed into scientific theory since the time of Charles Darwin.

The Ancients

The possibility of evolution and even suggestions as to the mechanisms of evolution were conceived as philosophical ideas during the times of the ancient Greeks and Romans. Some scholars consider it likely that the Greek philosophers borrowed and adapted their evolutionary ideas from the Hindus.

GREEKS

The concept of evolution was first moved from the realm of philosophical musing to the realm of scientific inquiry by Anaximander of Miletus (610 – 546 BC) who advanced a theory on the aquatic descent of humankind in the sixth century B.C. (**Figure 14.1**) Anaximander proposed that the first animals lived in water during a wet phase of Earth's past and that the first land-dwelling ancestors of humankind must have been born in water, and only spent part of their life on land.

Figure 14.1 Anaximander was an early proponent of science and tried to observe and explain different aspects of the universe, with a particular interest in its origins, claiming that nature is ruled by laws, just like human societies, and anything that disturbs the balance of nature does not last long.

Often called the "father of evolutionary naturalism," the Greek philosopher Empedocles (493-435 BC) (**Figure 14.2**) argued that "chance alone was responsible for the entire process of the evolution of simple matter into humankind." Empedocles also maintained that living organisms originate through spontaneous generation and gradually evolve through the process of trial-and-error recombination of animal parts. Furthermore, he concluded that fitter beings were more likely to survive and pass their traits on to their offspring or what Darwin would later dub natural selection. Empedocles postulated four ultimate elements which make up all the structures in the world—fire, air, water, and earth. Empedocles called these four elements "roots" and this theory became standard dogma for the next two thousand years (**Figure 14.3**).

Empedocle's.

Figure 14.2 Empedocles was a citizen of Acragas (Agrigentum), a Greek city in Sicily

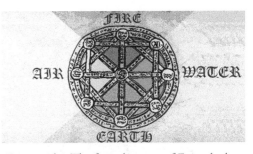

Figure 14.3 The four elements of Empedocles.

Other influential Greek philosophers, including Plato (428/427-348/347 BC) and Aristotle (384-322 BC) believed that the species of all things, not only living things, were fixed by divine design. Plato (**Figure 14.4**) promoted belief in *essentialism*, which is also referred to as the Theory of Forms. This theory holds that each natural type of object in the observed world is an imperfect manifestation of the ideal, form or "species" which defines that type and that every species has essential characteristics that are unalterable. However some historians of science have

Figure 14.4 Plato's real name is thought to have been Aristocles, after his grandfather. 'Plato' comes from the Greek for 'broad.' It is unclear if he got this nickname from his broad body, forehead, or style of teaching.

questioned how much influence Plato's essentialism had on natural philosophy by stating that many philosophers after Plato believed that species might be capable of transformation and that the idea that biologic species were fixed and possessed unchangeable essential characteristics did not become important until the beginning of biological taxonomy in the 17th and 18th centuries.

Aristotle (**Figure 14.5**), the most influential of the Greek philosophers for the Europeans during the Middle Ages, was a student of Plato and is also the earliest natural historian whose work has been preserved in any real detail. His writings on biology resulted from his research into natural history on and around the island of Lesbos, and have survived in the form of four books—*De anima*(*On the Soul*), *Historia animalium* (*History of Animals*), *De generatione animalium* (*Generation of Animals*), and *De partibus animalium* (*On the Parts of Animals*). Aristotle also delved into a number of other scientific areas besides animals including classification of living things, geology, optics, physics, and medicine.

Figure 14.5 Aristotle's works contain some remarkably astute observations and interpretations—along with sundry myths and mistakes reflecting the disjointed state of knowledge during his time.

Aristotle also proposed the *Scala Naturae* or the *Great Chain of Being* (literally "ladder [or stairway] of nature") (**Figure 14.6**). Aristotle saw the universe as an ultimately perfect place in which the relationship between the physical world, living beings, and the creator God could be visualized as a stairway. The stairway progressed upward with rocks and minerals on the bottom followed by plants, animals, humans, noblemen, the Sun and Moon, demons, and angels with God overarching all at the top. The Chain was a very rigid and static view of the natural world where species could never change and must remain fixed (no evolution). Aristotle believed that features of living organisms showed clearly that they must have had what he called a final cause, that is to say, that they had been designed for a purpose. He explicitly rejected the view of Empedocles that living creatures might have originated by chance. Unfortunately, Aristotle's inflexible and incorrect view of the natural world influenced and impeded Western thought and science for centuries after his death.

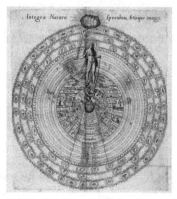

Figure 14.6 Within this system, everything that existed could be placed in order, from "lowest" to "highest," with Hell at the bottom and God at the top—below God, an angelic hierarchy marked by the orbits of the planets, mankind in an intermediate position, and worms the lowest of the animals. As the universe was ultimately perfect, the great chain of being was also perfect. There were no empty links in the chain, and no link was represented by more than one species. Therefore, no species could ever move from one position to another.

ROMANS

In the poem *De rerum natura* written around 60 B.C. the Roman Titus Lucretius Carus (99-55 BC) described the development of Earth in stages from atoms colliding in the void as swirls of dust, then early plants and animals springing from the early Earth's substance to a succession of animals, including a series of progressively less brutish humans, a description that sounds remarkably similar to the modern nebular origin theory of the solar system and evolution of life. This work describes the development of the cosmos, Earth, living things, and human society through purely naturalistic mechanisms, without any reference to divine involvement. *De rerum natura* would influence the cosmological and evolutionary speculations of philosophers and scientists during and after the Renaissance.

CHINESE

Ancient Chinese thinkers such as Zhuang Zhou (369 – 286 BC), a Taoist philosopher, expressed ideas on changing biologic species. Taoism explicitly denies the fixity of biological species and Taoist philosophers speculated that species had developed differing attributes in response to differing environments. Taoism regards humans, nature and the heavens as existing in a state of "constant transformation" known as the *Tao*, in contrast with the more static view of nature typical of Western thought of the time.

In line with earlier Greek thought, the 4th-century bishop and theologian, Bishop Augustine of Hippo (354-430 AD) (modern day Algeria), wrote that the creation story in the Book of Genesis should not be read too literally. In his book *De Genesi ad litteram* (*On the Literal Meaning of Genesis*), he stated that in some cases new creatures might have come about through the "decomposition" of earlier forms of life. For Augustine, "plant, fowl and animal life are not perfect... but created in a state of potentiality," unlike what he considered the theologically perfect forms of angels, the firmament, and the human soul. Augustine's idea that forms of life had been transformed "slowly over time'" prompted Father Giuseppe Tanzella-Nitti, Professor of Theology at the Pontifical Santa Croce University in Rome, to claim that Augustine had suggested a form of evolution.

Middle Ages

Although Greek and Roman evolutionary ideas died out in Europe after the fall of the Roman Empire, they were not lost to Islamic philosophers and scientists. In the Islamic Golden Age of the 8th to the 13th centuries, philosophers explored ideas about natural history. These ideas included transmutation from non-living to living: "from mineral to plant, from plant to animal, and from animal to man."

With the rise of Christianity and the decline and fall of the Roman Empire, the Catholic papal state assumed authority on all things political, social, philosophical, and scientific. The universe, the planet, and living things were explained solely in religious terms, and evolutionary thought was considered heretical and even rebellious. In Europe, a spiritual view of the natural world developed that held species to be unconnected, unrelated, and unchanging since the moment of their creation. Humans were not considered part of the natural world in this view but were considered instead to be above and outside it. And Earth itself was thought to be unchanging and so young—perhaps only 6,000 years old—that species would not have had time to change. Species were considered to have been divinely created and unchanged since the moment of their creation whereas humans were considered to be above and outside the natural world. Thus a shroud of scientific stagnation descended on Europe, a situation that would last for centuries.

As evolutionary ideas died out or were suppressed in European Christian society, they continued to be pronounced in the Eastern world as Islamic scholars embraced and advanced the works of those such as Aristotle and Galen, the great Roman physician. In the Islamic Golden Age of the 8th to the 13th centuries, philosophers explored ideas about natural history. These ideas included transmutation from non-living to living: "from mineral to plant, from plant to animal, and from animal to man."

Renaissance—Rebirth of Reason

The dawn of the age known as the Renaissance in Europe in the 14th century brought a refreshing questioning of the ideas and beliefs of antiquity in all aspects of human endeavor—culture, art, medicine, and especially science—as well as a healthy skepticism about the validity of those ancient values. From the English physician William Harvey who discovered the true nature of human blood circulation to the Dutch microscopist Anton von Leeuwenhoek who first viewed the unimaginable complexity of life in a drop of pond water, naturalism was again on the rise. This new generation of naturalists envisioned life as machines and found that they could apply the same principles to life as they used in physics to invent machines.

The clergy of the time worried that this mechanistic approach to the study of life smacked of atheism and heresy. However, many of the naturalists themselves believed that they were actually on a religious mission. In fact, a number of them were both naturalists and theologians. By studying the intricate structures of an eye or a feather, these naturalists believed they could better appreciate God's benevolent design, an approach that came to be known as *natural theology*. The most significant scientific achievement of the era was not a single discovery but the gradual development of a process for divining natural truths that would come to be called the *scientific method*.

A Scientific Revolution

The period of the 16th through 18th centuries was a time of scientific revolution during which new ideas and knowledge in physics, astronomy, biology, medicine, and chemistry transformed ancient and medieval views of the natural world and laid the foundations for modern science. Highlights of the advance of evolutionary thought during this time would include:

Figure 14.7 Georges-Louis Lecler Comte de Buffon. It has been said that "Truly, Buffon was the father of all thought in natural history in the second half of the 18th century."

1700s:

- Georges-Louis Lecler Comte de Buffon (1770-1788) (**Figure 14.7**) published a large four-volume encyclopedia of the natural world, *Historie Naturelle* over the course of his working life, and continued in eight more volumes over two decades after his death. The books cover what was known of the "natural sciences" at the time, including what would now be called material science, physics, chemistry, and technology as well as the natural history of animals. In these massive tomes, Buffon toyed with the idea of common ancestry for old world primates and humans. Although he believed in natural change, Buffon could not come up with a mechanism to explain it.

- Carolus Linnaeus (1707-1778) (**Figure 14.8**) was chief among a rising number of taxonomists, an increasingly important endeavor, during the mid-eighteenth century. Linnaeus, like other taxonomists of his time, believed in the fixity of species, that each species had an "ideal" form. He also believed in the *scala naturae*, a sequential ladder of life where the simplest beings occupy the lowest rungs and the most complex and spiritual beings—angels, humans, and then God—occupy the two highest rungs.

• Erasmus Darwin (1731-1802), Charles Darwin's grandfather, proposed one of the first formal theories on evolution in his work, *Zoonomia* (*Laws of Organic Life*) in which he suggested that "all warm-blooded animals have arisen from one living filament... with the power of acquiring new parts" in response to stimuli, with each round of "improvements" being inherited by successive generations. His writings on both botany and zoology contained comments and footnotes that suggested the possibility of evolution. He based his conclusions on changes in animals during development, animal breeding, by humans, and the presence of vestigial structures—anatomical structures that apparently functioned in an ancestor but have since lost most or all of their functionality in a descendant, such as the human appendix. Like Buffon, Erasmus Darwin thought that species might evolve, but he could offer no mechanism by which that could happen.

Figure 14.8 Linnaeus developed the binomial system of nomenclature and a system of classification for living things we still use today.

1800s:

• Jean Baptiste de Lamarck (1744-1829) attempted to explain the process of evolution in his book, *Philosophie Zoologique*. Lamarck (**Figure 14.9**) proposed characteristics an organism acquired during its lifetime could be passed on to its offspring. Lamarck's theory of the *inheritance of acquired characteristics* came to be known as "Lamarckism." One oft-stated example is how the giraffe acquired its long neck. Then (and now) it was believed that giraffes originally had relatively short necks that somehow grew longer over time. Lamarckism explained this phenomenon by contending that the giraffe's neck grew progressively longer as generations of giraffes stretched their necks reaching ever higher for food and then passed this stretched neck on to their offspring. Over the centuries, this stretching and passing along of longer and longer necks resulted in the modern giraffe's exceptionally long neck. Playing a central role in Lamarck's theory of the inheritability of acquired characteristics was *orthogenesis*, the hypothesis where life has an innate tendency to progress from simple forms to ever higher, ever more complex and perfect forms.

Figure 14.9 Lamarck did not believe that all living things shared a common ancestor but thought simple forms of life were created continuously by spontaneous generation.

Charles Darwin not only praised Lamarck for supporting the concept of evolution and bringing it to popular attention, but he also accepted the idea of use and disuse and believed the inheritance of acquired characteristics to be not only plausible but likely. However, with the dawn of Mendelian genetics and a better understanding of the true nature of inheritance, genetic experimentation simply did not support the concept that purely "acquired traits" were inherited. Eventually, Lamarck and his ideas were ridiculed and discredited. An ironic turn of events, however, may have given Lamarck the last laugh. An emerging field of genetics has shown that Lamarck may have been at least partially correct all along. We now know that reversible and heritable changes (phenotypes)—Lamarck's "acquired traits"—can occur without a change in DNA sequence (genotype) and that such changes may be induced in response to environmental factors.

Geneticists have known for several decades that some extra element associated with the tightly-wound spirals of DNA inside each cell was necessary to designate which genes get transcribed. Those designator molecules were found to be methyl groups, a common structural component of organic molecules. Since these methyl groups are attached to the genes, residing beside but not within the double-helix DNA code, this field was dubbed epigenetics (Gr. *epi* = over, outer, above).

Originally epigenetic changes or DNA methylation were believed to occur only during embryonic development. However, pioneering studies have demonstrated that not only could methylation occur in adulthood due to infection, diet, chemicals, and environmental exposure but that epigenetic change is inheritable and can be passed down from parent to child, generation after generation. For example, when female mice are fed a diet rich in methyl groups, the fur pigmentation of subsequent offspring is not only permanently altered but also inheritable and passed to future generations much like a mutation in a gene. Evolutionary biologists have long pondered the apparent speed at which human evolution, in particular, seems to be progressing. Epigenetics may hold the key not only to understanding human evolution but to directing it as well.

- Thomas Robert Malthus (1776-1834) was an English cleric and scholar whose book *An Essay on the Principle of Population* predicted a grim future in which population would increase geometrically, doubling every 25 years, but food production would only grow arithmetically resulting in famine and starvation unless births were controlled. Of course, Malthus (**Figure 14.10**) could not possibly have imagined first steam-powered and then diesel-powered machines, chemical fertilizers, or irrigation, the combination of which staved off Malthus' dire predictions.

Figure 14.10 The views of Malthus became influential, and controversial across economic, political, social and scientific thought. The pioneers of evolutionary biology Charles Darwin and Alfred Russel Wallace were both influenced by his works.

- Baron Georges Cuvier (1776-1832) was a distinguished zoologist that used comparative anatomy to develop a system of classifying animals. He also founded the science of paleontology, the study of fossils. Curvier was a staunch advocate of the fixity of species and special creation, but his studies revealed that the assemblage of fossil varieties changed suddenly between different layers of sediment (strata) within a geographical region.

- The development of biological evolutionism was accompanied by an increasing understanding of geologic change and the true age of Earth. In his revolutionary work, *Principles of Geology* published in three volumes in 1830-33, the English geologist Charles Lyell (1797-1875) established the principle of geologic uniformitarianism. The central argument in *Principles* was that "the present is the key to the past." Highly controversial for its time, uniformitarianism proposed that Earth was and continued to be shaped by slow-acting natural forces over very long time periods. From his geological studies, Lyell (**Figure 14.11**) concluded that the age of Earth must be measured in millions of years. These ideas were in direct conflict with the prevailing belief in **catastrophism** which held that huge geologic changes occurred planet-wide over very short periods of time (thousands of years) as implied by Biblical chronology.

Figure 14.11 Lyell's books had a great influence on Charles Darwin. Darwin took Volume I with him when he left for the Galapagos, and he received Volume II when he arrived in South America.

- In 1859 Charles Darwin (1809-1882) publishes *On the Origin of Species by Means of Natural Selection, or the Preservation of Favoured Races in the Struggle for Life.* Darwin's theories and published works stand among the greatest intellectual achievements of all time as they continue to influence scientific, religious, and social thought with ramifications extending to the daily existence of each one of us.

Box 14.1
A Journey of Body and Mind

After having been driven back twice by heavy southwestern gales, Her Majesty's ship Beagle, a ten-gun brig, under the command of Captain Robert Fitzroy, R.N., sailed from Devenport on the 27ᵗʰ of December, 1831.

—from an account by Charles Darwin

Figure 14.12

So began a five-year round the world voyage of survey, exploration, and collection that while modest in scope, produced repercussions whose importance reverberates to this day. Aboard the *Beagle* as the ship's naturalist was young Charles Darwin (**Figure 14.13**). When he sailed away, Darwin was a young university graduate filled with a life-long passion for plants, animals and all things natural and an immense interest in all the sciences, especially geology. Darwin had been stunned and elated to receive an invitation to accompany the voyage as he had longed for some time to travel and explore the natural history of tropical lands. His father, Robert Darwin, however, had serious misgivings about this "wild scheme" and considered it reckless, dangerous, and unbefitting of the future clergyman he hoped Charles would become. Darwin sadly declined the offer but with only a month to go before sailing, Charles and his uncle, Josiah Wedgwood, convinced his father to allow Charles to make the voyage. In fact, Robert Darwin paid for all of Charles's expenses—no small consideration, since it was an unpaid position.

Figure 14.13 Charles Darwin at age 51, one year after publishing one of the most revolutionary and controversial theories of all time.

The original plan called for a two-year mission but the voyage eventually stretched to five years (1831-1836). The British government backed the voyage and sent them off with the primary purpose of surveying the coastline and charting the harbors and islands of South America to produce better maps and protect

British interests in the Americas. It was understood that secondary to the main mission, Darwin was to make scientific observation and collect specimens. Spending nearly two-thirds of his time ashore, Darwin explored the South America wilderness of Brazil, Argentina, and Chile as well as the remote Galapagos Islands. He filled dozens of notebooks with painstaking observations of plants, animals, and geologic formations as well as collecting and crating over 1,500 specimens, hundreds of which had never been seen before in Europe (**Figure 14.14**).

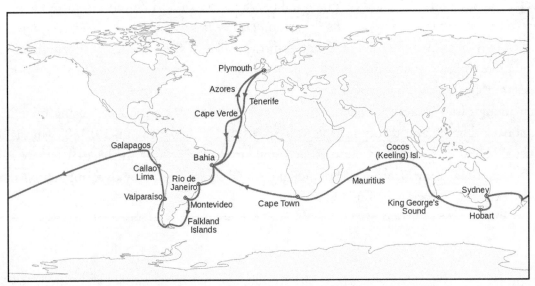

Figure 14.14 Map of the circumnavigation voyage of the HMS Beagle.

By the time he returned, Darwin was an established naturalist, known for the remarkable specimens he had sent ahead and he had grown from a mere observer into a probing theorist. By any measure, Darwin's labors were hugely successful but more importantly, the trip gave him a lifetime of experiences to ponder, and it planted the seeds of a theory he would mull over for the rest of his life.

Darwin was presented Volume I of Lyell's *Principles* by Captain Fitzroy just before the *Beagle* departed. He received Volume II while in South America. These works had a major impact both on Darwin's interpretation and understanding of the geologic world and the fossils he encountered on his voyage and in the development of his own theory of organic evolution. On his return, not only did Darwin and Lyell become friends but Lyell would be the first scientist of standing to endorse Darwin's theory of evolution once it was published.

Darwin was barely off the ship before his next great journey—a journey of the mind—began. For the next six years, Darwin would contemplate and question the biological wonders he had seen. He quietly gathered evidence from every possible source and sought out new ideas to support a notion regarding the transmutation of species that was gaining form and clarity in his thinking.

By the late summer of 1842, Darwin felt ready to commit an outline of his theory to paper. This rough sketch of his reasoning and arguments was a condensed version of his future masterwork, yet he kept his ideas secret for nearly two decades more. Why? Darwin feared the ridicule of other scientists, especially his friends and mentors and he knew his ideas would be very controversial and seen as an attack on religion and established society. He considered it prudent to wait and amass as much compelling evidence as possible to convince the world of the plausibility and possibility of such a radical idea.

By 1856 Darwin felt the time was right and he began working to compile the findings from his vast research into a large thesis on the origin of species. Though his plans were to compile four volumes, it was not to be. Darwin had been corresponding with and receiving bird specimens from Alfred Russel Wallace, an English naturalist working in Malaya. In 1858, he received a manuscript from Wallace that summarized the main points of the very theory Darwin had struggled to piece together over the past two decades. Darwin was stunned. Darwin worked feverishly over the next twelve months in his preparation of an "abstract" to summarize his planned larger, multi-volume treatise. In November 1859, *On the Origin of Species by Means of Natural Selection, or the Preservation of Favoured Races in the Struggle for Life* was finally published. It took but one day for all 1250 first-printing copies to sell out forcing the publisher to quickly rush 3,000 more copies to print. As Darwin suspected and feared, the book created instant controversy and misunderstandings that continue to this day.

Even though frail in health, Darwin entered a very productive period producing book after book on evolutionary thought and theory for the next 23 years. He died on April 19, 1882, and is buried in Westminster Abbey. Darwin's ideas and theories stand among the greatest intellectual achievements of all time as they continue to influence scientific, religious, and social thought with ramifications that extend to the daily existence of each one of us.

- In 1866 Gregor Mendel's plant hybridization research is published in a small book entitled *Experiments on Plant Hybridization*. Mendel's work served as the basis for modern genetics. Mendel's work and the geneticists that followed him finally provided evolutionary biologists with the "holy grail" of evolutionary theory—the mechanism that drives evolution.

1900s:

- Mendel's work is "rediscovered" by several geneticists and when applied to the development of evolutionary theory not only proves Lamarck's theory of the inheritability of acquired characteristics incorrect but also provides the mechanism by which Darwin's natural selection progresses.

- In the years immediately following Darwin's death, evolutionary considerations splintered into a number of different interpretations. Although the scientific community generally accepted that evolution had occurred, many disagreed that it had happened as explained by Darwin. The ideas of the various disagreeing camps were brought together in the 1930s when Darwinian natural selection and Mendelian inheritance were combined to form *neo-Darwinism* or *Modern Evolutionary Synthesis* as it later became known. The Modern Synthesis holds that the processes responsible for small-scale or micro-evolutionary changes can be extrapolated indefinitely to produce large-scale or macro-evolutionary changes leading to major changes and innovation in body form. By the 1950s, natural selection acting on genetic variation was virtually the only acceptable mechanism of evolutionary change.

- In the 1940s and 1950s, the identification of DNA as the genetic material by Oswald Avery and colleagues and the articulation of the structure of DNA by James Watson and Francis Crick revealed the true nature of genes and the genetic material. These discoveries launched the era of molecular biology and transformed our understanding of evolution into a molecular process. Since then, the role of genetics in evolutionary biology has become increasingly central.

- In the 1960s Niles Eldredge and Stephen Jay Gould propose the *theory of punctuated equilibrium,* which holds species remained persistently unchanged phenotypically for long periods of time with relatively sudden and brief periods of speciation resulting in phenotypic change ("evolution by jerks and creeps.) Evolution can also occur rapidly when a species moves into an environment it had not previously occupied, a process known as **adaptive radiation**. Adaptive radiation can occur when animals find themselves in an environment with unoccupied niches, such as a newly formed lake or oceanic island. Through adaptive radiation, some introduced animal species have become serious environmental problems in many parts of the world.

2000s:

- Increasing scrutiny of the tenets of modern evolutionary synthesis has resulted in the combining of the fields of phylogenetics, paleontology, and comparative developmental biology into a new discipline known as evolutionary developmental biology (or evo-devo as it is sometimes known). Developmental biologists compare the developmental processes of different organisms in an attempt to determine the ancestral relationship between organisms and how the developmental processes themselves evolved.
- Microbiology as an evolutionary discipline emerges. Increasingly, evolutionary researchers are taking advantage of our extensive understanding of microbial physiology and microbial genomics to answer evolutionary questions.

Today, evolutionary investigators of all biological disciplines continue to refine and advance our understanding of the origin and evolution of the multitude of species inhabiting this planet. Using this chapter as a foundation, you should be better prepared to understand the details of modern evolutionary theory that you will encounter in chapter 15.

CHAPTER 15

MODERN EVOLUTIONARY THEORY

Nothing in biology makes sense except in the light of evolution.

—Theodosius Dobzhansky

Introduction

As this tiny and fragile craft that is our home planet pirouettes around the nurse star we call Sol, it carries aboard it a living crew of such amazing diversity that one can only marvel at the variety and complexity of it all. When evolutionary biologists contemplate our fellow passengers on this tiny mote, questions about this glorious realm of creatures invariably spring to mind: Where did living things come from? (*origins*) How have living things changed since they first originated? (*evolution*) Although these questions have yet to be answered in detail, we do know with great certainty that once living things did appear, they flourished against the backdrop of ever-changing geologic conditions.

Questions of Evolution

Questions as to the evolution of living things can be answered by examining the tenets of modern evolutionary theory. Note that the word "theory" as it is applied to evolution in this chapter is used in an integrational sense to indicate a concept that unifies a set of ideas, not in the experimental sense to indicate an unproven hypothesis. Furthermore, the word *evolution* is defined here in its broadest sense to indicate change over time. Given that definition, no clear-thinking and rational person can deny that biological evolution (change) has occurred on this planet. To hold even a single fossil in one's hand is to possess overwhelming evidence that life on this planet has evolved over the course of geologic time.

Mechanisms of Evolution

Evolution is responsible for both the remarkable similarities we see across the spectrum of life and the amazing diversity of that life, but exactly how does evolution work? The modern theory of evolution may best be understood when viewed not as a single monolithic set of ideas but rather as the integration and melding of five different but mutually compatible subtheories or components. For evolutionists to fathom the total process of evolution, they must first understand the workings of each Darwinian subtheory.

1. **Perpetual change** This is the core on which the other theories rest. It states that the living world is constantly changing and, as it does so, organisms undergo transformation across generations throughout time.
2. **Common descent** The second theory states that all forms of life descended from a common ancestor through a branching of lineages. Studies of organismal anatomy, cell structure, and macromolecular structures confirm the idea that life's history has all the components of a branching evolutionary tree (**Figure 15.1**).

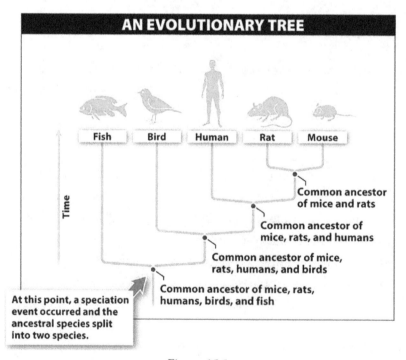

Figure 15.1

3. **Multiplication of species** The third theory states evolutionary processes produce new species by splitting and transforming older ones. Although evolutionists generally agree with this idea, there is still much controversy concerning the details of the process.

4. ***Gradualism*** The fourth subtheory states that the large differences in anatomical traits that characterize different species originate through the accumulation of many small incremental changes over long periods of time. Evolutionists today concede that although gradual evolution is known to occur, it may not fully explain the origin and development of all the structural differences between species.

5. ***Natural Selection.*** The fifth theory, Darwin's most famous, holds that all members of a species vary slightly from each other and thus have different structural, behavioral, and/or physiological traits. Those organisms with variation in traits that permit them to best exploit their environment will preferentially survive and pass these beneficial traits on to future generations. Over long periods of time, the accumulation of such favorable variations produce new organismal characteristics and eventually new species (**Figure 15.2**). This proposition has been popularly termed "survival of the fittest" but might be more accurately described as "survival of the best adapted and most reproductively fit."

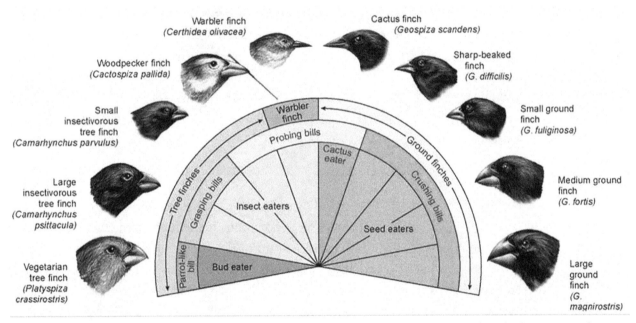

Figure 15.2 Ten species of Darwin's finches from Isla Santa Cruz, one of the Galapagos Islands. The correspondence between the beaks of the various finch species and their food source immediately suggested to Darwin that evolution had shaped them. About these finches Darwin said, 'Seeing this gradation and diversity of structure in one small, intimately related group of birds, one might really fancy that from an original paucity of birds in this archipelago, one species has been taken and modified for different ends.' Scientists have concluded that these 10 species derived from a single common ancestor that evolved into different species through the process of natural selection.

Evolution encompasses changes on two vastly different scales—from an increase in the frequency of a gene for colored spots on the feathers of a bird (**microevolution**) to something as grand in scale as the evolution of the entire bird lineage (**macroevolution**). Microevolution happens on a small scale and possibly over short time periods and is simply a change in the gene frequency of a single population. A relatable example would be the evolution of resistance to chemicals and antibiotics in certain organisms, such as insects becoming resistant to insecticides and bacteria becoming resistant to antibiotics. Taking long periods of time to occur, macroevolution encompasses the grandest trends, and transformations (speciation) in evolution and its patterns are what we see when we look at a tree of the history of life. Despite the scale on which it happens, evolution at both levels is driven by the primary mechanisms of natural selection mutation, genetic drift, gene flow, and speciation.

Natural Selection

Darwin's grand idea of evolution by natural selection is relatively simple but often misunderstood. For natural selection to occur certain conditions must be meant. There must be (1) variations in traits within a population, (2) differential reproduction within a population, and (3) inheritance.

Let us use an imaginary population of small fish to illustrate the mechanism of natural selection:

- **Variation in traits** Some of the fish are a bright silver color on top (dorsal side) while others are a darker grayish color.
- **Differential reproduction due to advantageous adaptation** No environment can support unlimited population growth, so not all individuals can reproduce to their full potential. If in the fish population, the bright colored fish are more easily seen by predators and thus tend to get eaten more often, the dark colored fish will survive in greater numbers and therefore produce more offspring.
- **Inheritance** The surviving dark colored fish pass this advantageous color trait (genes) on to their offspring. If you recall, this was the missing piece Darwin could not explain in his theory. He knew that offspring tended to possess the traits of their parents; he just lacked a mechanism to explain it all.

The more advantageous trait, dark coloration, allows the fish that possess the trait to have more offspring. As a result, the dark coloration genes become more common (higher gene frequency) in the population. If this process continues, eventually all (or at least most) of the fish in that population will be dark colored. The light-colored trait is likely to persist for a period of time. Fixation is the ultimate outcome of genetic drift, but there are other factors that may maintain the alternate trait in a population.

If you have variation, differential reproduction, and heredity occurring in a population, natural selection will result. A living example of what appears to be natural selection in action can be seen in the peppered moth (*Biston betularia*) of Britain. The moths come in two color variations or subspecies—light colored or *typica* and dark colored or *carbonaria*. Originally, the vast majority of peppered moths had the light coloration that effectively camouflaged them against the light-colored lichen-covered trees on which they rested. However, with the onset of the Industrial Revolution and the burning of massive quantities of coal, the lichens died, and the trees were blackened by soot. What had been an advantage now became a disadvantage as the typica moths became more visible against the trees (**Figure 15.3**). As predators found and ate more of the *typica* moths, the *carbonaria* moths flourished as their dark color now gave them an advantage.

Figure 15.3 Typica and carbonaria peppered moths (a) on an unpolluted lichen-encrusted tree and (b) on a sooty polluted tree."

With improved environmental standards, light-colored peppered moths have again become more common. These dramatic changes over relatively short periods of time have remained a subject of much interest and continued study and strongly implicate birds as the agent of selection in the case of the peppered moths.

Researchers have defined three general types of natural selection: directional selection, stabilizing selection, and disruptive selection (**Figure 15.4**).

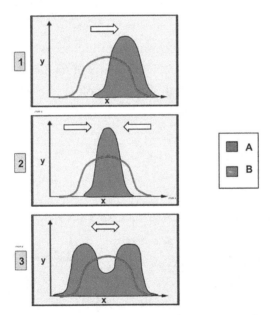

Figure 15.4 These graphs depict the different types of genetic selection. On each graph, the x-axis variable is the phenotypic trait and the y-axis variable is the number of organisms. Group A is the original population and Group B is the population after selection. Graph 1 shows directional selection, in which a single extreme phenotype is favored. Graph 2 depicts stabilizing selection, where the intermediate phenotype is favored over the extreme traits. Graph 3 shows disruptive selection, in which the extreme phenotypes are favored over the intermediate.

Directional selection occurs when an extreme phenotype is favored, and the distribution curve shifts toward one of the extremes. Over time, directional selection changes the frequency of a phenotype within a population. Such a shift can occur when a population is adapting to a changing environment. It now becomes clear that the peppered moth is undergoing directional selection. In pristine environments, the light-colored phenotype is selected whereas, in polluted environments, the dark-colored phenotype is favored.

Stabilizing selection results in a decrease of a population's genetic variance when natural selection favors an intermediate phenotype and selects against extreme variations. Human birth weight is an example of stabilizing selection (**Figure 15.5**). Over many years, hospital data have shown that human infants born within an intermediate range birth weight have a better chance of survival than those at either extreme. Thus, stabilizing selection reduces the variability in birth weight in human populations.

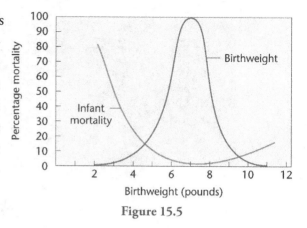

Figure 15.5

Disruptive selection (also called **diversifying selection**) is found when two or more extremes phenotypes are favored over the intermediate phenotype. Suppose there is a population of rabbits. The color of the rabbits is governed by two incompletely dominant traits: black fur, represented by "B," and white fur, represented by "b". A rabbit in this population with a genotype of "BB" would have a phenotype of black fur, a genotype of "Bb" would have gray fur (a display of both black and white), and a genotype of "bb" would have white fur. If this population of rabbits occurred in an environment that had areas of black rocks as well as areas of white rocks, the rabbits with black fur would be able to hide from predators amongst the black rocks, and the rabbits with white fur likewise amongst the white rocks. The rabbits with gray fur, however, would stand out in all areas of the habitat, and would thereby suffer greater predation. As a consequence of this type of selective pressure, our hypothetical rabbit population would be disruptively selected for extreme values of the fur color trait: white or black, but not gray.

Mutation

A mutation is a change in the DNA of an organism, usually occurring because of errors in replication or repair. Evolutionists and geneticists have come to realize that mutation is the ultimate source of genetic variation.

Not all mutations serve evolution. *Somatic cell mutations* occur in the DNA of somatic (body) cells and are not passed on to offspring. Some somatic cell mutations are deleterious such as those at work in human cancer tumors, whereas others have proven useful in the development of navel oranges and red delicious apples. The only mutations that matter in evolution are *germ line mutations* because those mutations occur in reproductive cells—sperm or egg—and can be passed on to offspring (**Figure 15.6**).

Mutations may occur spontaneously at the molecular level in several forms, or they may be induced by mutagens such as certain chemicals and solar radiation. Mutations may be small-scale (**point mutations**) affecting only a few nucleotides of a single gene or large-scale affecting the structure of entire chromosomes. Mutations may be benefi-

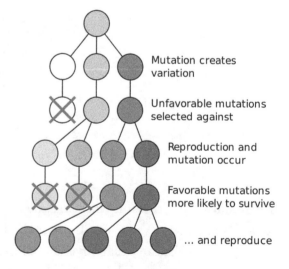

Figure 15.6

cial, neutral, or harmful for the organism, but they are always totally random. Whether a mutation happens or not is unrelated to how useful or detrimental that mutation could be.

In our previous example of a population of different colored fish, the bright silver-colored fish may have arisen from a germ line mutation. This mutation may have proven detrimental in some situations but favorable in others. Bright colored fish in shallow water on a cloudy day would be easily seen by predators, but the same fish in deeper water near the surface on a bright sunny day would very hard for predators to detect. Because the mutation (adaptation) was beneficial, natural selection would then maintain the mutation in the fish population.

Genetic Drift

By the simple matter of chance, in each generation, some individuals may leave behind a few more descendants that other individuals. These individuals passed on their genes due to a lucky accident, not because they were naturally selected. Unlike natural selection, genetic drift is a totally random process that doesn't work to produce new adaptations.

Returning to our imaginary fish population yet again, suppose fishermen looking for minnows (any small, shiny fish) to use as bait begin netting the mixed population of small fish in a single small lake (Lake A). They keep (remove) the bright colored ones but toss back the dark colored ones. Following this "catastrophe," genetic drift for at least one to possibly several generations could result in the removal of the unique genes carried by the bright colored fish.

Genetic drift acts more quickly to reduce variation in small populations than it does in large populations. This can create a **population bottleneck**, an evolutionary event in which a significant portion of a population is prevented from reproducing (**Figure 15.7**). Population bottlenecks result in increased sensitivity to genetic drift, increased inbreeding, and greatly reduced genetic variation in the remaining population. Suppose the fisherman seeking bait decided to keep all the small fish they netted from Lake A, regardless of color. If a substantial number of fish were removed from what was already a small population to begin with, a population bottleneck for that species in that lake would result, a catastrophe from which they would only slowly, if ever, recover.

The incidences of real population bottlenecks are increasing in many species, especially large birds and mammals. These are the realities conservation biologists struggle with as they attempt to prevent the extinction of large animals such as the California condor, the whopping crane, the elephant and the rhinoceros to name a few.

A special type of genetic drift known as the **founder effect** occurs when a new population is established by a very small number of individuals that splinter off from a larger population (**Figure 15.8**). Suppose the small lake (Lake A) containing our imaginary population of dark and bright colored

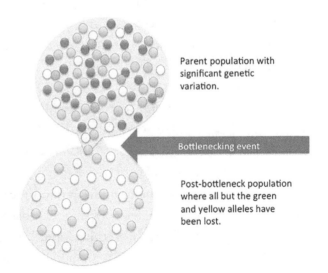

Parent population with significant genetic variation.

Bottlenecking event

Post-bottleneck population where all but the green and yellow alleles have been lost.

Figure 15.7 A bottlenecking event would be some sort of catastrophe that greatly reduces the original population.

fish of the same species is hit with a massive flood sweeping some of the fish from their ancestral lake. The founder effect would occur if a small group of the flood fish was to establish themselves downstream of their ancestral lake in a small pond(s) where their species had not existed before.

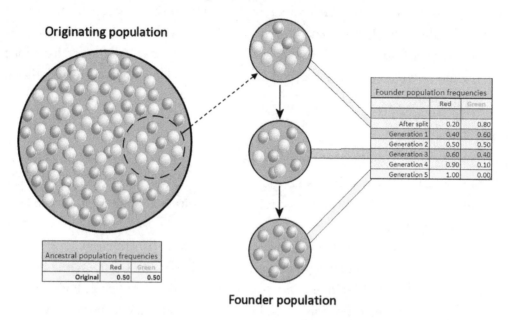

Figure 15.8 Representation of the Founder Effect. The colored balls represent the two alleles for a specific locus that are present in a hypothetical population; once a random subgroup of a population becomes separated from its ancestral population, the allele frequencies in the two groups' subsequent generations can diverge widely within a relatively short period of time as a consequence of a purely random selection of alleles for reproduction.

Apparently, humankind has not been immune to population bottlenecks in the past. Recent genetic research indicates that in our evolutionary path to modern *Homo sapiens*, our species may have experienced as many as four population bottlenecks and at one point there may have been as few as 15,000 individual humans on the planet. It seems probable that humanity has stared extinction in the face perhaps more than once but managed to avoid, at least for the time being, a fate that befell so many other animal species in the past. Incidences of population bottlenecks have increased in modern times, especially in large birds and mammals.

Genetic variation within natural populations was a puzzle to Darwin and his contemporaries. As Mendel's landmark genetic work emerged, however, and became integrated into evolutionary thought, the picture became clearer. In the early 1900s, it was thought that dominant genes must, over time, inevitably swamp recessive genes out of existence, a theory known as *genophagy* ("gene eating").

In 1908, an English mathematician, Godfrey Hardy, and a German physician, Wilhelm Weinberg, independently proposed a mathematical model that proved the theory of genophagy incorrect. The **Hardy-Weinberg principle** (or **Hardy-Weinberg equilibrium**) as it has come to be known is based on probability and concludes that gene frequencies are inherently stable but that evolution should be expected in all populations virtually all of the time.

Evolutionary biologists and population geneticists came to understand that evolution will not occur in a population if seven conditions are met:

1. Mutation is not occurring.
2. Natural selection is not occurring.
3. The population is infinitely large.
4. All members of the population breed.
5. All mating is totally random.
6. All matings produce the same number of offspring.
7. There is no migration in or out of the population.

In other words, if no mechanisms of evolution are acting on a population, evolution will not occur, and the gene pool frequencies will remain unchanged. In the real world, however, few if any of the above conditions exist. Hence, evolution is the inevitable result. The Hardy-Weinberg principle does not describe natural populations, but is an important tool for population geneticists, because the violation of one or more of the seven conditions causes the allele and/or genotype frequencies of a population to change in predictable ways. These predictions permit population geneticists to identify the factors that cause microevolution by measuring how the allele and genotype frequencies of a population are different than those of a population in Hardy-Weinberg equilibrium.

The Hardy-Weinberg principle defines evolution from the standpoint of population genetics as the sum total of the genetically inherited changes in the individuals who are the members of a population's gene pool. It is clear that the effects of evolution are felt by individuals, but it is the population as a whole that evolves. Genetically speaking, evolution is simply a change in the frequencies of genes in the gene pool of a population.

If a population is small, Hardy-Weinberg may be violated. Chance alone may eliminate certain members out of the proportion to their numbers in the population. In such cases, the frequency of a gene may begin to drift toward higher or lower values. Genetic drift may ultimately cause the frequency of the gene to represent 100% of the gene pool or, depending on the initial frequency of the alleles, completely disappear from the gene pool altogether. Genetic drift produces evolutionary change, but there is no guarantee that the new population will be fitter than the original one. Evolution by drift is aimless, not adaptive.

Gene Flow

Gene flow (also known as gene migration) results when genes are transferred from one population to another. If genes are carried to a population where those genes did not previously exist, gene flow can be an important source of genetic variation.

The most significant factor affecting the rate of gene flow between different populations is the mobility of the organisms comprising those populations. Animal populations are more mobile than plant populations and thus show a greater tendency to experience gene flow. **Immigration** (new individuals coming in) can add new genes to an established **gene pool** while **emigration** (individuals leaving a population) may result in the removal of genes. If two populations maintain a gene flow between them, this can lead to a combination of both their gene pools, reducing the genetic variation between the two groups (**Figure 15.9**).

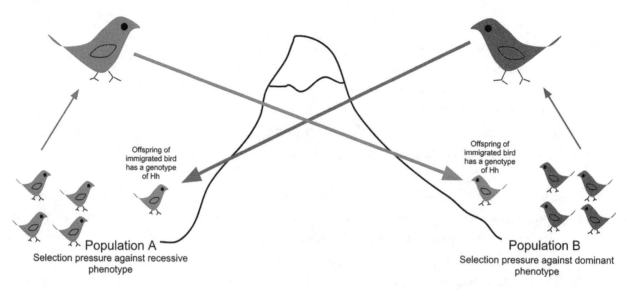

Figure 15.9 Population A has HH dominant genes whereas Population B has recessive hh genes. The population A immigrant introduces H genes into population B whereas the B immigrant introduces h genes into population A.

Barriers to gene flow are usually natural. They may include impassable mountains, open oceans, or vast tracts of deserts. Human development of the landscape has thrown up many artificial barriers. A multi-lane highway stretching for miles can be a major barrier to the movement of small animals. Human development and occupation can also fragment ecosystems into isolated islands and in the process seriously reduce the genetic variation of the creatures imprisoned on those islands. A great challenge facing conservation biologists is the maintenance or establishment of corridors or connections between these isolated fragments to maintain a healthy genetic variation through the facilitation of gene flow.

Although human constructions, occupation, and activity may pose barriers to other species, humans themselves are the most mobile and wide-ranging animal on the planet. Genetic analysis of humans of different geographical regions and races reveal tremendous rates of gene flow and a subsequent mixing of many human gene pools or the "dilution of the races" as it has been called.

Let us again consider our imaginary fish population. Suppose our mixed population of small fish—the dark-colored fish, representing the usual coloration of the species, and the bright-colored fish which arose through germ line mutation—being maintained by natural selection is found in a small lake (Lake A). A land barrier of a short distance separates Lake A from another small lake with a population of only the standard dark colored fish (Lake B). After torrential rains lasting for days, the lakes overflow and a few of the bright colored fish only from Lake A are washed into Lake B. This unintentional migration of bright colored fish has resulted in gene flow from Lake A to Lake B.

Box 15.1
Are Botanical Gardens Ecological Salvation or Evolutionary Stagnation?

Botanical gardens, the plant equivalent of zoos, are tranquil places filled with beautiful and often unusual plants. For over 100 years many botanical gardens have also served as a repository for rare and endangered plant species. Faced with the fact one of every eight plant species in the world is threatened with extinction, concerned botanists have been working diligently to conserve endangered species by propagating them in artificial habitats such as botanical gardens. But is such a strategy ecologically and evolutionarily/genetically sustainable in the long term?

How have we come to the point where the number of imperiled plants fills 750 pages and counting in a large bound book and accounts for 13% of all known plant species worldwide? Botanists cite two main reasons why so many plant species are threatened: the past and continuing destruction of large areas of habitat by human development, agriculture, and logging and invasion by introduced plant species that crowd out indigenous species. While endangered mammals and birds garner more public attention and sympathy, it is the Green World that forms the warp and woof of the natural world, the underpinnings on which eukaryotic life is based.

Ecologically, botanical gardens may be the only choice for sheltering and preserving many endangered species from total extinction. For example, in partnership the National Tropical Botanical Garden (NTBG) and the Plant Extinction Prevention Program (PEP) of Hawaii monitor and survey endangered plants on the islands and collect plants, seeds, and cuttings of these endangered plants. Plants grown from seeds or cuttings are established *ex situ* (out of native habitat) in NTBG gardens or *in situ* (in native habitat) in native restoration sites on various islands as well as on private land. The goal of such projects is two-fold: preserve the species with hopes of eventually reestablishing the species into its natural habitat.

Unfortunately, what may preserve a species from extinction in the short-term may result in evolutionary changes (perhaps for the worst) in the long-term. Many plant species have been brought to the brink of extinction for a combination of reasons. This forces these species into a population bottleneck seriously restricting the gene pool of those species. Botanists are then forced to gallantly battle against the total extinction of these species by protecting them, usually ex situ. By collecting the healthiest specimens of endangered plants for preservation, botanists further decrease the gene pool in the remaining wild population. With only a small number of individuals with small stilted gene pools maintained in botanical gardens and parks isolated from gene flow with others of their kind, genetic drift can cause these captive species to evolve rapidly. An endangered species protected in captivity may be lost genetically because it evolves into one or several species better adapted to survive in artificial humanly-maintained conditions than they are to living in the natural habitat of the original species.

The only way to preserve endangered species ecologically and genetically is to provide a sanctuary large enough to contain and maintain the pollinators and seed dispersers of the species and diverse enough to force natural selection pressures. That sanctuary would then need to be populated with enough individuals of the species to remain stable against genetic drift. Clearly, the best approach to solving this problem is to maintain vast tracts of native habitat in the first place rather than trying to recreate them artificially after they are gone, a monumental challenge given the relentless grind of human encroachment and development.

Speciation

When Darwin arrived at the Galapagos Islands, he was astounded by the variety of plants and animals found nowhere else in the world. Darwin wrote in his diary: "Both in space and time, we seem to be brought near to that great fact—that mystery of mysteries—the first appearance of new beings on this Earth."

The "mystery of mysteries" that captivated Darwin and evolutionary biologists since then is **speciation**, the process by which one species splits into two or more new species. Speciation not only explains the great diversity of life on Earth but also reveals the common ancestors from which new species have arisen. Speciation also forms a conceptual bridge between microevolution, changes over time in the allele frequencies in a population, and macroevolution, the broad pattern of evolution above the species level.

A species may be defined in three main ways: morphologically, phylogenetically, or biologically.

Morphological species concept In this concept, species are distinguished from each other by one or more distinct physical characteristics known as *diagnostic traits*. Morphological identification of species works reasonably well for large plants and animals as evidenced by the fact that many of the morphological species that Linnaeus defined have held up to 200 years of scrutiny. However, the morphological species concept breaks down at the microscopic level because many bacteria and other microorganisms do not have many measurable morphological traits.

Phylogenetic species concept In this concept, a phylogeny or evolutionary "family tree" is used to identify species based on a common ancestor. According to the phylogenetic species concept, a species is the smallest set of interbreeding organisms—usually a population—that shares a common ancestor. In phylogeny, a branch that contains all the descendants of a common ancestor is said to be monophyletic. Monophyly is the main criterion for defining species in the phylogenetic species concept (**Figure 15.10**). One advantage of the phylogenetic species concept over the morphological species concept is that it relies on genetic makeup rather than morphological traits. Thus, species of microorganisms and **cryptic species** (species that look almost identical morphologically) can be identified with great confidence.

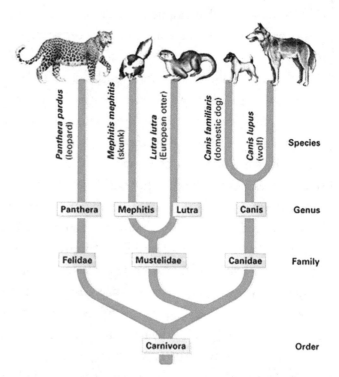

Figure 15.10 The species designation for select members of order Carnivora as determined by their phylogeny.

Biological species concept This concept relies primarily on reproductive isolation to identify different species. Biologically, a species may be defined as *the smallest cluster of organisms that possess at least one diagnostic character— morphological (structural), biochemical, or molecular— and that are reproductively isolated from their ancestral species.* That is, if organisms cannot mate and produce offspring in nature or if their offspring are sterile, they are defined as different species. The biological species concept often cannot be tested in nature, because many potential species do not overlap in their distribution and thus do not have an opportunity to demonstrate reproductive isolation.

What mechanism drives the biological speciation concept? The first step in the process is that a population of interbreeding individuals of the same species must be splintered into at least two separate groups. These groups must then be isolated in such a way as to cause a reduction or elimination of the gene flow between them. Over a long period of time, as different mutations and adaptations accumulate in each group, the groups may become so reproductively isolated from each other that they could no longer interbreed even if they were brought back together again. Thus by our definition, a new species would have arisen.

A population may be separated into groups through geographic isolation as rivers change course, mountain chains arise, canyons form, continents drift, and organisms migrate. However, even in the absence of a geographical barrier, reduced gene flow across a species' range can encourage speciation.

Groups may be reproductively isolated from each other through a number of different mechanisms:

- Two species might occupy such a large range that eventually those individuals at the far edges of that range stop breeding with each other thereby reducing the gene flow between them (*habitat isolation*).
- They may evolve different mating times or mating rituals (*temporal isolation*).
- A lack of "fit" between sexual organs may evolve. When animal genitalia or plant flora structures are incompatible, reproduction cannot occur (*mechanical isolation*). This seems to be especially true in insects where damselflies alone have nearly half a dozen different penis shapes, and male dragonflies have claspers that are suitable for holding only female dragonflies of their own species.
- Even if the gametes of two different species meet, they may not fuse to become a viable zygote (*gamete isolation*). In animals, the sperm of one species may not be able to survive in the reproductive tract of another species, or the egg may have receptors only for the sperm of its species. In plants, only certain types of pollen grains can germinate in order that sperm might reach the egg.
- A hybrid zygote with two different sets of chromosomes may not be viable (*hybrid inviability*). The developing embryo may not receive compatible instructions from the maternal and paternal genes thus preventing proper development.
- The hybrid zygote may develop into a sterile adult (*hybrid sterility*). Sterility of hybrids results from complication in meiosis that lead to an inability to produce viable gametes. On the rare occasion when two hybrids mate and produce offspring, the F_2 hybrids have a reduced fitness.

Suppose the bright-colored fish taken from Lake A were carted hundreds of miles, and after some had been used as bait, the rest were dumped into an isolated lake that contained no other fish of that species. Thus geographically and physically isolated, all gene flow between these fish and others of their species would be eliminated. If the founder effect kicked in and they become established, mutations, adaptations, and behavioral changes might accumulate over long periods of time resulting in these isolated bright-colored fish becoming an entirely new species.

There are two basic modes of speciation: *allopatric* ("Other Country") and *sympatric* ("Same Country").

Allopatric speciation is the eventual result of populations that have become separated ("Other Country") by a geographic or physical barrier. When populations of a species become geographically or otherwise isolated, microevolutionary processes, such as genetic drift and natural selection alter the gene pool of each population independently. If the differences between the groups become large enough, reproductive isolation may occur, resulting in the formation of new species (**Figure 15.11**).

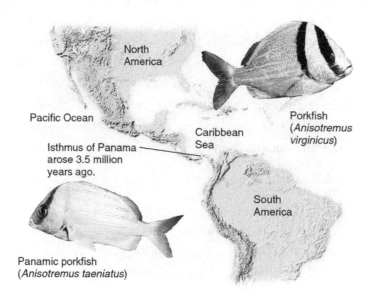

Figure 15.11 The formation of the Isthmus of Panama about 3.5 million years ago split an ancestral fish population into two. Since that time, different genetic changes have occurred in the two populations because of their geographic isolation; these changes eventually led to the formation of different species. The porkfish (*Anisotremus virginicus*) is found in the Caribbean Sea, and the Panamic porkfish (*Anisotremus taeniatus*) is found in the Pacific Ocean.

Sympatric speciation is speciation that occurs in the absence of geographic or physical separation ("Same Country"). Sympatric speciation is difficult to observe in nature because no physical barrier prevents mating between populations, as in allopatric speciation. Most examples of sympatric speciation in nature involve divergence of diet, microhabitat, or both. In these cases, a new species evolves when a population becomes specialized to live in a new microhabitat.

Cichlid fish found in a small Nicaraguan lake provide one of many examples of sympatric speciation. Initially, the lake was colonized by the Midas cichlid (*Amphilophus citrinellus*) that occupied rocky shallow water areas along the shoreline (**Figure 15.12**). Over time, a population of the Midas cichlid adapted to living and feeding in an open water habitat and evolved into the arrow cichlid (*Amphilophus zaliosus*). The partitioning of the lake habitats and changes in body size, jaw morphology, and tooth size and shape to accommodate different diets led to sympatric speciation of the cichlids.

Figure 15.12 Lake Cichlids. (a) Midas cichlid, (b) Arrow cichlid.

One of the criticisms leveled at the modern theory of evolution by the misguided or ill-informed is that if evolution is dynamic and ongoing, why have no new species appeared? In truth, molecular and genetic analysis reveals that a few new species of both plants and animals have come into being in modern times. These *speciation events* as they are dubbed are occurring, but at what rate? Unfortunately, there seems to be a lack of interest among biologists in pursuing this question. For one thing, the biological community considers this a settled question, so few researchers have bothered to look closely at the issue. Second, most biologists accept the idea that speciation takes a very long time (relative to human life spans). Thus, the number of speciation events that actually occur during the course of a researcher's life would be very small, if any.

Evidence of Evolution

At the very core of evolutionary theory is the basic idea that life has existed in some form for billions of years and that life has changed drastically over time. The history of living things is documented through multiple lines of evidence that detail the story of life through time.

Evidence from the Fossil Record

Humans have attempted to understand and interpret fossils since the days of the ancient Greeks and Romans. More scientific views of fossils began to emerge during the Renaissance, but fossils were still largely misunderstood and misinterpreted. For example, in China, the fossil bones of ancient mammals were thought to be "dragon bones" and useful in medicine when ground into powder. In the West, the presence of fossilized sea creatures high up on mountainsides was seen as proof of the Biblical flood.

In 1799, an English canal engineer named William Smith noted that in undisrupted layers of rock, fossils occurred in a definite sequential order, with more modern-appearing ones closer to the top. Because bottom layers of rock logically were laid down earlier and thus are older than top layers, the sequence of fossils could also be given a chronology from oldest to youngest.

When Darwin wrote the *Origin of Species*, the oldest animal fossils known were those from the Cambrian Period, about 540 million years old. The scarcity and relatively young age of fossils and especially the rarity of intermediate fossil forms held implications that worried Darwin concerning the validity of his theories. Since Darwin's time, the fossil record has been pushed back to 3.5 billion years before the present, and many of the gaps in the fossil record have been filled in by the research of paleontologists.

Given the low likelihood of environmental conditions suitable for fossil formation and the even lower likelihood of finding fossils after they have formed, it is not surprising that there are gaps in the fossil record. These gaps are being slowly filled in by the discovery of transitional forms that show the intermediate states between and ancestral form and its descendants.

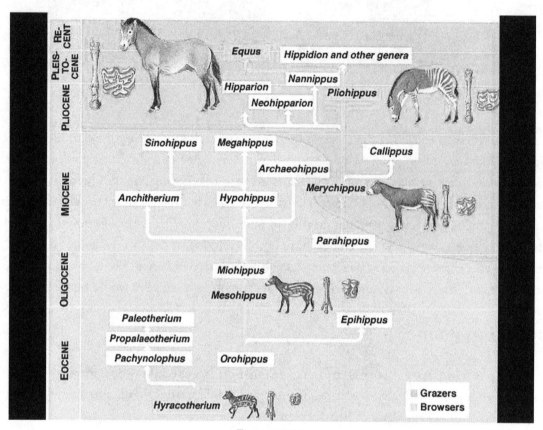

Figure 15.13

One of the most studied examples of transitional forms concerns the evolution of horses (**Figure 15.13**). Modern horses, zebras, donkeys, and asses (Equidae) represent just one tiny twig on an immense family tree that spans tens of millions of years. These species, all classified in the genus Equus, are the last living descendants of a long lineage that have produced 34 genera since its origin in the Eocene period, approximately 55 million years ago.

❖ *Hyracotherium*, the first equid, appeared some 55 mya and flourished in North America and Europe. Often referred to as the "dawn horse," this prehorse averaged two feet (60-cm) in length and eight to 14-inches (20cm) high at the shoulder and weighed about 50 pounds. It had four-hoofed toes on each front foot and three-hoofed toes on each hind foot. Each toe had a pad on its underside, similar to those of a dog. It had a short face with eye sockets in the middle and a short *diastema* (the space between the front teeth and the cheek teeth). The skull was long, having 44 low-crowned teeth. Although it had low-crowned teeth, the beginnings of the characteristic horse-like ridges on the molars can be seen. *Hyracotherium* is believed to have been a browsing herbivore that ate primarily soft leaves as well as some fruits, nuts and plant shoots. The dawn horse was succeeded by *Orohippus*, which differed from *Hyracotherium* primarily in dentition.

❖ *Mesohippus* lived some 30 to 40 mya and, like many fossil horses was common in North America. *Mesohippus* had longer legs than its predecessor standing about 24 inches (60 cm) tall and weighing around 75 pounds. It had also lost a toe and stood predominantly on its middle toe, although the other two were also used. The face of *Mesohippus* was longer and larger than earlier equids. It had a slight *facial fossa*, or depression, in the skull. The eyes were rounder and were set wider apart and farther back than in *Hyracotherium*. Its teeth were low-crowned and contained a single gap behind the front teeth, where the bit now rests in the modern horse. In addition, it had another grinding tooth, making a total of six. *Mesohippus* was a browser that fed on tender twigs and fruit. The cerebral hemisphere, or cranial cavity, was notably larger than that of its predecessors; its brain was similar to that of modern horses.

❖ *Merychippus*, considered a proto-horse, was endemic to North America during the Miocene some 20 mya. It stood about 35 in (89 cm) tall and weighed up to 500 pounds. The muzzle was longer, the jaw deeper, and the eyes wider apart than any other horse-like animal to date. The brain was also much larger, making it smarter and more agile. *Merychippus* was the first equine to have the distinctive head of today's horses. The foot was fully supported by ligaments, and the middle toe developed into a hoof, which did not have a pad on the bottom. In some *Merychippus* species, the side toes were larger, whereas, in others, they had become smaller and only touched the ground when running. Its teeth were beginning to form a series of tall crests with higher crowns.

❖ *Pliohippus* arose during the middle Miocene around 15 mya. Sometimes called the "grandfather of the modern horse," these equines stood about six feet high at the shoulder and weighed about 1,000 pounds. It was the first singled toe, or hoofed, horse although it did have two much-reduced side toes that would have been barely visible. The teeth of *Pliohippus* were larger, taller and more complexly folded than those of earlier horses indicating a greater dependence on grazing than browsing for food. The skull was marked by deep pre-orbital fossae (deep depressions in front of the eyes) not found in the skull of modern *Equus*.

The horse spread into South America, as well as Asia, Europe, and Africa. In the last two million years, *Equus* emerged as the large, magnificent creature we admire today. About 8,000 years ago, *Equus* became extinct in the New World and was not to return until the Spanish brought horses to the Western Hemisphere in the 1,400's.

The changes that occurred in the horse lineage may be understood as adaptations to changing global climates. During the lat Miocene and early Oligocene epochs (20 to 25 mya), grasslands and steppes became widespread across the globe. As horses adapted to these habitats, the need to run faster to escape predators became increasingly important. The flexibility provided by multiple toes and shorter limbs that had provided so advantageous for ducking through forest vegetation was no longer beneficial. At the same time, horses were eating grasses and other coarse vegetation that favored skull and teeth better suited for grinding such materials.

Investigations of the history of any group of living things such as the horse reveal that early notions of evolution as a progressive, guiding force, consistently pushing living organisms in a single direction are wrong. Rather, the fossil record demonstrates that even though overall trends have been evident in a variety of characteristics and adaptations, evolutionary change has been far from constant and uniform through time. Instead, rates of evolution have varied widely, and when changes happen, they often occur simultaneously in different lineages with a group of creatures. And when a trend exists, exceptions are not uncommon.

Evidence from Morphology—Homology and Analogy

In nature, organisms often look like one another for different reasons. Two beetles of the same species might look alike because they inherited their yellow and black spots from their parents. On the other hand, two flying animals, such as a bird and a bat, also appear similar to each other. The similarity between the beetles is inherited, but the similarity between the bird and the bat is not.

The skeletons of tetrapods are strikingly similar, despite the different lifestyles of these animals and the different environments they inhabit. Bone by bone, the similarity of these animals can be observed in every part of the body, including the limbs, yet a person throws a ball, a whale swims, an alligator shuffles, and a bird flies. Each of these animals is framed of bones that are different in detail but similar in general structure and relation to each other (**Figure 15.14**).

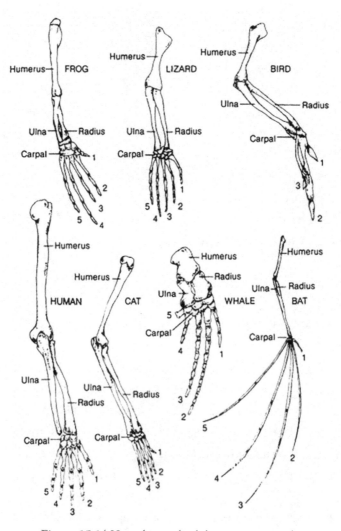

Figure 15.14 Homologous limb bones in tetrapods.

Evolutionary biologists regard such similar traits and structures as being **homologous** and have concluded that such similarities are best explained by descent from a common ancestor. Comparative anatomists investigate such homologies, working out relationships from degrees of similarity. Their conclusions provide valuable inferences into the details of evolutionary history, inferences that can be tested by comparison with the sequence of ancestral forms in the fossil record and through cladistic analysis. Different animals end up with the same sort of limb inherited from a common ancestor, just as cousins might inherit their hair color from their grandfather (**Figure 15.15**).

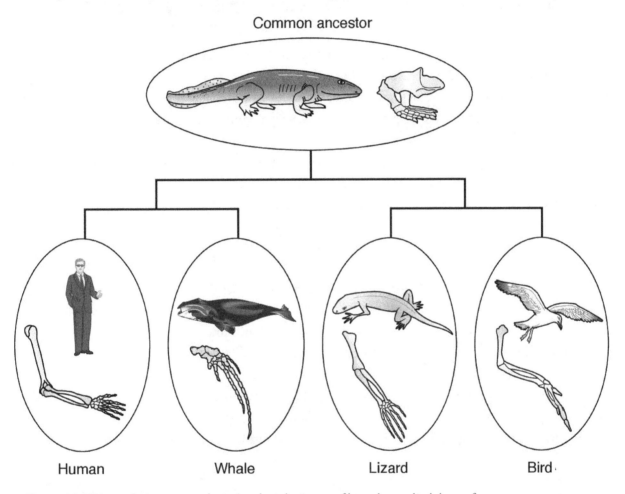

Figure 15.15 An evolutionary tree depicting the inheritance of homologous limb bones from a common ancestor.

Over time, a trait or structure inherited from a common ancestor may be changed and adapted taking on a different structure and function. Consider the teeth of the beaver and the elephant. The gnawing front teeth of the beaver (**Figure 15.16**) and the tusks of the elephants (**Figure 15.17**) are both basically incisor teeth. Although inherited in a basic form from an ancestor common to the beaver and the elephant, these teeth have been greatly modified by evolutionary mechanisms into the seemingly dissimilar teeth we see today in the beaver and the elephant.

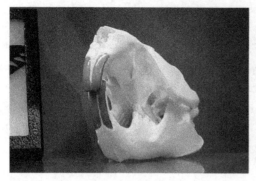

Figure 15.16   Figure 15.17

At first glance, many traits and structures that seem dissimilar are homologous but some that seem homologous are not. The wings of a bird and those of a bat might appear to be homologous but are, in fact, **analogous**. Analogous traits or structures are those that evolve independently in organisms that are not closely related as a result of having adapted to similar environments or niches through a process known as **convergent evolution**.

Examples of convergent evolution abound in the animal kingdom. Several mammal groups have independently evolved prickly protrusions of the skin or spines, such as echidnas (monotremes), hedgehogs (insectivores), and porcupines (rodents). Claws and long, sticky tongues that allow them to open the nests of ants and termites are the analogous structures that mark mammals loosely grouped as "anteaters" including four species of true anteaters, eight species of pangolin, four species of echidna, the African aardvark, and the Australian numbat.

In birds, the Little Auk of the North Atlantic and the diving petrels of the southern oceans are remarkably similar in appearance and habits. Both Old World vultures and New World vultures have featherless necks and heads, search for food by soaring and feed on carrion. However, Old World vultures are in the eagle and hawk family and use eyesight to find food, whereas the New World vultures are related to storks and use the sense of smell and sight to discover food.

Convergent evolution can also be found in plants. Numerous nitrogen-deficient plants have become carnivorous: flypaper traps such as sundews (**Figure 15.18**) and butterworts, spring traps like the Venus fly trap (**Figure 15.19**), and pitfall traps (**Figure 15.20**) in order to capture and digest insects to obtain scarce nitrogen.

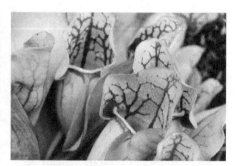

Figure 15.18 Fly-paper trap plants such as this sundew (*Drosera capensis*) use sticky secretions to trap insect

Figure 15.19 Modified leaves of the Venus flytrap (*Dionaea muscipula*) serve as insect-collecting traps

Figure 15.20 The slippery edge around the opening of the pitcher plant (*Sarracenia purpurea*) causes insects to slide down to their doom in liquid at the base of the pictcher.

Convergent evolution is even seen in humans. People in most populations around the world stop producing lactase, the enzyme that digests milk, some time in childhood. Individuals in African and European populations that raise cattle, however, produce lactase throughout their lives. DNA analysis reveals that the retention of lactase production into adulthood is the result of different mutations in Africa and Europe, which indicates that those populations have independently acquired this adaptation.

Evidence from the Molecular Record

Evolutionary biologists believe that all living creatures have descended from simple cellular organism that arose about 2.5 billion years ago. Some of the characteristics of that earliest organism have been preserved in all living things alive today. By comparing the genomes (sequence of all the genes) of different groups of animals, we can specify the degree of relationship among the groups more precisely than by any other means.

All organisms store their biological history in their DNA. Therefore, evolutionary changes over time should involve a continued accumulation of genetic changes in the DNA. As a result, organisms that are more distantly related should have accumulated a greater number of these genetic differences than organisms that are more closely related. Thus, the genome of gorillas, which the fossil record indicates diverged from humans between 6 and 8 million years ago, differs from the genome of humans by 1.6% of its DNA. However, the genome of chimpanzees, which diverged about 5 million years ago, differs by only 1.2%. These echoes of the evolutionary past also show up protein sequences, such as hemoglobin, the oxygen-binding component of red blood cells. Humans and macaques, both primates, show less difference in their hemoglobin proteins than to more distantly related mammals, such as dogs. Nonmammalian vertebrates, such as birds, reptiles, and frogs, differ from primates even more.

Genes evolve at different rates because although mutation is a random event, some proteins are more tolerant of changes in their amino acid sequence that are other proteins. For this reason, the genes that encode these more tolerant, less constrained proteins evolve faster. The average rate at which a particular kind of gene or protein evolves has given rise to the concept of a **molecular clock**. Molecular clocks run rapidly for less constrained proteins and slowly for more constrained proteins, though they all time the same evolutionary events.

The concept of a molecular clock is useful for two purposes. It determines evolutionary relationships among organisms, and it indicates the time in the past when species started to diverge from one another. It should be noted, however, that the idea of a molecular clock ticking away in proteins is a controversial concept that is not totally embraced and supported by all biologists.

Another interesting line of evidence supporting evolution involves sequences of DNA known as pseudogenes. **Pseudogenes** (or "junk DNA") are remnants of genes that no longer function, but continue to be carried along in DNA as excess baggage. Pseudogenes also change over time as they are passed from ancestors to descendants, and in the process offer an especially useful way of reconstructing evolutionary relationships. Because they perform no function, the degree of similarity between pseudogenes must simply reflect their evolutionary relatedness. The more remote the last common ancestor of two organisms, the more dissimilar their pseudogenes will be.

The evidence for evolution from molecular biology is overwhelming and is growing steadily. In some case, this molecular evidence transcends paleontological evidence. Take whales for example. Anatomical and paleontological evidence indicates that the whale's closest living land relatives seem to be the even-toed hoofed mammals (cattle, sheep, camels, goats, etc.). Recent analysis of some milk protein genes have confirmed this relationship and have suggested that the closest land-bound living relative of the whales may be the hippopotamus.

Evidence from Biogeography

Darwin in his time and biologists of today are fascinated by the inconsistency and variability in the distribution of plants and animals around the world. Why do the Galapagos Islands of South America and the Cape Verde Islands off Africa have strikingly different fauna and flora, despite having similar environmental conditions? Why are there polar bears but no penguins in the Arctic when the Antarctica has penguins, but no polar bears? In *Origin of Species*, Darwin states: "In considering the distribution of organic beings over the face of the globe, the first great fact that strikes us is, that neither the similarity nor the dissimilarity of the inhabitants of various regions can be wholly accounted for by climatic and other physical conditions."

Biogeography is the science that attempts to understand why (and sometimes, more importantly, why not) plants and animals exist where they do. All creatures are adapted to the abiotic and biotic factors of their habitat. Thus, one might assume the same species would be found in a similar habitat in a similar geographic area, e.g. Africa and South America. That is not the case. Plant and animal species are discontinuously distributed around the planet. This discontinuity of distribution can be seen in African and South American fauna. Africa has short-tailed (Old World) monkeys, elephants, lions and giraffes, whereas South America has long-tailed monkeys, cougars, jaguars, and llamas.

Before humans arrived 40,000 to 60,000 years ago, Australia had more than 100 species of kangaroos, koalas, and other marsupials, but none of the more advanced terrestrial placental mammals such as wolves, lions, bears, horses. Land mammals were entirely absent from the even more isolated islands that make up Hawaii and New Zealand. However, each of these places had a great number of plant, insect, and bird species that were found nowhere else in the world. The most likely explanation for the existence of Australia's, New Zealand's, and Hawaii's mostly unique biotic environments is that the life forms in these areas have been evolving in isolation from the rest of the world for millions of years.

All creatures are adapted to the abiotic and biotic factors of their habitat. One might assume, therefore, that the same species would be found in a similar habitat in a similar geographic area, for instance, in Africa and South America. That is not the case. Plant and animal species are discontinuously distributed around the planet. This discontinuity of distribution can be seen in African and South American fauna. Africa has short-tailed (Old World) monkeys, elephants, lions and giraffes while South America has long-tailed monkeys, cougars, jaguars, and llamas. The flora of North and South America show a similar discontinuity. Deserts in North and South America have native cacti, but deserts in Africa, Asia, and Australia have succulent native euphorbs that resemble cacti but are very different, even though in some cases cacti have done very well (for example in Australian deserts) when introduced by humans.

The discontinuity of species distribution is even more striking when it comes to islands. Biogeography divides islands into two categories. *Continental islands* are islands like Great Britain, and Japan that have at one time or another been part of a continent. *Oceanic islands* like the Hawaiian Islands and the Galapagos Islands, on the other hand, are islands that have formed in the ocean and have never been part of any continent.

Oceanic islands have distributions of native plants and animals that are unbalanced in ways that make them distinct from the flora and fauna found on continents or continental islands. Save for bats and seals, oceanic islands do not have native terrestrial mammals, amphibians, or fresh water fish. In some cases oceanic islands have terrestrial reptiles but most do not.

Starting with Charles Darwin, many scientists have conducted experiments and made observations that have shown that the types of animals and plants found, and not found, on such islands are consistent with the theory that these islands were colonized accidentally by plants and animals that were able to reach them. Such accidental colonization could occur by air, such as plant seeds carried by migratory birds, or bats and insects being blown out over the sea by the wind, or by floating from a continent or other island by sea. Many of the species found on oceanic islands are endemic to a particular island or group of islands, meaning they are found nowhere else on earth. However, they are clearly related to species found on other nearby islands or continents.

Geologic and Environmental Influences

It would be nearly impossible to understand the evolutionary patterns which shaped life on this planet without considering the canvas of geologic and environmental changes on which those patterns have been and continue to be painted.

The supercontinent Pangaea and the Carboniferous forests once found there no longer exist. Where did they go? What happened to them? Pangaea was eventually wrenched asunder by the slow but powerful process known as **continental drift**. Powerful movements in the hot putty-like upper mantle break the cool brittle crust into various sizes pieces known as *continental plates*. These tightly interlocked plates "float" on the mantle below them. As the mantle continues to churn, the plates are forced apart in some places, forced together in other places, and slip side by side in others. Over geologic periods of time, these movements change not only the shape but also the location of the continents (**Figure 15.21**). The North American plate broke from the remains of Pangaea, headed north, and became cooler and drier. As a result, the Carboniferous forests and the animals that inhabited the plate disappeared and in their place arose the plants and animals of present-day North America. Continental drift also solves mysteries which had previously puzzled scientists such as the fact that beneath miles of ice on Antarctica

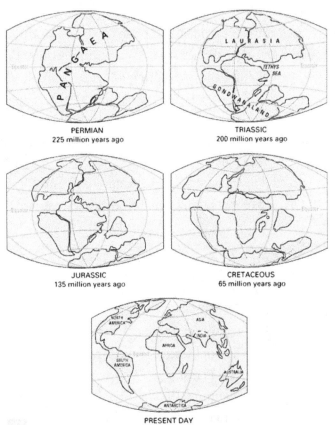

Figure 15.21 The powerful dance of the continents known as continental drift slowly makes and remakes the face of Earth.

we find the fossilized remains of tropical plants and animals and that Australia has many unique species of plants and animals found nowhere else on the planet.

As the continents changed, so did the flora and fauna occupying those continents. Some types of plants and animals became extinct yet others adapted and evolved into new species to fill the void. As a result, continental drift has caused the continents to become the burial ground for some species (extinction) but an evolutionary birth place (origins) for others.

The second major environmental influence on plant evolution has been ice, ice in the form of continental glaciers, polar ice sheets miles thick periodically forming and retreating in a series of cycles known as the *ice ages*. Starting around 2.4 billion to 2.1 billion years ago, ice ages have come and gone at least five times. Within each long-term ice age, there are pulse of extra cold temperatures (*glacial periods*) mixed with intermittent warm periods (*interglacial periods*). The "ice age" colloquially refers to the most recent glacial period which peaked at the Last Glacial Maximum approximately 20,000 years ago, in which extensive ice sheets lay over large parts of the North American and Eurasian continents. Earth is currently in a warm interglacial period known as the Holocene, which began 10,000 to 50,000 years ago.

As continental glaciers up to 2 miles thick slowly ground over the land, they obliterated all flora in their path, stripped away soil down to bare rock, and carved out many strange geologic features such as the Great Lakes. In fact, the tremendous weight of these glaciers deformed the crust in numerous places. When the glaciers retreated (melted) a barren, cold, dry landscape challenged plants and animals to adapt and evolve or perish.

All available evidence supports the central conclusions of evolutionary theory that life on Earth has evolved and that species share common ancestors. Biologists do not argue about these conclusions. What they are debating and attempting to discover are the details of the process:

1. Does evolution proceed slowly and steadily or does it happen in fits and starts?
2. Why are some clades (branches on the evolutionary tree) very diverse, whereas others are sparsely populated?
3. How does evolution produce new and complex adaptations?
4. Are there trends in evolution and if so, what processes generate these trends?

In Summary

- Darwin's theory of evolution consists of five components:
  1. Perpetual change
  2. Common descent
  3. Multiplication of species
  4. Gradualism
  5. Natural selection
- Evolution may be defined as change over time.
- Evolution happens on two levels: the genetic level (microevolution) and the taxonomic level (macroevolution).

- Evolution is driven by the mechanisms of mutation, genetic drift, gene flow, natural selection, and speciation.
- For natural selection to occur there must be variable traits within a population, differential reproduction within a population, and inheritance.
- There are three types of natural selection: directional selection, stabilizing selection, and disruptive selection.
- A mutation is a change in the DNA of an organism.
- Somatic cell mutations occur in somatic (body) cells whereas germ line mutations occur in germ (reproductive) cells.
- Genetic drift is a variation in the relative frequency of different genotypes in a small population, owing to the chance disappearance of particular genes as individuals die or do not reproduce.
- A population bottleneck is an event that drastically reduces the size of a population.
- A special type of genetic drift known as the founder effect occurs when a new population is established by a very small number of individuals that splinter off from a larger population.
- The Hardy-Weinberg principle (or Hardy-Weinberg equilibrium) states that both allele and genotype frequencies in a randomly-mating population remain constant—and remain in this equilibrium across generations—unless a disturbing influence is introduced.
- Genes flow into or out of the established gene pool of a population as individuals come in (immigration) or leave (emigration).
- Speciation not only explains the great diversity of life on Earth but also reveals the common ancestors from which new species have arisen.
- A species may be defined in three main ways: morphologically, phylogenetically, or biologically.
- There are two basic modes of speciation: allopatric ("Other Country") and sympatric ("Same Country").
- Evidence for evolution may be found in the fossil record, the homology and analogy of body structures, the molecular record, and biogeography.
- Biogeography is the science that attempts to understand why (and sometimes, more importantly, why not) plants and animals exist where they do.
- Evolutionary patterns and trends have happened against a backdrop of geologic and environmental change.

Review and Reflect

1. *In Your Own Words* Write you own personal definition of biological evolution.
2. *Foundation of Evolutionary Theory* The modern theory of evolution is said to rest on five pillars or components. Name and explain those pillars. (Reference Chapter 14)
3. *Mechanisms of Natural Selection* On your large private island there are many species of animals and plants. You plan on introducing a population of an animal species not found on the island in an attempt to observe natural selection in action. What three things will need to happen to the introduced species for natural selection to occur?

4. **Selection Types** Use Figure 15.4 to determine the type of selection occurring in each of the following situations. Name the type of selection and explain how you arrived at that conclusion.

Situation 1: Fossil records that show that the size of the black bears in Europe decreased during interglacial periods of the ice ages, but increased during each glacial period.

Situation 2: Suppose that in an imaginary population of rabbits the color of the rabbits is governed by two incompletely dominant traits: black fur, represented by "B," and white fur, represented by "b". A rabbit in this population with a genotype of "BB" would have a phenotype of black fur, a genotype of "Bb" would have gray fur (a display of both black and white), and a genotype of "bb" would have white fur.

If this population of rabbits occurred in an environment that had areas of black rocks as well as areas of white rocks, the rabbits with black fur would be able to hide from predators amongst the black rocks, and the rabbits with white fur likewise amongst the white rocks. The rabbits with gray fur, however, would stand out in all areas of the habitat, and would thereby suffer greater predation.

Situation 3: Robins typically lay four eggs.

5. **You Mutant** Mutations may be harmful, helpful, or neutral. Which type of mutation—somatic cell mutation or germ line mutation—would have the greatest effect on the population in which the mutation appears?

6. **Hardy and Weinberg Speak** Let us suppose that a trait in beetles, such as spots, is determined by the inheritance of a gene with two alleles—S and s. Fifty years ago, the frequency of S in a small population of beetles was determined to be 92% while the frequency for s was found to be 8%. Today we find S at 90% and s at 10%.

   A. Has evolution occurred in this population of beetles? Defend your answer using the Hardy-Weinberg equilibrium.
   B. How might a population bottleneck and resulting genetic drift have played a role in the observed change in gene frequency in the beetles?

7. **In the Blood of Cheetahs**. Clocked at 70 miles per hour while pursuing prey, the cheetah is by far the fastest and most agile member of the cat family. Superbly adapted for hunting, cheetahs are even more successful than lions in capturing prey. Yet the cheetah population is dwindling with estimates placing their numbers at anywhere from 1,000 to 20,000. What is happening? Why should the population of an animal with such extraordinary capabilities dwindle in this way? Is this an example of a population bottleneck?

It has been proposed that at some time in the cheetah's recent evolutionary history, an infectious disease or a natural catastrophe must have reduced its population, and a population bottleneck resulted. This, in turn, caused a reduction in the variety of alleles (gene pairs) in the population. The genetic drift resulting from the reduction of a population results in a surviving population that over time is no longer representative of the original population.

Biologists began to investigate the population bottleneck hypothesis by studying cheetahs at a breeding and research center in Africa. At this center, as with others, there was little success in breeding cheetahs in captivity; only a few reproduced and over 37% of the infant cheetahs died.

Biologists first examined semen from 18 male cheetahs and a similar number of male domestic cats. They found that the concentration of sperm in male cheetah semen was only one-tenth as great as the concentration of sperm in male domestic cats. Furthermore, 71% of the cheetah sperm had abnormal shapes with flagella that were either bent or coiled or sperm heads that were too large or too small. By comparison, only 29% of the sperm from domestic cats was abnormally shaped.

A. What conclusions can be drawn from this study?

B. Why were domestic cats included in this study?

C. The formation of sperm is directed by genes. What can you conclude about the genes that control sperm formation in the cheetahs studied?

When inbreeding occurs naturally or is carried out deliberately in a population, the number of heterozygous alleles (Aa) decline while the number of homozygous alleles (aa) increase. This results in a lack of genetic variability that can be detrimental to a population. For example, if an environmental change occurs, a lack of genetic diversity can reduce the population's chance of adapting to the change. Suppose a group of cheetahs all carried a gene that made them susceptible to the fatal disease *feline infectious peritonitis* (FIP).

D. What might happen if cheetahs were exposed to this disease?

To test how genetically similar cheetahs were biologists took blood samples from both cheetahs and domestic cats then isolated various proteins. They determined whether each protein was the product of heterozygous or homozygous genes. Of 52 proteins tested, all of the cheetah proteins were found to be the products of homozygous genes. In domestic cats, a healthy mix of both heterozygous and homozygous proteins was found.

E. Does the protein evidence indicate that the cheetahs were seriously inbred?

F. Does the fact that the cheetahs are inbred support the population bottleneck hypothesis?

Next biologists studied the rejection rates of skin grafts on cheetahs. Usually, an animal rejects a skin graft unless it is from a close relative with a similar genotype. Domestic cats usually reject skin grafts from unrelated animals in 10 to 12 days. Biologists grafted three patches of skin onto a group of cheetahs: one patch from the cheetah itself, one patch from another cheetah, and one patch from a domestic cat.

G. Why were the cheetahs given grafts of their own skin?

H. From what you have learned so far, predict what you think was the outcome of the skin graft experiment?

I. Does the skin graft experiment support the population bottleneck hypothesis? Explain.

8. **Define it** Biologists define a species in three ways: morphologically, phylogenetically, or biologically. Compare and contrast each of these different ways of defining a species.

9. **Gilligan's Island** Oceanic islands have distributions of native plants and animals that are unbalanced in ways that make them distinct from the flora and fauna found on continents or continental islands. Save for bats and seals, oceanic islands do not have native terrestrial mammals, amphibians, or freshwater fish. In some cases oceanic islands have terrestrial reptiles but most do not. Propose a hypothesis or hypotheses to explain this discrepancy.

10. **Continental Grind** Explain what role continental drift has played in the evolution of living things.

Create and Connect

1. **Prove It** You have been taught and probably accept that living things on this planet evolve, but can you prove it? Using any organism of your choice, design an experiment to determine if that organism evolves. HINT: The working definition of evolution we have used in this book is simply that evolution is change over time.

2. **A Classroom Controversy** How the concept of biological evolution should be taught in schools is a raging controversy that impacts society on a number of levels—educationally, legally, religiously, scientifically, and legislatively. What are your feelings and beliefs on this controversy? Write a position paper in which you take a stand on this issue and defend the stand you take.

   Guidelines
   A. Write a position paper, NOT an opinion paper. Defend your position with as many facts, figures, quotes, and pertinent information as possible.
   B. When completed your paper should detail: (1) your personal belief about the origin of life in general, (2) the origin of humans in particular (3) what should or should not be taught in public schools regarding these concepts, (4) who—persons or state or federal agencies—should decide through laws, rules, or policies what should or should not be taught in school regarding these concepts, (5) who—persons or state or federal agencies—should enforce any laws, rules, or policies established as to the teaching of these concepts, and (6) what punishment should there be for anyone—individual or school district—that violates established laws, rules, or policies.
   C. Your work will be evaluated not on the "correctness" of your position but the quality of the defense of your position.

3. **Playing Defense** In the century and a half since he proposed it, Darwin's theory of evolution by natural selection has become nearly universally accepted by biologists, but it has been the source of controversy among some members of the general public. Following are the principle objections critics raise to the teaching of evolution as biological fact. Investigate each of these arguments and respond to each objection as if you were a scientist defending the modern theory of evolution.

A. *Evolution violates the Second Law of Thermodynamics argument.* "The Second Law of Thermodynamics tells us that there is a natural and irreversible tendency of any complex system to degenerate into a more disordered state. Things become more disorganized due to random events, not more organized."

B. *The irreducible complexity argument.* "Many biological systems are too complex to have arisen by natural selection. If you take a part away from an organism and it stops functioning, then it must be irreducibly complex and cannot have evolved." (Two of the most often used examples of irreducible complexity are protein structure and the mammalian eye.)

C. *Evolution cannot be experimentally tested argument.* "Where are the experiments showing new species evolving from other species?"

D. *The intelligent design/special creation argument.* "Living things are too complex to have produced by a random process in a series of steps. The existence of a clock is evidence of the existence of a clockmaker. Living things had to have been created by a diving power."

E. *Evolution is a theory argument.* "Evolutionists even call it the 'theory of evolution' and a theory is nothing more than a guess."

HISTORY OF LIFE ON EARTH

It is often said that all the conditions for the first production of a living organism are present, which could ever have been present. But if (and Oh! what a big if!) we could conceive in some warm little pond, with all sorts of ammonia and phosphoric salts, light, heat, electricity, etc., present, that a protein compound was chemically formed ready to undergo still more complex changes, at the present day such matter would be instantly devoured or absorbed, which would not have been the case before living creatures were formed.

—Charles Darwin

Introduction

In this corner of the Milky Way galaxy billions of years ago was a giant cloud of molecular gas and dust called a **nebulae**. Then, about 4.57 billion years ago, something happened that caused the cloud to collapse. This could have been the result of a passing star or shock waves from a supernova, but the end result was a gravitational collapse at the center of the cloud.

From this collapse, pockets of dust and gas began to collect into denser regions. As the denser regions pulled in more and more matter, conservation of momentum caused it to begin rotating, while increasing pressure caused it to heat up. Most of the material ended up in a ball at the center while the rest of the matter flattened out into a disk that circled it. The ball at the center formed the Sun, and the rest of the material formed into the protoplanetary disc. The planets, moons, and all other objects in the solar system formed from the materials in the disc. And Earth came to be some 4.6 billion years ago (give or take 500 million years). Around 45 million years after the planets began to form, the Moon formed. Probably a large planetoid, about the size of Mars, crashed into Earth (**Figure 16.1**). Little bits of hot rock splashed off during the crash and orbited around Earth. Eventually, these bits joined, cooled off, and became the Moon.

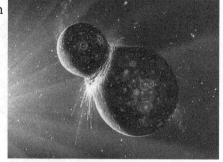

Figure 16.1

Marking Time

To understand the origin and development of life one must also understand the geologic history of Earth as well. The stretch of geologic history or **geologic time** is commonly referred to as *deep time,* and it is metered in hundreds of millions to billions of years. Can we measure deep time? Yes. Will we ever fully comprehend the immensity of it? No. However, to better understand the history of life on Earth, we must learn to cope with it.

The age of Earth is calculated to be roughly 4.6 *billion* years. Although we cannot possibly grasp the enormity of that amount of deep time, we can cast it against time scales we can fathom. Let us first shrink all geologic time into 100 years (a century). On that scale, each year represents 46 *million* actual years. Each month corresponds to 3.8 million years, and each day signifies 127,000 years. An hour is 5,300 years; a minute is 88 years, and each second is a year and a half. By that time scale, the average human lifespan lasts only a minute or less.

Let us compress our time scale even more and squeeze all geologic time into one year (**Table 16.1**). Finally, if we condense just the 3.5 to 3.8 billion years of geologic time that has past since the first appearance of life on Earth into a single minute, we would have to wait about 50 seconds just for multicellular life to appear, another four seconds for vertebrates to invade the land, and then only in the last 0.002 seconds would *Homo sapiens* arise.

Table 16.1	
Geologic Time Compressed into 1 Year	
Time	**Event**
January 1- 12:00 AM	First forms from a planetary nebulae
February 25 -4:41 PM	First primitive living cells appear
July 17- 9:54 PM	First eukaryotic cells appear
September 3- :39 PM	First multicellular organism (algae) appear
November 8- 4:35 PM	Marine worms and jellyfishes appear
November 21- 7:40 PM	First fish appear
December 2- 3:54 AM	First land animals (amphibians) appear
December 5- 5:50 PM	First reptiles appear
December 12- 9:42 PM	First crocodiles appear
December 13- 8:37 PM	First dinosaurs appear
December 14- 9:59 AM	First mammals appear
December 28- 9:31 PM	First monkeys appear

December 31- 5:18 PM	First hominids appear
December 31- 10:30 PM (2.7 million ago)	First *Homo sapiens* appear
December 31- 11:46 PM (420,000 years ago)	Domestication of fire
December 31- 11:59:20 PM (20,000 years ago)	Invention of agriculture
December 31- 11:59:56 PM (20,000 years ago)	Roman Empire; birth of Christ
December 31- 11:59:58 PM (500 years ago)	Renaissance in Europe; emergence of the experimental method in science
December 31- 11:59:59 PM	Widespread development of science and technology; first steps in space exploration; advanced global communication

Reading the Record

How do we know what happened when? What is the evidence to support scientific beliefs about the age of Earth and the sequence of events in the history of life on earth? Several methods are employed to determine past dates and events:

Relative dating is the method of determining the relative order of past events such as the age of an object in comparison to another, without necessarily determining their absolute age. In the nineteenth century, even before the theory of evolution was formulated, geologists sought to correlate rock strata worldwide. They discovered that even though strata of rock contain different sediments worldwide, each stratum of the same age contained certain **index fossils**. Index fossils can identify deposits made at approximately the same time in different parts of the world. For example, if a particular fossil of a species of trilobites has been found over a wide range geographically and for a limited time period, all strata around the world that contain this fossil must all be the same relative age.

Absolute dating is assigning an actual date to fossils using radiometric techniques. Radiometric techniques make use of the fact that certain elements in rocks, fossils, and living things are naturally radioactive. Many elements have naturally occurring *isotopes*, varieties of the element that have different numbers of neutrons in the nucleus. Some isotopes are stable, but some are unstable or radioactive. Some radioactive isotopes *decay* (break down) almost instantaneously back into their stable atomic form; others take tens of thousands to millions or even billions of years to decay. Since the rate of decay of an isotope (known as its *half-life*) is predictable, we can use these radioactive isotopes to date rocks

Imagine you have been given charcoal briquettes like those used for outdoor grilling except yours have been very precisely formed and weighed. Let us also suppose that those briquettes (representing the initial amount of radioactive isotopes in a rock or fossil.) will reduce to ash at a known rate in a specific temperature of X° C You then begin to fire those briquettes at X° C. After allowing the briquettes to "decay" for a time, you stop the firing process. By carefully weighing the amount of ash (representing the now stable isotopes in a rock or fossil) and comparing it to the weight of the briquettes before firing began, you can use the rate of "decay" of the briquettes to determine how long they had been fired. By comparing the amount of radioactive isotopes in a present day rock or fossil to the amount of radioactive isotopes that was present when that rock or fossil formed, the rate of decay of that radioactive isotope can be used to determine the approximate age of that rock or fossil.

Radiocarbon or carbon-14 (C-14) is probably one of the most widely used and best known absolute dating methods. However, because C-14 has a half-life of only 5,730 years, it can only be used to accurately date carbonaceous materials up to 60,000 years old. Other isotopes, such as potassium-argon dating and uranium-lead dating, can be used to date fossils and rocks millions to even billions of years old.

Molecular clocks allow us to use the amount of genetic divergence between species to extrapolate backward to estimate dates.

Stratigraphy uses the position of rock layers and the fossils within those layers to determine *relative* age. Reading from bottom to top, the **Law of Superposition** tells us that the lower a **strata** (layer) of rock or fossil is found, the older it is. That is, as you drill down, the older the rock layers and fossils.

Hadean Earth

The Hadean (from the Hebrew word "Hell") period between 4.6 billion and about 4.0 billion years ago is an informal designation within the Precambrian. This period was characterized by Earth's initial formation—from the accretion of dust and gases and the frequent collisions of larger planetismals—to the

Figure 16.2 The early period of Hadean Earth was a violent and unstable time not conducive to the formation of life.

stabilization of its core and crust and the development of its atmosphere and oceans. The planet was a hellish place in the early Hadean period with a partially molten surface, high volcanism, and continual bombardment with comets, meteors, and asteroids (**Figure 16.2**).

The surface of Hadean Earth was incredibly unstable during the early part of the Hadean period (**Figure 16.3**). Convection currents in the mantle brought molten rock to the surface and caused cooling rock to descend into magmatic seas. Heavier elements, such as iron, descended to become the core, whereas lighter elements, such as silicon, rose and became incorporated into the growing crust. Although no one knows when the first outer crust of the planet formed, some scientists believe that the existence of a few grains of zircon dated

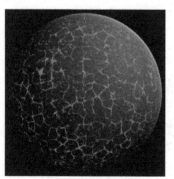

Figure 16.3 Artistic rendering of Hadean Earth.

to about 4.4 billion years ago confirm the presence of stable continents, liquid water, and surface temperatures that were probably less than 100 °C (212 °F).

Since Hadean times, nearly all of this original crust has subducted from the movements of tectonic plates, and thus few rocks and minerals remain from this time. The oldest rocks known are the faux amphibolite volcanic deposits of the Nuvvuagittuq greenstone belt in Quebec, Canada estimated to be 4.28 billion years old. The oldest minerals are the aforementioned grains of zircon, which were found in the Jack Hills of Australia.

Considerable debate surrounds the timing of the formation and initial composition of the atmosphere. Although many scientists contend that the atmosphere and the oceans formed during the latter part of the Hadean period, the discovery of the zircon grains in Australia provides compelling evidence that the atmosphere and ocean formed before 4.4 billion years ago. The early atmosphere likely began as a region of escaping hydrogen and helium. It is thought that ammonia, methane, and neon were present sometime after the crust cooled, and volcanic outgassing added water vapor, nitrogen, and additional hydrogen. Some scientists believe that ice delivered by comet impacts could have supplied the planet with additional water vapor. Later, it is thought, much of the water vapor in the atmosphere condensed to form clouds and rain that left large deposits of liquid water on Earth's surface.

Origin of Life

We do not know nor will we ever know exactly how the first living organisms arose from the warm primordial soup of Hadean oceans. Observations and experiments in chemistry, geology, physics, paleontology, microbiology and other branches of science have led to an ever-growing body of evidence that life originated 3.5-4 billion years ago from nonliving, inorganic matter in a series of four phases:

Phase 1: *Organic monomers* In the 1920s, biochemist Alexander Oparin and geneticist J. B. S. Haldane independently proposed a hypothesis about the chemical origin of life. The *Oparin-Haldane hypothesis*, sometimes called the *primordial soup hypothesis*, proposes that the first stage in the origin of life was the evolution of simple organic compounds, or *monomers*, from the inorganic compounds present in Earth's early atmosphere—water vapor (H_2O), hydrogen gas (H_2), methane (CH_4), and ammonia (NH_4). The hypothesis holds that because of its composition, this early atmosphere was a reducing (electron adding) environment, in which organic compounds could have formed from simpler molecules. Because a reducing atmosphere would not have required as much energy to drive chemical reactions as it would today, it would have been much easier to form the carbon-rich molecules from which life evolved. The energy to power this synthesis could have come from lightning, heat from volcanoes, radioactive isotopes, and solar radiation, especially UV radiation.

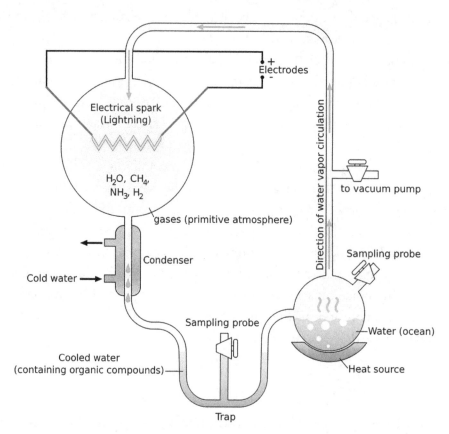

Figure 16.4 Gases that were thought to be present in the early Earth's atmosphere were admitted into the apparatus, circulated past an energy source (electric spark), and cooled to produce a liquid that could be withdrawn. Upon chemical analysis, the liquid was found to contain various small, organic molecules that could serve as monomers for large, cellular polymers.

In 1953, Stanley Miller, working with Harold Urey at the University of Chicago, attempted to test the Oparin-Haldane hypothesis by simulating conditions in the laboratory as they were thought to have existed in early oceans under a reducing atmosphere (**Figure 16.4**). For his experiment, Miller placed a mixture of methane (CH_4), ammonia (NH_4), hydrogen gas (H_2), and water (H_2O) in a closed system, heated the mixture (warm ocean), and circulated it past an electric spark (simulated lightning). After a week's run, Miller discovered a variety of amino acids, and other organic acids had formed.

Other researchers have achieved similar results as Miller using other, less-reducing combinations of gas dissolved in water and there is some indication that small pockets of early atmosphere, such as those near the openings of volcanoes, may have been reducing. In 2008, a group of investigators used modern equipment to reanalyze molecules that Miller had saved from one of his experiments. This 2008 study found a greater variety of organic molecules than Miller reported, including all 20 amino acids that had formed under conditions simulating a volcanic eruption.

If early atmospheric gases did react with one another to produce small organic compounds, there would have been neither oxidation (no free oxygen present then) nor decay (no bacteria existed) to destroy them as would happen today. Rainfall would have washed them into the ocean, where they would have accumulated for hundreds of millions of years forming a warm, thick primordial soup or "warm, little pond" in Darwin's words.

Are there other possibilities? In the late 1980s, biochemist *Günter Wächtershäuser* proposed his *iron-sulfur world* hypothesis in which organic compounds were produced in deep-sea *hydrothermal vents*, areas on the sea floor where superheated water and minerals flow from the mantle into the ocean (**Figure 16.5**). According to the iron-sulfur world hypothesis, dissolved gases emitted from the vents, such as carbon monoxide (CO), ammonia, and hydrogen sulfide, would pass over warm iron and nickel sulfide minerals that are present at thermal vents. The iron and nickel sulfide would act as catalysts for the formation of ammonia and even monomers of large organic molecules that occur in cells.

Figure 16.5 A venting black smoker emits jets of particle-laden fluids. The particles are predominantly very fine-grained sulfide minerals formed when the hot hydrothermal fluids mix with near-freezing seawater. These minerals solidify as they cool, forming chimney-like structures. "Black smokers" are chimneys formed from deposits of iron sulfide, which is black. "White smokers" are chimneys formed from deposits of barium, calcium, and silicon, which are white.

Another possibility is the *panspermia theory*, which suggests that life did not originate on Earth, but was transported here from somewhere else in the universe. It may have been as simple as organic molecules which seeded the early ocean with the chemical origin of life or actual bacterium-like organisms formed the ancestral types that let to all other life forms.

Phase 2: *Organic Polymers* Within a cell, organic monomers join to form polymers in the presence of enzymes. An example would be the synthesis of protein polymers from amino acid monomers. The problem is that enzymes are proteins, but there were no proteins present on early Earth. Several hypotheses have been advanced to deal with this discrepancy.

- **Iron-Sulfur World Hypothesis**. Research shows that organic molecules will react and amino acids will form peptide bonds in the presence of iron-nickel sulfides under conditions found at thermal vents. The iron-nickel sulfides have a charged surface that attracts amino acids and binds them together into proteins. Researchers have also produced polymers by dripping solutions of amino acids or RNA nucleotides onto hot sand, clay, or rock (*Clay hypothesis*). The polymers formed spontaneously without the help of enzymes or ribosomes.

- **Protein Interaction World Hypothesis**. The sudden appearance on early Earth of a large self-copying molecule such as RNA was exceedingly improbable. Energy-driven networks of small molecules afford better odds as the initiators of life. Could proteins have been those small molecules? Research has shown that amino acids polymerize abiotically when exposed to dry heat. This suggests that once amino acids were present in the ocean, they could have collected in shallow puddles along the water's edge. There the heat of the sun could have caused them to form *proteinoids*, small polypeptides that have some catalytic properties.

  Laboratory simulations reveal that when placed in water, proteinoids form **microspheres**, structures composed only of protein that have some properties of a cell. Perhaps some newly formed polypeptides had enzymatic properties and that some may have been more enzymatically active than others. If a certain level of enzyme activity provided an advantage over others, this would have allowed natural selection to shape the evolution of these first organic polymers. The protein interaction world hypothesis assumes that proteins arose before RNA or DNA. Thus, the genes that encode for proteins followed the evolution of the first polypeptides.

- **RNA World Hypothesis**. RNA can act as both genes and enzymes (ribozymes) and to serve as a substrate. These properties could offer a way around the "chicken-and-egg" problem. (Genes require enzymes; enzymes require genes.) Furthermore, RNA can be transcribed into DNA, in reverse of the normal process of transcription. Some viruses today have RNA genes; therefore, the first genes could have been RNA and could have carried out the processes of life commonly associated with DNA and proteins. Others have proposed that polypeptides and RNA evolved simultaneously. Therefore, the first true cell would have contained RNA genes that could be replicated because of the presence of proteins. This eliminates the chicken-and-egg paradox of whether proteins or RNA came first, but requires that the evolution of two events would have to happen at exactly the same time, a huge improbability.

Phase 3: *Protocells* The first true cells could not have arisen directly from organic polymers and RNA; there had to be an intermediate structure or *protocell (protobiont)*. The theoretical protocell is a self-organized spherical aggregate of lipids containing two or more *RNA replicases* that are capable of making copies of each other (**Figure 16.6**). Protocells were capable of growth, replication, and evolution. Growth was achieved by the addition of *micelles* (small spheres of fatty acid) to the outer membrane. Concurrent with growth was the replication of the RNA. The membrane becomes unstable and eventually divides, forming two daughter protocells, with the RNA replicases randomly divided between them. Occasionally mutant RNA replicases appear that have new, different catalytic activity. If these mutant replicases allow the protocell to grow or divide faster, this would give them an advantage over other protocells that could be enhanced by natural

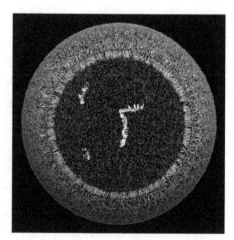

Figure 16.6 An illustration of a protocell composed of a fatty acid membrane encapsulating RNA replicase.

selection (evolution). Although a working version of a protocell has yet to be achieved in a laboratory setting, the goal appears to be well within reach.

The contents of protocells and true cells must be protected from the outside environment. This is reflected in the *membrane-first hypothesis* that contends the first cell had to have a plasma membrane before any of its other parts. It is possible that micelles were the first stage in the formation of both plasma membranes and the vesicles found in the cytoplasm of true cells.

Phase 4: *Self-replication* Living cells carry on protein synthesis in order to produce the enzymes that allow DNA to replicate. The central dogma of genetics states that DNA directs protein synthesis and that information flows from the DNA to RNA to protein. Whether proteins or RNA came first or whether they evolved simultaneously, self-replication was the line that had to be crossed for the first true cells to appear.

History of Life

At the very core of evolutionary theory is the basic idea that life has existed in some form for billions of years and that life has changed drastically over time. The history of living things is best documented through the fossil record, and **paleontology** is the branch of science that studies this record to learn more about the history of life and ancient climates.

A fossil (L., *fossus*, having been dug up) is any remains, impressions, or traces of living organisms of past geological ages that have been preserved in Earth's crust. Paleontological studies of fossils across geologic time and how they formed are vital to our understanding of the evolutionary relationships (phylogeny) between taxa (taxonomic groups). Large specimens (**macrofossils**) are more commonly found and displayed, although tiny remains (**microfossils**) are far more prevalent in the fossil record.

The great majority of fossils are founded embedded in or recently eroded from sedimentary rock. A number of different mechanisms can account for fossil formation:

- ***Permineralization*** For permineralization to occur, the organism must become covered by sediment very soon after death, sealing it and preventing decay. (Sudden freezing, rapid desiccation (drying) or coming to rest in an anoxic (no oxygen) environment such as the bottom of a lake also suffices.) The degree of decay of the remains before being sealed determines the fineness of detail that will be found in the fossil. Mineral-rich groundwater slowly trickling down through the buried remains permeates the empty spaces of the dead organism and gradually, organic structures are replaced molecule by molecule by rock-like minerals. One example of fossilization by permineralization that most people are familiar with is petrified wood. Astonishingly, petrified wood can exhibit the original structure of the wood in all its detail, down to the microscopic level.

- ***Molds, Replacement, and Compression*** When all that is left of the original organism is an organism-shaped hole in the rock, a *mold fossil* (or *typolite*) has formed. If this mold

Figure 16.7)

depression is later filled in with other minerals, a *cast fossil* has formed through the replacement of the original materials by new, unrelated ones. In some cases, replacement occurs so gradually and on such a minute scale that even microstructural features are preserved despite the total loss of the original organic material (**Figure 16.7**).

If the remains of an organism become pressed into soft mud or sand which then turns into rock, a *compression fossil* has formed. Usually, only plant stems and leaves form such depressions with only a general outline of the external structure of the plant part; internal anatomy is never preserved in compression fossils (**Figure 16.8**).

Figure 16.8 Compressed in mud but frozen in rock, the molded remains of this ancient plant speak to us across time.

- *Traces* Trace fossils are the indirect evidence of the presence and activity of some ancient organism including the remains of trackways, burrows, footprints, eggs and eggshells, and even fecal droppings (known as **coprolites**).

- *Resins* In past geologic times as now, some species of trees excreted a sticky, syrupy material known as **resin**. Often, other organisms such as fungi, bacteria, seeds, leaves, insects, spiders, and rarely small invertebrates (collectively known as **inclusions**) became trapped in and covered by this sticky ooze. Over time, the resin hardened into *amber*, preserving in exquisite detail any inclusions entombed within (**Figure 16.9**).

Figure 16.9 To hold or even see a fossil is to transcend millions of years of time to a prehistoric world we can scarcely imagine.

Scroll of Time

The history of the planet and the history of life are inexorably intertwined. Geologic and atmospheric conditions of early Earth were not suitable to the type and kinds of life as we recognize them today. The fossil record and other evidence suggest that as Earth changed so did living things.

Instead of trying to view all of geologic time as a single span, scientists find it more meaningful and useful to break deep time down in a hierarchal system known as the *geologic time scale*. The geological time scale (GTS) is a system of chronological dating that relates geological strata (**stratigraphy**) to time and is used by geologists, paleontologists, and other earth scientists to describe timing and relationships of geologic and biologic events that have occurred during Earth's history. The geologic time scale can be expressed either as a table (**Figure 16.10**) or in clock form (**Figure 16.11**).

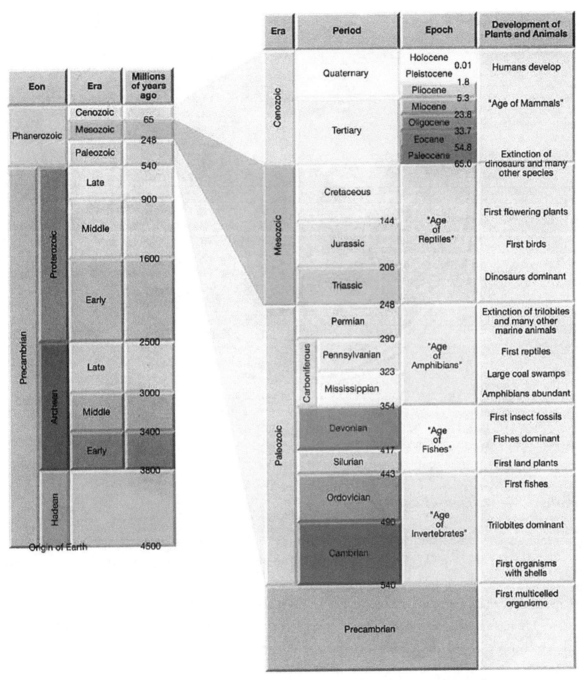

Figure 16.10 The geologic timescale was derived from the accumulation of data from the age of fossils in strata all over the world.

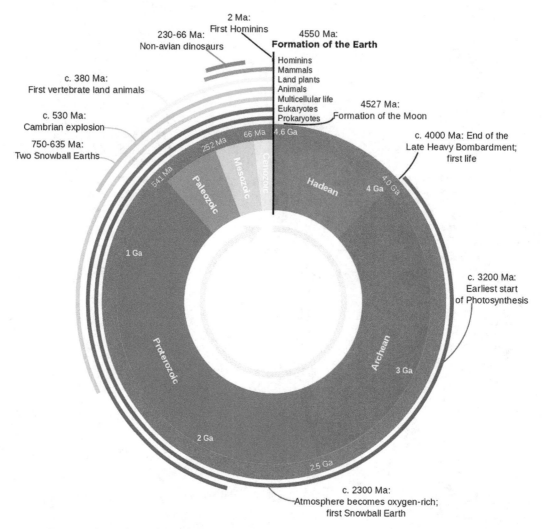

Figure 16.11 The geological clock, a projection of Earth's 4,5 Ga history in clock form. ("MA" = a million years [Megayear] ago; "GA" = a billion years [Gigayear] ago).

Precambrian—Archaean Eon

The Precambrian eon was a very long period of time, comprising about 87% of the geologic timescale. During the informal Hadean time of the Precambrian the planet formed and cooled, the early atmosphere formed, and the ocean developed.

As the end of Hadean time ushered in the Archaean eon, about 3.8 billion years ago, Earth was still about three times as hot as it is today, but no longer hot enough to boil water. Most of the planet was covered with a warm ocean with deep basins. By the end of the Archaean, plate tectonic activity may have been similar to that of the modern Earth and evidence suggests the assembly and destruction of one and perhaps several super continents (**Figure 16.12**). The Archaean atmosphere contained mainly nitrogen and carbon dioxide but was nearly completely lacking in oxygen.

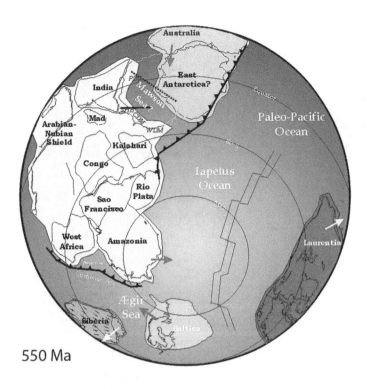

Figure 16.12 Positions of the continents toward the end of the Proterozoic eon.

The first living cells known as prokaryotes appeared early in the Archaean eon some 3.8 billion years ago. Prokaryotes are simple cells with no membrane-bound nucleus or organelles. Prokaryotes are able to survive in the most inhospitable of environments—high salinity and pH, high temperatures, and even in environs lacking in oxygen. The first identifiable fossils are those of complex prokaryotes. Chemical fingerprints of prokaryotes are found in sedimentary rocks from Greenland, dated at 3.8 billion years ago. The oldest prokaryote fossils have been discovered in Western Australia and resemble present day cyanobacteria. These fossils were found in fossilized stromatolites. *Stromatolites* or *stomatoliths* are layered rock that form when certain prokaryotes bind thin films of sediments together (**Figure 16.13**). Early prokaryotes resembling present day cyanobacteria began to undergo photosynthesis and gradually add oxygen to the Archaean atmosphere.

Figure 16.13 Living stromatolites in shallow, warm water off western Australia.

Precambrian—Proterozoic Eon

The Proterozoic eon began 2.5 billion years ago. During this time, stable continents continued to assemble and drift, and evidence suggests that about 1.6 to 1 billion years ago oxygen began to build-up in the

Figure 16.14 An artist's reconstruction of life in the Proterozoic ocean 630 million years ago.

atmosphere. This first "pollution crisis" was a global catastrophe for most anaerobic bacteria groups but made possible the explosion of eukaryotic forms possible.

Fossils of prokaryotic bacteria and archaeans from this time are abundant, and around 2.1 billion years ago eukaryotic cells appear as fossils as well. The appearance of eukaryotic cells during this time is the foundation on which all life forms have developed. And once eukaryotic cells did develop, a profusion of life forms followed, such as protists, marine algae, and soft-bodied marine invertebrates (**Figure 16.14**).

Phanerozoic Eon—Paleozoic Era

During the Paleozoic there were six major continental land masses; each of these consisted of different parts of the modern continents. For instance, at the beginning of the Paleozoic, today's western coast of North America ran east-west along the equator, while Africa was at the South Pole. These Paleozoic continents experienced tremendous mountain building along their margins, and numerous incursions and retreats of shallow seas across their interiors. Large limestone outcrops, like the one pictured here, are evidence of these periodic incursions of continental seas.

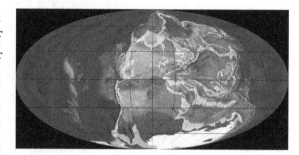

Figure 16.15 Position of the land masses during the late Permian Period of the Paleozoic Era.

The Paleozoic Era is bracketed by the times of global super-continents. The era opened with the breakup of the world-continent Pannotia and closed with the formation of Pangea, as Earth's continents came together once again (**Figure 16.15**).

Figure 16.16 Cambrian Period Fauna. (a)Trilobites; (b) Brachipods; (c) Hyolithids

Two great animal faunas dominated the seas during the Paleozoic. The "Cambrian fauna" typified the Cambrian period oceans; although members of most phyla were present during the Cambrian, the seas were dominated by trilobites, brachiopods, hyolithids, and archaeocyathids (**Figure 16.16**). Ordovician

Figure 16.17 An artist's reconstruction of life in the Ordovician ocean some 450 million years ago.

rock strata reveal the "Ordovician fauna" to be characterized by numerous and diverse trilobites types and conodonts (extinct chordates resembling toothed eels). In addition, stalked blastoid echinoderms, bryozoans (small filter feeders with tentacles), corals, crinoids, as well as many kinds of brachiopods, snails, clams, and cephalopods appeared for the first time in the geologic record in tropical Ordovician environments (**Figure 16.17**). Remains of ostracoderms (jawless, armored fish with a spade-shaped head) from Ordovician rocks comprise some of the oldest vertebrate fossils (**Figure 16.18**). By the end of the Ordovician Period, life was no longer confined to the seas as nonvascular plants, such as mosses and liverworts, had begun to colonize the land.

The end of the Ordovician Period and the beginning of the Silurian Period is marked by the first of what are believed to be six major extinctions events. The Ordovician-Silurian mass extinction is considered the third largest such event in Earth's history. Almost all major taxonomic groups were affected during this extinction event. Extinction was global during this period, eliminating 49-60% of marine genera and nearly 85% of marine species. Brachiopods, bivalves, echinoderms, bryozoans, and corals were particularly affected. This event had two peak dying times separated by hundreds of thousands of years. The cause of this extinction event is thought to be glaciation and a subsequent lowering of ocean levels as the supercontinent Gondwanaland moved over the South Pole.

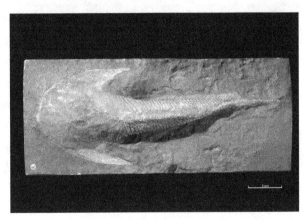

Figure 16.18 Ostracoderms were small, primitive, jawless fish covered in an armor of bony scales. With their mouth permanently open, they slowly moved over the bottom of Ordovician seas.

The early Silurian Period (444-416 million years ago) saw the end of the glaciation of the late Ordovician Period and a substantial rise in the levels of the ocean. This meant rich shallow sea ecosystems with new ecological niches. Silurian fossils show evidence of extensive reef building and the first signs that life beginning to colonize the new estuary, fresh water, and terrestrial ecosystems. Earth's climate underwent a relative stabilization, ending the previous pattern of erratic climatic fluctuations.

In the oceans, there was a widespread radiation of crinoids, a continued proliferation and expansion of the brachiopods, and the oldest known fossils of coral reefs. This time period also marks the wide and rapid spread of jawless fish, along with the important appearances of both the first known freshwater fish and the appearance of jawed fish. Other marine fossils commonly found throughout the Silurian record include trilobites, graptolites, conodonts, corals, stromatoporoids, and molluscs.

During the Silurian Period, the climate was generally warm and stable, in contrast to the glaciers of the late Ordovician and the extreme heat of the Devonian. A warm, stable climate provided for one of the most significant developments to take place during the Silurian Period: the arrival of the first plants to colonize the land. Lichens were probably the first photosynthetic organisms to cling to the rocky coasts of the early continents. When organic matter from decaying lichens joined the action of erosion to wear away rock, the first real soil began to build up in shallow, protected estuaries. Nonvascular spore-forming bryophytes such moss, hornworts and liverworts first appeared in the late Ordovician. A major advance in plant history occurred with the appearance of the first vascular spore plants. With the advent of a vascular system, plants could grow much taller and larger than the small, short bryophytes that preceded them.

Figure 16.19 *Cooksonia* lacked true leaves. This suggests that the stalk developed to disperse spores and was not itself photosynthetic

The first known plant to have an upright stalk, and vascular tissue for water transport, was the *Cooksonia* of the mid-Silurian deltas. This little plant was a few centimeters high with a branched structure with small bulbous tips (**Figure 16.19**).

During the Devonian Period (416-359 million years ago) the supercontinent Gondwana occupied most of the Southern Hemisphere, although it began significant northerly drift during the Devonian Period. Eventually, by the later Permian Period, this drift would lead to a collision with the equatorial continent known as Euramerica, forming Pangaea.

The Devonian Period was a time of extensive reef building in the shallow water that surrounded each continent and separated Gondwana from Euramerica. Reef ecosystems contained numerous brachiopods, numerous trilobites, and horn corals. Placoderms (the armored fishes) underwent wide diversification and became the dominant marine predators. Cartilaginous fish such as sharks and rays were common by the late Devonian. Devonian strata also contain the first fossil ammonites (**Figure 16.20**).

By the mid-Devonian, the fossil record shows evidence that there were two new groups of fish that had true bones, teeth, swim bladders, and gills. The ray-finned fish were the ancestors of most modern fish. Like modern fish, their paired pelvic and pectoral fins

Figure 16.20 Ammonites were early cephalopods. The closest living relative to ammonites is the present day *Nautilus*.

were supported by several long thin bones powered by muscles largely within the trunk. The lobe-finned fish were more common during the Devonian than the ray-fins but largely died out. (The coelacanth and a few species of lungfish are the only lobe-finned fishes left today.) Lobe-finned fishes are the accepted ancestors of all tetrapods (four-limbed terrestrial animals). The radiation and appearance of many new species of fish during the Devonian and Silurian Periods has led to these periods being called "the Age of Fish."

Plants, which had begun colonizing the land during the Silurian Period, continued to make evolutionary progress during the Devonian. Lycophytes, horsetails, progymnosperms, and ferns grew to large sizes and formed Earth's first forests. This rapid appearance of so many plant groups and growth forms has been called the "Devonian Plant Explosion."

On land, the millipedes, centipedes, and arachnids that appeared in the Silurian continued to diversity during the Devonian Period and we find the fossils of the first insects, such as *Rhyniella praecusor*,

Figure16.21 *Tiktaalik* was probably mostly aquatic and would be found "walking" on the bottom of shallow water estuaries. It had a fish-like pelvis, but its hind limbs were larger and stronger than those in front, suggesting it was able to propel itself outside of an aquatic environment. It had a crocodile-like head, a moveable neck, and nostrils for breathing air."

a flightless hexapod (insect) with antennae and segmented body between 412 and 391 million years old. This period also ushered in the first tetrapods (four limbs). The earliest known tetrapod is *Tiktaalik rosae*, a fossil creature considered to be the link between the lobe-finned fished and early amphibians (**Figure16.21**).

The Devonian Period closed with yet another mass extinction event. The latest research suggests a series of distinct extinction pulses through an interval of some three million years in which around 83% of all species on land and in the ocean perished. Life in the shallow seas was the hardest hit, especially corals who did not return to their former abundance until new types of coral evolved over 100 million years later.

The Carboniferous Period (359-299 million years ago) was yet another rebound from a mass extinction event. The Carboniferous takes its name from the large underground coal deposits that date to it. Formed from lush prehistoric vegetation, the majority of these deposits are found in parts of Europe, North America, and Asia. Carboniferous coal was produced by bark-bearing trees that grew in vast lowland swamp forests. Vegetation included giant club mosses, tree ferns, great horsetails, and giant lycopod trees. Over millions of years, the organic deposits of this plant debris formed the world's first extensive coal deposits—coal that humans are still burning today (**Figure 16.22**).

Figure 16.22 An artist's rendering of a Carboniferous forest.

The growth of these forests removed huge amounts of carbon dioxide from the atmosphere, leading to a surplus of oxygen. Atmospheric oxygen levels peaked around 35 percent, compared with 21 percent today. Increased oxygen levels may account for deadly poisonous centipedes some six feet (two meters) in length crawled in the company of mammoth cockroaches and scorpions as much as three feet (one meter)

Figure 16.23 Fossil imprint of a giant dragonfly compared to a present day dragonfly (upper right).

long. Most impressive of all were dragonflies that grew to the size of seagulls. One exquisitely detailed fossil of a dragonfly that died 320 million years ago shows it had a wingspan of 2.5 feet (0.75 meters) (**Figure 16.23**). There was a great radiation of arthropods in general, especially insects.

Amphibians, the dominant land vertebrates, began to diversify and developed so many different types so rapidly that the Carboniferous and Permian periods are known as "the Age of Amphibians." One branch of Amphibia evolved into reptiles with the development of the amniotic egg. The amniotic (shelled) egg prevents eggs laid on land from drying out. The oldest shelled egg fossils are 280 million years old.

The Permian Period (299-251 million years ago) brought the Paleozoic Era to an end. By the end of the early Permian, the two great continents of the Paleozoic, Gondwana and Euramerica had collided to form the supercontinent Pangaea (**Figure 16.24**). Because Pangaea was so immense, the interior portions of the continent had a much cooler, drier climate than had existed in the Carboniferous. As a result, the giant swamp forests of the Carboniferous began to dry out and were being replaced by the first seed-bearing vascular plants, the gymnosperms.

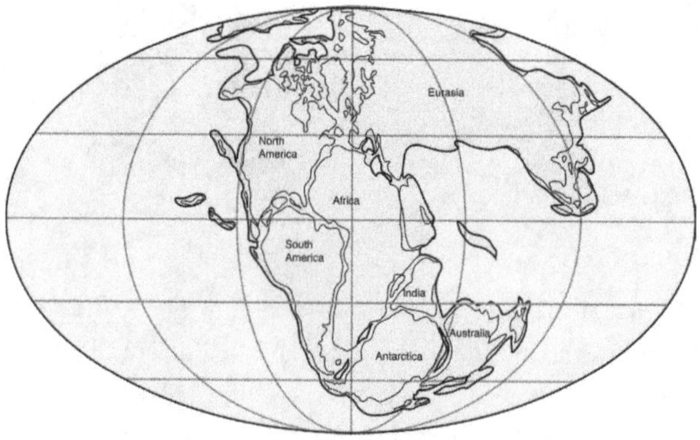

Two important groups of animals dominated the Permian landscape: Synapsids and Sauropsids. Synapsids

Figure 16.24 Pangaea and the position of what would become today's continents within that ancient supercontinent.

had skulls with a single temporal opening and are thought to be the lineage that eventually led to mammals. Sauropsids had two skull openings and were the ancestors of the reptiles, including dinosaurs and birds.

In the early Permian, it appeared that the Synapsids were to be the dominant group of land animals. The group was highly diversified. The earliest, most primitive Synapsids were the Pelycosaurs, which included an apex predator, a genus known as *Dimetrodon*. This animal had a lizard-like body and a large bony "sail" fin on its back that was probably used for thermoregulation (**Figure 16.25**).

In the late Permian, Pelycosaurs were succeeded by a new lineage known as Therapsids. These animals were much closer to mammals. Their legs were under their bodies, giving them the more upright stance typical of

Figure 16.25 *Dimetrodon* is mistaken for a dinosaur more often than any other prehistoric reptile, but the fact is that this creature lived tens of millions of years before the first dinosaurs ever evolved.

quadruped mammals. They had more powerful jaws and more tooth differentiation. Fossil skulls show evidence of whiskers, which indicates that some species had fur and were endothermic. The Cynodont ("dog-toothed") group included species that hunted in organized packs (**Figure 16.26**). Cynodonts are considered to be the ancestors of all modern mammals. At the end of the Permian, the largest Synapsids became extinct, leaving many ecological niches open. The second group of land animals, the Sauropsid group, weathered

Figure 16.26 The Cynodont was a link between reptiles and mammals. About 1m (3 feet) in length, cynodonts likely ate insects and small reptiles and is thought to have lived in burrows.

the Permian Extinction more successfully and rapidly diversified to fill them. The Sauropsid lineage gave rise to the dinosaurs that would dominate the Mesozoic Era.

The Permian Period of the Paleozoic Era closed with the most severe extinction event ever recorded. Known as the Permian-Triassic (P-Tr) extinction event or more commonly as the Great Dying, this event served as the boundary between the Permian and Triassic geologic periods as well as the Paleozoic and Mesozoic eras. This global catastrophe took out up to 96% of all marine species with 70% of terrestrial vertebrate species becoming extinct. It is the only known mass extinction of insects. Some 57% of all taxonomic families and 83% of all genera became extinct. Because so much biodiversity was lost, the recovery of life took significantly longer than after any other extinction event, possibly up to 10 million years.

There is evidence for one to three distinct pulses, or phases, of extinction. Suggested mechanisms for include one or more large impact events, massive volcanism, coal or gas fires and a runaway greenhouse effect triggered by the sudden release of methane from the sea floor. Other contributing factors may have been sea-level change and increasing aridity due to a shift in ocean circulation driven by climate change.

Phanerozoic Eon—Mesozoic Era

The Triassic Period (251-200 million years ago) that opened the Mesozoic Era was both a time of transition and recovery from the P-Tr extinction event. The creatures of the Triassic can be considered to belong to one of three groups: survivors of the P-Tr extinction event, new groups that flourished briefly, and new groups that went on to dominate the Mesozoic world.

The oceans had been massively depopulated by the Permian Extinction when as many as 95 percent of extant marine genera were wiped out by high carbon dioxide levels. Fossil fish from the Triassic Period are very uniform, which indicates that few families survived the extinction. The mid- to late Triassic Period shows the first development of modern stony corals and a time of modest reef building activity in the shallower waters near the coasts of Pangaea.

Plants and insects did not go through any extensive evolutionary advances during the Triassic. Due to the dry climate, the interior of Pangaea was mostly desert. In higher latitudes, gymnosperms survived and conifer forests began to recover from the extinction. Mosses and ferns survived in coastal regions. Spiders, scorpions, millipedes and centipedes survived, as well as the newer groups of beetles. The only new insect group of the Triassic was the grasshoppers.

Early in the Triassic, a group of reptiles, the order Ichthyosauria, returned to the ocean. Fossils of early ichthyosaurs are lizard-like and clearly show their tetrapod ancestry. Their vertebrae indicate they probably swam by moving their entire bodies side to side, like modern eels. Later in the Triassic, ichthyosaurs evolved into purely marine forms with dolphin-shaped bodies and long-toothed snouts (**Figure 16.27**).

The Mesozoic Era is often known as the "Age of Reptiles." Two groups of animals survived the P-Tr extinction: Therapsids, which were mammal-like reptiles, and the more reptilian Archosaurs. However, by the mid-Triassic, most of the Therapsids had become extinct, and the more reptilian Archosaurs were clearly dominant leading to one lineage of Archosaurs evolving into dinosaurs. By the late

Figure 16.27 Ichthyosaurs ("fish lizards") were reptile inhabitants of the seas, living at the same time as dinosaurs trod the land and pterosaurs glided overhead.

Triassic, a third group of Archosaurs had branched into the first pterosaurs, flying reptiles about the size of a present day crow with wing membranes attached to long hind legs (**Figure 16.28**).

The first mammals evolved near the end of the Triassic Period from the nearly extinct Therapsids. Scientists have some difficulty in distinguishing where exactly the dividing line between Therapsids and early mammals should be drawn. Early mammals of the late Triassic and early Jurassic were very small, rarely more than a few inches in length. They were mainly herbivores or insectivores and therefore were not in direct competition with the Archosaurs or later dinosaurs.

At the beginning of the Triassic, most of the continents were concentrated in supercontinent known as Pangaea. The climate was generally very dry over much of Pangaea with very hot summers and cold winters in the continental interior. A highly seasonal monsoon climate prevailed nearer to the coastal regions. Although the climate was more moderate farther from the equator, it was warmer than today with no polar ice caps. Late in the Triassic, seafloor spreading in the Tethys Sealed to rifting between the northern and southern portions of Pangaea, which began the separation of Pangaea into two continents, Laurasia and Gondwana, which would be completed in the Jurassic Period (**Figure 16.29**).

Again, as one period (Triassic) ends, and another (Jurassic) begins, life is rocked by another extinction event, the Triassic-Jurassic (T-R) extinction event 201 million years ago. At least half of the species now known to have been living on Earth at that time became extinct. This event vacated terrestrial ecological niches, allowing the dinosaurs to assume the dominant roles in the Jurassic period. This event happened in less than 10,000 years and occurred just before Pangaea started to break apart.

Great plant-eating dinosaurs roamed Earth feeding on lush ferns, palm-like cycads and gymnosperms. Smaller but vicious carnivores stalked the great herbivores. This was the Jurassic Period, 199.6 to 145.5 million years ago,

Figure 16.28 Pterosaurs ("flying lizards") were an extremely successful group of reptiles. They flourished all through the age of dinosaurs, a period of more than 150 million years. Over time the relatively small early pterosaurs evolved into a wide variety of species. Some had long, slender jaws, elaborate head crests, or specialized teeth, and some were extraordinarily large.

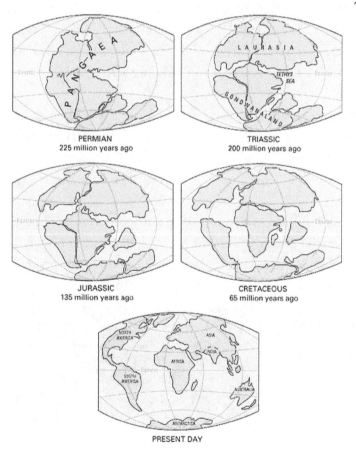

PERMIAN
225 million years ago

TRIASSIC
200 million years ago

JURASSIC
135 million years ago

CRETACEOUS
65 million years ago

PRESENT DAY

Figure 16.29

a 54-million-year chunk of the Mesozoic Era. The Jurassic was a golden age for the large herbivorous dinosaurs known as the sauropods. These were preyed upon by large theropods.

But there was more to life than dinosaurs! In the seas, the fishlike ichthyosaurs were at their height, sharing the oceans with the plesiosaurs, giant marine crocodiles, and modern-looking sharks and rays. Jurassic cephalopods included the ammonites, with their coiled, external shells (upper left), and the belemnites, close relatives of modern squid but with heavy, calcified, bullet-shaped, partially internal shells. Numerous turtle species could be found in lakes and rivers, and the first birds appear.

The warm, humid climate of the Jurassic allowed lush jungles of tree ferns, ferns, cycads, and ginkgoes to cover the land (**Figure 16.30**). In these landscapes the first flowering seed plants appear and creeping about in this foliage, no bigger than rats, were a number of early mammals.

Figure 16.30

The Cretaceous Period (145-66 million years ago) saw no great extinction or burst of diversity separating the Cretaceous from the Jurassic Period that had preceded it. In some ways, things went on as they had. Dinosaurs both great and small moved through forests of ferns, cycads, and conifers. Ammonites, belemnites, other molluscs, and fish were hunted by great marine reptiles, and pterosaurs and birds flapped and soared in the air above. The Cretaceous saw the appearance of new creatures, such as placental mammals, and the diversification and radiation of creatures such as flowering plants and many types of insects that had first appeared in earlier periods.

The most famous of all mass extinctions marks the end of the Cretaceous Period, about 65 million years ago. This was the great extinction in which the dinosaurs died out. The other lineages of marine reptiles — the ichthyosaurs, plesiosaurs, and mosasaurs — also were extinct by the end of the Cretaceous, as were the flying pterosaurs, but some, like the ichthyosaurs, were probably extinct a little before the end of the Cretaceous. But many groups of organisms, such as flowering plants, gastropods and pelecypods (snails and clams), amphibians, lizards and snakes, crocodilians, birds, and mammals survived the Cretaceous-Tertiary boundary, with few or no apparent extinctions at all.

Phanerozoic Eon—Cenozoic Era

The Cenozoic Era is the most recent of the three major subdivisions of life history. The Cenozoic spans only about 65 million years, from the end of the Cretaceous Period and the extinction of the dinosaurs to the present day. The Cenozoic is sometimes called the Age of Mammals because the largest land animals have been mammals during that time. This is a misnomer for several reasons. First, the history of mammals began long before the Cenozoic began. Second, the diversity of life during the Cenozoic is far wider than mammals. The Cenozoic could be called the "Age of Flowering Plants" or the "Age of Insects" or the "Age of Teleost Fish" or the "Age of Birds" just as accurately.

Classically, the Cenozoic Era is divided into the Tertiary and Quaternary periods. Another scheme, dividing the Cenozoic into the Paleogene and Neogene periods, is gaining popularity. The new system divides the epochs differently. In any case, we are currently living in the Holocene Epoch.

Early in the Cenozoic, the planet was dominated by relatively small fauna, including small mammals, birds, reptiles, and amphibians. From a geological perspective, it did not take long for mammals and birds to greatly diversify in the absence of the large reptiles that had dominated during the Mesozoic. Mammals came to occupy almost every available niche (both marine and terrestrial), and some also grew very large, attaining sizes not seen in most of today's terrestrial mammals. Climate-wise, Earth had begun a drying and cooling trend, culminating in the glaciations of the Pleistocene Epoch. The continents also began looking roughly familiar at this time and moved into their current positions (Figure 16.29).

In Summary

- The sun and planets, including Earth, formed from a nebulae in this area of the Milky Way galaxy.
- The geologic and biologic history of the planet is measured in geologic time.
- Past dates and events are determined in several ways:
  - Relative dating
  - Absolute dating
  - Molecular clocks
  - Stratigraphy
- Hadean time is an informal designation within the Precambrian denoting the formation of Earth.
- It is believed that life originated 3.5-4 billion years ago from nonliving, inorganic matter (abiogenesis) in a series of four phases:
  - Phase 1. *Organic monomers*
  - Phase 2. *Organic Polymers*
  - Phase 3. *Protocells*
  - Phase 4. *Self-replication*
- Paleontology is the branch of science that studies this record to learn more about the history of life and ancient climates.
- A fossil is any remains, impressions, or traces of living organisms of past geological ages that have been preserved in Earth's crust.
- A number of different mechanisms can account for fossil formation:
  - Permineralization
  - Molds, Replacement, and Compression
  - Traces
  - Resins
- Scientists break deep time down in a hierarchal system known as the geologic time scale.
- The geological time scale (GTS) is a system of chronological dating that relates geological strata (stratigraphy) to time, and is used by geologists, paleontologists, and other earth scientists to describe timing and relationships of geologic and biologic events that have occurred during Earth's history.

- The geologic time scale divides geologic time into four eras:
  - Precambrian
  - Paleozoic
  - Mesozoic
  - Cenozoic
- Each era of the geologic time scale is characterized by certain geologic and biological events.

Review and Reflect

1. *You Have to Begin Somewhere* Detail the evolution of the first protocells and then explain how those protocells are thought to have evolved into eukaryotic cells.

2. *Can You Dig It?* Where should the young paleontologist pictured in **Figure 16.31** begin digging to find the oldest fossils? Explain your reasoning.

3. *Two Worlds* Explain why scientists think RNA World came before DNA World in the history of life.

4. *Hadean World in a Bottle* Describe Miller and Urey's ancient atmosphere device and explain what hypothesis they were trying to test with that device,

Figure 16.31

5. *What a Grind* What role have plate tectonics and continental drift played in the history of life on Earth? Explain

6. *Fossilize Rover* Imagine your beloved dog has died. If you wished to memorialize her by turning her into a fossil, detail what procedure you would follow.

7. *Monkey Business* Use the concept of evolution and continental drift to explain why New World monkeys can hang by their tail (prehensile tail) while Old World monkeys cannot.

8. *Dinosaur Myths* There is a popular notion that the dinosaurs were driven to extinction by the impact of a very large extraterrestrial object, such as comet or asteroid, and that within a few weeks to a few months all the dinosaurs were gone. Is this idea accurate? Investigate and explain.

9. *Life Goes Kaboom!* Around 542 million years ago most major animal phyla appeared in a relatively short time from an evolution standpoint. This period is known as the Cambrian Explosion. List and explain the factors that led to the Cambrian Explosion.

Create and Connect

1. *How Old is Papa?* Fossils are often dated by radiometric dating techniques. A dispute has arisen over the age of a piece of fossil bone nicknamed the "Papa Femur." Some contend the bone is over 100,000 years old while others maintain it is less than 20,000 years old. Using radiocarbon-14 dating techniques, you have established that the Papa Femur contains 12.5 units of C-14. You

know that such bone contained 100 units of C-14 when the organism was alive and that C-14 has a half-life of 5,730 years. Which camp in this dispute is correct about the age of the Papa Femur? Defend your answer with the appropriate calculations.

Radiometric dating techniques are based on certain assumptions and have some limitations. Investigate and explain the limitations of radiometric dating techniques and the assumptions on which these techniques are based.

2. *Travel Back* You are going on a temporal expedition with the famous time-traveling paleontologist, Dr. Buckaroo Chronos. You will be traveling first to a Carboniferous forest and then to a Jurassic forest. Your job is to detail what type of flora and fauna you will encounter in each type of forest. May the force be with you!

3. *Box It Up* Oops! The author forgot to include a boxed reading in this chapter. Study some of the boxed readings in other chapters of this book and then write an appropriate boxed reading the author might include in this chapter in the next edition of the book.

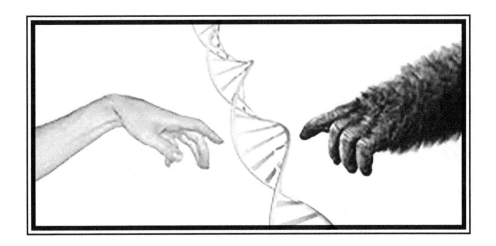

CHAPTER 17

THE HUMAN CONDITION: RISE OF THE CULTURAL APE

We are just an advanced breed of monkey on a minor planet of a very average star. But we can understand the Universe. That makes us something very special.

—Stephen Hawking

Introduction

Humans may well be the most unusual species we will encounter on our journey through the realms of life. Humans are creatures with an almost alien-sized brain that can write poetry, raise giant cities, travel in space, and even contemplate and explore their own mind; an advanced technological being wrapped in the body of a primate, a naked primate at that. This duality of mind and body distinguishes humanity for we what we are biologically—a speaking cultural ape.

As a species that is capable of contemplating itself, we have always sought, and **anthropology** continues to seek answers to certain questions about ourselves: What and where are our origins? How are we anatomically different than our closest primate kin? What does it mean to be "human"? The field of anthropology unites many other disciplines as it searches both the past and present to understand the human condition (**Figure 17.1**).

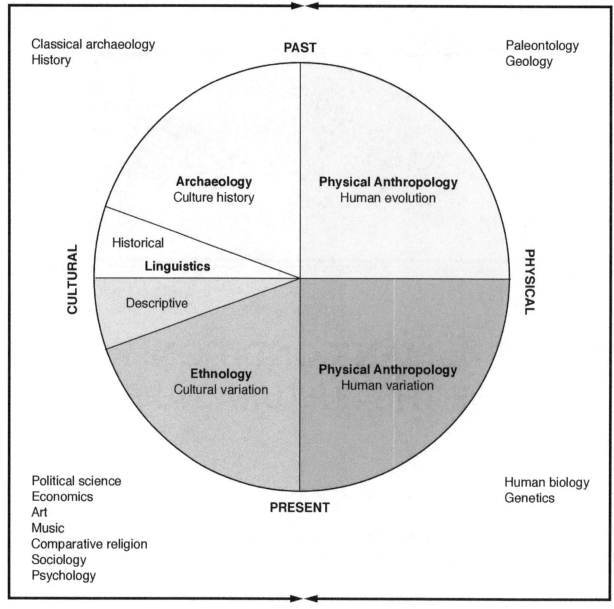

Figure 17.1 Anthropology and related disciplines in the study of humankind.

Rise of Primates

To comprehend our biological nature, we need to understand our place in the hierarchy of animals. As previous chapters have revealed, humans are clearly vertebrate in morphology and placental mammals by our reproductive patterns. On one branch of the placental mammal portion of the tree of life, we find the primates, our closest mammalian kin (**Figure 17.2**). Primates, especially humans, are remarkably recent animals on the evolutionary stage. Primate-like mammals (or proto-primates) first arose in the early Paleocene epoch about 65 million years ago. At that time, the world was a very different place than it is today. The continents were in other locations, and they had somewhat different shapes. North America was still connected to Europe, but not yet to South America. India was not yet part of the Asia but headed towards it, and Australia lay close to Antarctica. Most land masses had warm tropical or subtropical climates. The flora (plants) and fauna (animals) of the time would be unrecognizable to us since most of the plants and animals that are familiar to us had not yet evolved. From fragmentary fossil evidence (mostly from North Africa) it appears there lived a group of creatures at that time resembling modern squirrels and tree shrews that ranged from chipmunk-size to rat-size that were primarily insectivores. These were the proto-primate mammals (Plesiadapiformes). Their numbers declined into the Eocene, and they were extinct by the end of the epoch, possibly as a result of competition with rodents that first appear and radiate in the late Paleocene.

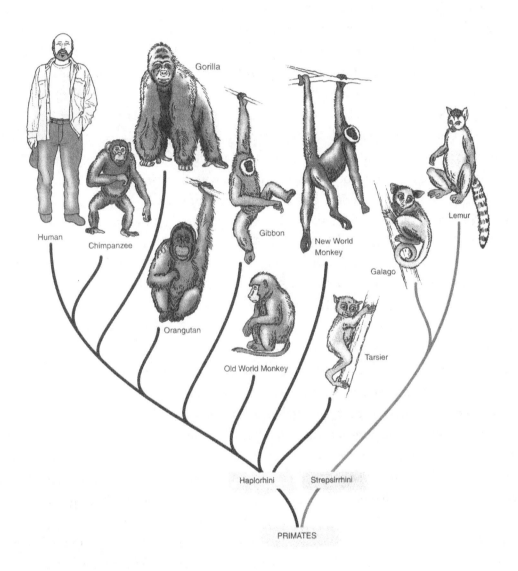

Figure 17.2 Primate Family Portrait. This tree depicts the possible phylogeny of primates. The green box at the base of the tree represents ancestral protoprimates.

The first true primates (Euprimates) appear in the fossil record during the early Eocene epoch some 55 million years ago in North America, Eurasia, and Northern Africa (Figure 17.2). These creatures were somewhat squirrel-like in size and appearance, but apparently they had grasping hands and feet that were efficient in manipulating objects and climbing trees. It is likely that they were developing effective stereoscopic vision as well. From this ancestral type developed the **Strepsirrhini** (or traditionally **prosimians**—lemurs, aye-aye, bush babies, potto, and lorises. The Eocene epoch was a period of maximum prosimian adaptive radiation with at least 60 genera developing. This is nearly four times greater prosimian diversity than today. These primates lived in what would become North America, Europe, Africa, and Asia, and it was during this time that they reached the island of Madagascar. The Eocene prosimians flourished as a result of a combination of lack of competition from monkeys and apes since these more highly developed primates had not yet evolved, and because of a favorable climate. With the transition of the Eocene epoch into the Oligocene epoch some 35 million years ago, temperatures cooled, and monkeys appeared for the first time resulting in many prosimian species becoming extinct. Today most modern prosimians either live in locations where monkeys and apes are absent, or they are normally active only at night when most of the anthropoids are sleeping

From the prosimians arose the **Haplorhini** (or **anthropoids**)—tarsiers, tamarins, baboons, monkeys, apes, and humans (Figure 17.2). Monkeys were first of the anthropoid lineage. Appearing during the Oligocene or slightly earlier, the ancestral monkeys ranged from the size of a large squirrel to a large domestic cat. They most likely were fruit and seed-eating forest tree-dwellers. Compared to the prosimians, early monkeys had fewer teeth, a less fox-like snout, larger brains, and more forward-looking eyes.

The Oligocene was an epoch of major geological changes with resulting regional climate shifts. By the beginning of the Oligocene, North America, and Europe had drifted apart becoming distinct continents. The Great Rift Valley system of East Africa also was formed during the Oligocene along a 1200 mile volcanically active fault zone. The collision of India with Asia that had begun in the Eocene continued into the Oligocene resulting in the progressive growth of the immense barrier we call the Himalayan Mountains and the forcing upward of the Tibetan Plateau beyond. These and other major geological events during the Oligocene are thought to have triggered global climatic changes. The cooling and drying trend that had begun in the late Eocene accelerated, especially in the northern hemisphere. This, in turn, resulted in the general disappearance of primates from these northern areas.

Geologic change continued apace into the Miocene epoch as continental drift ground slowly on, creating new mountain chains that altered local weather patterns. In addition, the progressive global cooling and drying trend continued. Growing polar ice caps reduced sea levels exposing previously submerged coastal lands. As a result of this and tectonic movement, a land connection was reestablished between Africa and Asia providing a migration route for primates and other animals between these continents. Much of the continuous East African and South Asian tropical forests were broken into a mosaic of forests and vast dry grasslands because of the climate changes. This resulted in new selection pressure being applied to the primates of the time.

Early in the Miocene epoch about 25 million years ago the first apes appear apparently having evolved from monkeys. Under pressure to adapt to less forest and more grassland, African apes became largely terrestrial, exploiting grains, tubers, and dead animals for food. Under these conditions, an upright posture became a distinct advantage as it provided a better view of predators while leaving the hands free for gathering food, caring for the young and using tools. This was the beginning of the Hominidae lineage that would lead eventually to orangutans, gorillas, chimpanzees, and a number of different species of *Homo*.

Diversity and Classification of Primates

The order Primate (L. *primas,* = excellent or noble) embraces 230-270 species aligned into two suborders and contains animals as primitive as the insect-eating aye-aye to the highly complex human. The development of sharp vision and depth perception, grasping hands, and a tremendous brain capacity has given primates a unique combination of specialized talents.

Geographically, primates are almost totally confined to the tropical latitudes, although the Barbary macaque (*Macaca sylvanus*) is found in northern Africa and the Japanese macaque (*Macaca fuscata*) occurs on both main islands of Japan. Humans, with cleverness of invention and aptitude for technology, have managed to dwell or at least visit everywhere on earth, including the greatest depths of the ocean, and have even ventured off the planet into space and walked on the moon.

Primatologists have traditionally classified primates into two suborders, prosimians—lemurs, lorises, bush babies, and tarsiers—and anthropoids—monkeys, baboons, gibbons, orangutans, gorillas, chimpanzees, and humans. Recent DNA and chromosomal analysis strongly support dividing living primates into one lineage comprising lemurs and lorises and another lineage composed of anthropoids (monkeys and apes) and tarsiers. Lemurs and lorises comprise the suborder Strepsirrhini—moist dog-like muzzle between nose and lip—whereas anthropoids and tarsiers comprise the suborder Haplorhini—dry skin or fur between nose and lip.

Strepsirrhines retain a greater number of earlier mammalian features (*e.g.* claws, long snout, lateral-facing eyes) than do anthropoid primates. With the exception of large-bodied species in Madagascar, an island that separated from Africa during the late Cretaceous before anthropoids evolved, strepsirrhines are small bodied and nocturnal (**Figure 17.3**).

Figure 17.3 Members of the Suborder Strepsirrhini. (a) Ring-tailed lemur (*Lemur catta*) and (b) Loris (Lorisidae).

Haplorhines are mostly larger than strepsirrhines and are generally diurnal rather than nocturnal. Haplorhines possess a shortened face, forward-directed eyes, and a larger, more complex brain (**Figure 17.4**).

Figure 17.4 Members of the Suborder Haplorhini. (a) Gorilla (*Gorilla*), (b) Orangutan (*Pongo*), and (c) Chimpanzee (*Pan*).

The classification of primates:

Domain Eukarya
    Kingdom Animalia
        Phylum Chordata
            Subphylum Vertebrata (Craniata)
                Class Mammalia
                    Order Primates
                        Suborder Strepsirrhini
                            Family Cheirogaleidae—dwarf and mouse lemurs
                            Family Lemuridae—Lemurs
                            Family Lepilemuridae—Sportive lemurs
                            Family Indriidae—Wooly lemurs and allies
                            Family Daubentoniidae—Aye-aye
                            Family Lorisidae—Lorises, pottos, and allies
                            Family Galagidae—Galagos
                        Suborder Haplorhini
                            Family Tarsiidae—Tarsiers
                            Family Callitrichidae—Marmosets and tamarins
                            Family Cebidae—Capuchins and squirrel monkeys
                            Family Cecopithecidae—Mandrils, baboons, and macaques
                            Family Hylobatidae—Gibbons
                            Family Hominidae
                                Subfamily Ponginae—Orangutan
                                Subfamily Homininae
                                    Tribe Gorillini—Gorillas
                                    Tribe Hominini—Chimpanzees and humans

Until recently humans were classified in a family separate from the gorilla, chimpanzees, and the orangutan. Molecular evidence, however, reveals that chimpanzees and humans are as similar to each other as many sister species. (That is, they are as similar as different species within the same genus.) The primate classification system used here keeps the chimps and humans in different genera, but in the same subfamily (Homininae). The term "hominin" is now used to refer to the chimpanzees, humans, and extinct members of direct human lineage, whereas the term "hominid" is used in reference to the entire family Hominidae—orangutans, gorillas, chimps, and humans.

General Characteristics of Primates

The traits and tendencies found in primate groups are:

- Shortening of the face accompanied by reduction of the snout
- The general retention of five functional digits on the fore and hind limbs. Enhanced mobility of the digits, especially the thumb and big toes, which are opposable to the other digits.
- A semi-erect posture that enables hand manipulation and provides an optimal position preparatory to leaping.
- Claws modified into flattened and compressed nails to support the sensitive digital pads on the last phalanx of each finger and toe.
- A shoulder joint allowing a high degree of limb movement in all directions and an elbow joint permitting a rotation of the forearm.
- A reduced snout and olfactory apparatus
- A reduction in the number of teeth compared to primitive mammals.
- Teeth and digestive tract that are adapted to an omnivorous diet
- A complex visual apparatus with high acuity and a trend toward development of forward-directed binocular eyes and color perception. Well-developed hand-eye motor coordination.
- A large brain relative to body size, in which the cerebral cortex is particularly enlarged.
- Only two mammary glands (some exceptions).
- Typically, only one young per pregnancy associated with prolonged infancy and pre-adulthood.

Primates are eclectic in their food choices. Most species eat a wide array of foods ranging from insects and other small animals to fruits, flowers, and foliage. However, different species occupying the same habitat differ considerably in the time of feeding, the levels of the forest from which they feed, the type of food eaten, and how far they range to find food.

The shape and structure of the primate body are adapted and suited to its environment. Some walk on arms (forelimbs) and legs (hind limbs) nearly the same length; others that **brachiate** (swing) through the trees with extra-long arms and reduced legs. In humans, the arms are shorter, and the legs elongated to accommodate upright bipedal (two legs) locomotion.

The contrast in locomotion is reflected in the shape of the hands and feet. Some brachiating monkeys lack a thumb, but in apes and humans, the thumb is well developed, agile, and opposable to provide a precise but powerful grip. The feet of lemurs are long and narrow, and those of the apes are broader and adept at grasping. Humans have a foot with reduced toes and an arch designed to carry the weight of the body on the heel and balls of the feet.

The structure and length of the tail vary considerably. Tails are retained in some strepsirrhines, such as lemurs, but lost in the slow-moving potto and lorises. Tails are present in most monkeys. Some Central and South American monkeys (New World monkeys) have a **prehensile** (grasping) **tail**, but none of the African and Asian monkeys (Old World monkeys) demonstrate such ability with their tail. Apes and humans have no tail at all.

Whether we like it or not, there can be no doubt that modern humans are ape-like primates. However, we are certainly much more. How did we become *Homo sapiens sapiens*? What does it mean to truly be human? Let us begin to attempt to answer those questions by examining the saga of human evolution. As we do so, it will become apparent that the human ape has gone through a series of changes and advances since the human lineage diverged from the chimpanzee lineage. Over the course of his evolutionary history, the human ape evolved first into a bipedal speaking tribal ape then into a hunting social ape, an intellectual tool-making ape, an agricultural ape, and finally a cultural ape. Some of these stages developed simultaneously, and all blended to form the human species we have become today.

Rise of the Speaking Bipedal Tribal Ape

In his book *The Descent of Man and Selection in Relation to Sex* (1871), Charles Darwin proposed the notion that humans and apes shared a common ancestry. Such an idea was not necessarily original to Darwin considering that Linnaeus recognized that humans are primates well before Darwin made the evolutionary connection. Darwin was, however, the first to subject such an idea to scientific scrutiny. During Darwin's time, virtually no fossil evidence linking humans with apes existed. This forced Darwin to formulate his hypotheses mostly on anatomical comparisons between humans and apes.

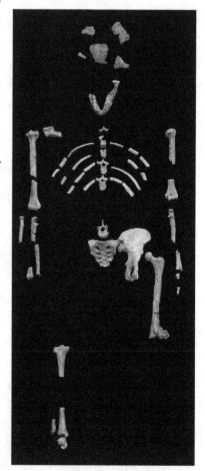

The idea of a common ancestry between humans and apes was met with indignation by the Christian Victorian world of that time. Since then the fires of controversy surrounding this idea have not dimmed but have, if anything, grown brighter. Today in the whole of biology there is nothing more controversial or hotly debated than is the origin of humans. And this debate has spilled over into many different venues ranging from religious pulpits to science classrooms, and even into legislative chambers and courtrooms.

Today the concept that humans and apes have a common ancestral past is supported by two main bodies of evidence. On the one hand, we have the data of DNA analysis and comparative biochemical and cytological studies of modern humans and apes that can be used to construct a reverse calendar of evolutionary events. On the other hand, we have fossil evidence that has been accumulating since the publication of Darwin's book. The early search for hominin fossils was driven by a desire to find a connection in lineage between humans and apes, a "missing link." First, there was the discovery of two Neanderthal skeletons in the 1880s. Then, in 1891, Eugene Dubois uncovered Java man (*Homo Erectus*). Between 1967 and 1977 (labeled the "golden decade" by paleoanthropologist Donald C. Johnson) many spectacular discoveries were made in Africa. Fortunately, important Hominin fossils continue to be discovered, but unfortunately, most hominin fossils are far from complete and may consist of only shards of bone, a single tooth, or piece of skull (**Figure 17.5**).

Figure 17.5 Pieces of bone of specimen AL 288-1 more commonly known as Lucy. These bones represent about 40% of the skeleton of an individual female *Australiopithecus afarensis.*

Based on best current evidence, one possible evolutionary sequence in human evolution is illustrated in **Figure 17.6**. Some pathways are debated, and not all fossil species are shown, especially some of the different species currently identified as *Australopithecus*. There is also considerable dispute concerning many overlapping species, especially the possible overlap between *Homo habilis* and *Homo erectus*. It could well be that the two are continuing examples of the same species. The same dispute exists with *Homo erectus*, *Homo sapiens* (archaic), and *Homo sapiens sapiens*.

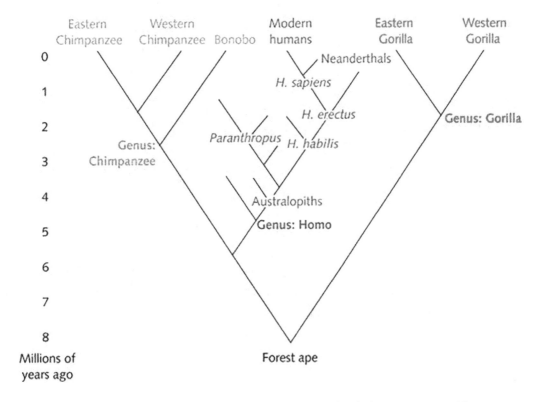

Figure 17.6 The possible phylogenetic tree of hominins. The phylogenies presented here are just a few of the working hypotheses from different authorities. Phylogenies and cladograms are constantly being revised as more fossil and genetic evidence becomes available.

Humans are more closely related to the African apes than to orangutans and even more closely related to chimpanzees than to gorillas. According to recent molecular dating, the divergence between the lineages leading to humans and chimpanzees occurred about 6.3 million years ago, and so the earliest species in the tribe Hominini, to which humans belong, would have appeared shortly thereafter (**Table 17.1**).

Table 17.1

Evolutionary Details of the Main Groups of Hominins

Lineage[a]	Approximate Time (Years ago)	Adaptations, Behavior and Habitats	Fossil and Archaeological Evidence
Hominin ancestors	8 to 5 My	Relatively large-bodied apes distributed in Central and Eastern Africa across forest-woodland mosaics	No fossil evidence yet, but when found, expected to be a group or groups ancestral to humans and chimpanzees
Australopithecines	4 to 2 My	Bipedal on the ground, occasionally arboreal Open savanna, and mosaic grasslands and woodland habitats Fibrous plant diet that may also have included meat[b]	Extensive fossils in Eastern and Southern Africa Large teeth and jaws
Homo habilis	Pliocene-Pleistocene 2 to 1.5 My	Improved bipedalism Tools to procure and process food. Habitats in drier areas indicating larger home ranges Scavenging and active animal hunting	Skeletal changes and increase in brain size Early stone tools
Homo erectus	Early-Mid Pleistocene 1.5 to 0.5 My	Entry into new habitats and geographical zones Definite preconception of tool form Manipulation of fire Increased level of activity and skeletal stress	Fossils found in formerly unoccupied areas of Africa, and outside Africa Development of a stone tool industry Archaeological hearths Increased cranial and postcranial development
"Archaic" *Homo sapiens*	Mid Pleistocene 500 to 150 thousand years	Geographical divergence and ecological adaptations More complex tools	Old World distribution with some distinct regional morphologies Bifacial axes: Acheulean-Mousterian stone tool industries
H. sapiens neanderthanlensis	Late Pleistocene 150-35 thousand years	Large and robust individuals More social complexity and development of ritual Increasingly sophisticated tools	Massive cranial and postcranial development Intentional Burial of the dead Increase number of stone-tool types
H. sapiens sapiens	Late Pleistocene to Present	Decreased levels of activity and skeletal stress Expansion of technology Development of complex cultures Increase in population size	Appearance of "anatomically modern" humans From Upper Paleolithic (Aurignacian) stone tools to satellite communication Beginnings and expansion of agriculture

Point(s) of Origin

Where did modern humans originate? Roughly 100,000 years ago, the world was occupied by a morphologically diverse group of hominins. In Africa and the Middle East there was *Homo sapiens*; in Asia *Homo erectus*, and in Europe, *Homo neanderthalensis*. However, by 30,000 years ago this diversity vanished, and humans everywhere had evolved into the anatomically and behaviorally modern form—*Homo sapiens sapiens*. There is great deliberation between two schools of thought regarding the nature of this transformation: one stresses multiregional continuity, and the other suggest a single origin for modern humans.

Multiregional Continuity Model This model contends that after *Homo erectus* left Africa and dispersed into other parts of the Old World, regional populations slowly evolved into modern humans. The emergence of Homo sapiens was not restricted to any one area but occurred throughout the entire geographic range where *Homo erectus* dispersed. Since their original dispersal, natural selection in regional populations is responsible for the regional variants (races) we see today.

Out of Africa Model this model asserts that *Homo sapiens* evolved relatively recently in Africa and the Middle East and migrated into Eurasia replacing all other hominin species, including *Homo erectus*.

At this time, the question of the origin(s) of modern humans remains unanswered. There is molecular evidence supporting both models, but a lack of fossil specimens muddies the waters. For the future, both strategic fossil discoveries and refinement of genetic information will hopefully decide the matter. Some investigators are becoming more open to a somewhat intermediate position in which there was an African origin with some mixing of populations.

Earliest Hominins

Three African fossil species—*Sahelanthropus tchadensis, Orrorin tugenensis,* and *Ardipithecus kadabba*—compete for the designation of earliest hominin. These three forms, which were added to the known hominin fossil record within the past decade, date from around the period of the human (hominin) and chimpanzee (panin) divergence and show a mix of hominin features including evidence that suggests bipedal locomotion. Perhaps the most surprising aspect of these finds is the diversity of forms that appeared around the divergence. Clearly, the traditional view of a single lineage leading progressively toward the human form is an oversimplification.

Figure 17.7 An artist's reconstruction of *Sahelanthropus tchadensis*

Sahelanthropus tchadensis (**Figure 17.7**), discovered in 2001 in Central Africa, has been dated between 7 and 6 million years old, suggesting that it may have lived slightly before the time when the chimpanzee and human lineages diverged. The nearly complete cranium and fragments of lower jaw and teeth reveal a protohominin mixture of ape and hominin features.

Figure 17.8 An artist's reconstruction of *Orrorin tugenensis*

Orroin tugenensis (**Figure 17.8**) is represented so far by lower jaw fragments and lower limb fragments including three femora that show clear indications that it was a habitual biped when on the ground. The upper limb fragments also show arboreal traits shared with chimpanzees, indicating that this species may have spent a fair amount of time in trees.

Ardipithecus kadabba (**Figure 17.9**) at 5.6 million to 5.8 million years ago is slightly younger than *Orrorin*. This third contender for the title of basal hominin is relatively well represented by limb remains but less well by cranial remains. It was clearly a bipedal hominin and shares cranial and dental features with later Australopithecines along with more primitive features such as fairly large canine teeth.

Figure 17.9 An artist's reconstruction of *Ardipithecus kadabba*

Australopithecines

Ardipithecus ramidus, dated to 4.4 million years old, may well be the earliest true hominin. The remains are incomplete with most *A. ramidus* fossils being teeth. Because limb fossils have not been found, questions about size and type of locomotion remain unanswered. Other fossilized animals found with the ramidus fossils would suggest that ramidus was a forest dweller.

Australopithecus anamensis was named in 1995 after its discovery in Allia Bay in Kenya. Living between 4.2 and 3.9 million years ago, its body showed advanced bipedal features, but the skull closely resembles the ancient apes.

Australopithecus afarensis lived between 3.9 and 3.0 million years ago. It retained the apelike face with a sloping forehead, a distinct brow ridge over the eyes, flat nose, and a chinless lower jaw. It stood between 107 cm and 152.4 cm (3 feet 6 inches and 5 feet tall and was fully bipedal. Its build (ration of weight to height) was about the same as the modern human, but its face and head were proportionately much larger, and the thickness of its bones showed that it was quite strong. The discovery of a nearly complete *A. afarensis* dubbed "Lucy" (Figure 23.6) is one of the most famous hominin discoveries of all times.

Australopithecus africanus was similar to *A. afarensis* and lived between 3 and 2 million years ago. It was also bipedal but was slightly larger in body and brain size, although the brain was not advanced enough for speech. This hominin was an herbivore and ate tough, hard to chew plant materials. The shape of the jaw was like modern humans.

Australopithecus robustus lived between 2 and 1.5 million years ago. It had a body similar to *A. africanus*, but a larger and more massive skull and teeth. Its huge face was flat with large brow ridges but no forehead. Brain size most likely could not support speech capabilities.

Australopithecus boisei existed at about the same time as *A. robustus* with an even more massive face but about the same size brain. Many authorities believe that *robustus* and *boisei* are variants of the same species. We should regard the Australopithecines as a group in which considerable evolutionary change was occurring, exemplifying rapid radiation of bipedal tropical apes (**Figure 17.10**).

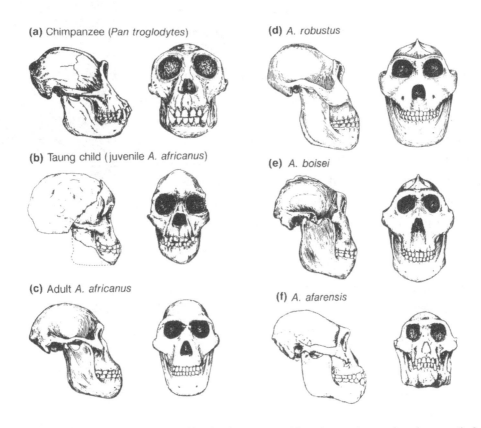

Figure 17.10 A comparison of the skulls of a chimpanzee (a) and several australopithecines (b-f).

Origin of Homo

Homo habilis is often referred to as the "handy man" because tools were found with its fossil remains. These associated artifacts indicate that these new hominins were engaged in making regularly patterned stone tools known as **Oldowan tools**. Australopithecine species had probably earlier embarked on using simple stone tools, bones, and sticks. This is likely because chimpanzees engage in simple tool use and modification. In the words of Thomas Carlyle: "Man is a tool-using animal. Without tools, he is nothing, with tools he is all."

Habilis existed between 2.4 and 1.5 million years ago (**Figure 17.11**). The brain size in earlier fossil specimens was about 500 cc but rose to 800 cc toward the end of this species existence. The increased brain size shows evidence that some speech may have developed. *Habilis* was about 152.4

Figure 17.11 An artist's reconstruction of *Homo habilis*.

cm (5 feet) tall and weighed about 45.4 kg (100 pounds). Some paleoanthropologists believe that *habilis* is not a separate species and should be considered as either a later *Australopithecus* or an early *Homo erectus*. It is possible that the smaller early examples are in one species and the later larger examples in the other.

Homo erectus lived between 1.8 million and 300,000 years ago and was a successful species for a million and a half years (**Figure 17.12**). Early examples had a brain size of 900 cc on the average, but the brain grew steadily during its reign reaching almost the same size as modern human, at 1200 cc. This species had speech, developed tools, and weapons, captured and kept fire, and learned to cook food. Though proportionately the same, *erectus* was sturdier in build and much stronger than modern humans. Only the head and face differed much from our own. Traveling out of Africa into China and Southeast Asia, *erectus* turned to hunting for food and developed clothing suitable for northern climates. These hominins became geographically widespread and existed for a very long period of time. There seem to be a number of variants in this species, and if we knew all the distinctions among different *H. erectus*

Figure 17.12 An artist's reconstruction of *Homo erectus.*

groups, it might be possible and prudent to mark off new species. However, these distinctions encompass anatomical and behavioral traits that we cannot always discern solely through fossils.

Archaic *Homo sapiens* (also *Homo sapiens heidelbergensis*) were types that may have provided a bridge between *erectus* and *Homo sapiens sapiens* during the period 500,000 to 200,000 years ago. The archaic *sapiens* represents a diverse group of skulls found with features intermediate between the two. Brain size averaged about 1200 cc and skulls are more rounded and with smaller features. The skeleton displays a stronger build than modern humans but was well proportioned.

Homo sapiens neanderthalensis (also *Homo neanderthalensis*) lived in Europe and the Mideast between 150,000 and 35,000 years ago. Neanderthals coexisted with the various archaic species and early *Homo sapiens sapiens*. Their brain size was slightly larger than that of modern humans at about 1450 cc, and the braincase was longer and lower than that of modern humans, with a marked bulge on the back of the skull. Like *erectus*, they had a protruding jaw and receding forehead with a weak chin. Neanderthals lived mostly in cold climates, and their body proportions were similar to those of modern cold-adapted peoples: short in height and solid of build (**Figure 17.13**). Males averaged about 168 cm (5 feet 6 inches) in height. Their bones were thick and heavy and show signs of powerful muscle attachments indicating that

Figure 17.13 An artist's reconstruction of *Homo neanderthalis.*

Neanderthals would have been extraordinarily strong by modern standards. Their skeletons also reveal that they endured brutally hard lives. A large number of tools and weapons have been found; most more advanced that those of *erectus*. Neanderthals appear to have been formidable hunters and are the first people known to have buried their dead, complete with flowers and artifacts. They also apparently used herbal medicines. Plaque on Neanderthal teeth yields traces of chamomile and yarrow, two anti-inflammatory drugs.

Neanderthals disappeared abruptly around 25,000 years ago with Gibraltar seemingly their last refuge. Although they are no longer with us as a species, traces of Neanderthals are with us still if nothing more than as segments of DNA within the human genome. Ancient liaisons between humans and Neanderthals as recently as 37,000 years ago in Europe reverberates today in the genomes of billions of modern Asian and European humans. In fact, recent research indicates that the DNA of living Asians and Europeans is, on the average, 2.5 percent Neanderthal.

Homo sapiens sapiens first appears about 195,000 years ago and for the last 24,000 years or so has been the sole survivor of genus *Homo* to walk the planet. Modern humans are characterized by a brain size of about 1,350 cc, a flat forehead, small or absent eyebrow ridges, a prominent chin, and a *gracile skeleton* (slender bones).

The long-term trend toward smaller molars and decreased robustness of the skeleton can be discerned even within the last 100,000 years. The face, jaw, and teeth of Mesolithic humans 10,000 years ago were about 10% more robust than our own. Upper Paleolithic humans 30,000 years ago were about 20% to 30% more robust than their modern counterparts in Europe and Asia. Interestingly, some modern humans, such as aboriginal Australians, have tooth sizes more typical of archaic *sapiens*. The smallest tooth sizes are found in those areas where food-processing techniques have been used for the longest time, a probable example of natural selection that has occurred within the last 10,000 years.

Box 21.1
Are Humans Still Evolving or Are We Ancient History?

For quite some time, the consensus among not only the general public but the world's preeminent biologists has been that human evolution is over. The late Stephen Jay Gould, eminent Harvard paleontologist and evolutionary theorist, once stated: "Natural selection has almost become irrelevant to us.....There have been no biological changes. Everything we've called culture and civilization we've built with the same body and brain." This view has practically become doctrine, and the doctrine states that modern *Homo sapiens* emerged around 50,000 years ago and that our bodies and brains were mostly sculpted during the long period when we were hunters and gatherers.

In other words, the modern human skull houses a Stone Age brain, and to suggest otherwise was nothing short of blasphemy. At the risk of being branded heretics, however, some researchers are challenging this long-held view and contend that not only is evidence beginning to mount that humans are still evolving, but that human evolution may possibly be occurring at a faster and faster pace.

The main evidence for continuing human evolution is etched in the human genome in the form of millions of rare gene variants. Furthermore, the data indicates that over the past 10,000 years human evolution has occurred a hundred times more quickly than in any other period in our species history. In fact, human evolution may be in overdrive as the result of the rapid increase in rare gene variations and DNA methylation. Humans have five times as many rare gene variants as would be expected (About 61 million instead of 12 million). On average, every duplication of the human genome includes 100 new errors. Thus as our population ballooned from as estimated 5 million individuals just 10,000 years ago to more than 7 billion today, our DNA was given many opportunities to accumulate mutations and variations. Unfortunately, evolution hasn't had time to weed out the harmful mutations so that an estimated 80 percent are probably deleterious to us.

These mutations relate to virtually every aspect of what it means to be a functioning human—our skin and eye color (apparently no one on earth had blue eyes just 10,000 years ago), our brain, our life span, our digestive system, our skeleton and bones, our immunity to pathogens, and even sperm production.

Many of these DNA variants seem to be unique to a continent of origin, a finding that has provocative implications. University of Utah anthropologist Henry Harpending states, "It is unlikely that human races are evolving away from each other. We are getting less alike, not merging into a single mixed humanity." Harpending continues, "We aren't the same as people even a thousand or two thousand years ago. Almost every trait you look at is under strong genetic influence." Chiming in is John Hawks, University of Wisconsin at Madison paleoanthropologists and bone specialist, "You don't have to look hard to see that teeth are getting smaller, skull size is shrinking, and overall stature is getting smaller." Some changes are similar in many parts of the world, but other changes, especially over the past 10,000years, are distinct to specific ethnic groups. In Europeans, the cheekbones slant backward, the eye sockets are shaped like aviator glasses, and the nose bridge is high. Asians have cheekbones facing more forward, very round eye orbits, and a very low nose bridge. Australians have thicker skulls and the biggest teeth, on average, of any human population today. Hawks wonders, "It beats me how leading biologists could look at the fossil record and conclude that human evolution came to a standstill 50,000 years ago."

Although populations of humans may be evolving apart in modern times, DNA studies conducted by genetic anthropologists show we all share a common female ancestor (mitochondrial Eve) who lived in Africa about 140,000 years ago. In addition, all living men share a common male ancestor (genetic Adam) who lived in Africa about 60,000 years ago. Around 50,000 years ago humans began pouring forth from Africa eventually spreading around the world.

Such information is divined by genetic anthropologists through the preparation of *haplotype maps*. Haplotype mapping makes use of the fact that genetic variants in the male Y chromosome and maternal mitochondria are often inherited together in segments called haplotypes, with each haplotype representing a unique mutation away from the original ancestral male and female.

As our ancestors spread across the face of the planet and came to occupy niches as diverse as the frigid Arctic Circle, to dry deserts and steamy rainforests, the cultural and demographic shifts necessary to survive those inhospitable conditions sparked a transformation in both the body and brain of our species that still continues. Over the past few centuries, and accelerating ever more quickly in the past 50 years, a steady stream of human innovations has begun to drastically speed up processes that were, until very recently, the sole providence of nature. It appears that our technology has created ways of accelerating change (genetic engineering) and new habitats (modern cities and space stations), essentially fracturing our species and transforming our future as a species.

The next time you look in the mirror consider that the image you see is not so much one of a Stone Age relic but rather that of a work in progress; a creature some experts believe is no more than a generation or two away from the emergence of *Homo evolutus*: a hominid that controls its own evolution.

Mark of Humankind

Based on the evolutionary evidence such as it is, a modern human seems to be defined anatomically by two main characteristics: bipedalism and a large brain with developed sites supporting speech and intelligence.

Bipedalism

Although paleoanthropologists have established bipedalism as a long-standing feature in hominin lineages, and although the questions of why and how bipedalism originated have been discussed for over a century, the origin of bipedalism remains a matter of controversy. What we do know is that bipedalism evolved very early in hominin evolution and that it represents a key innovation leading to a number of adaptations we associate with modern humans. Three different arenas may have exerted selection pressures that influenced bipedalism, each of which may have been influenced by the others.

Improved Food Acquisition Beginning with the Late Miocene and extending through Pliocene-Pleistocene times, there were periodic decreases in global temperatures, marked by the onset of ice-sheet formation in Antarctica and the northern hemisphere. As ice locked up water, various terrestrial areas became relatively dry, and open environments such as woodlands and grasslands replaced rain forests in many tropical regions.

Early hominins lived in a patchy environment of mixed woodland and savanna (relatively dry grassland and bushland with occasional trees) that provided seasonal food supplies. Their habitat dictated an omnivorous diet, demanding relatively more time devoted to searching for food over longer distances. An upright stance and bipedal striking would have enhanced long-distance foraging by enabling them to carry food gathered in different places.

Bipedalism may have arisen as a byproduct of adaptations that reduced forelimb involvement in quadrupedal support and movement. As hands became increasingly specialized for grasping, manipulating and carrying food, tools, weapons, and offspring, selection occurred for an upright stance and for transferring locomotion to hind limbs.

Improved Predator Avoidance Bipedalism enhances height, and so improves a hominin's ability to see over tall grass and obstructions and to wade in deeper water to pursue game or seek protection from predators. Bipedalism fosters the use of manual weapons such as stick-wielding and stone throwing, which extends the reach of hominins beyond the teeth, claws, and other defenses of animal competitors, predators, and prey. Bipedalism also allows hominins to carry tools and weapons from place to place as well as to move offspring from one camp to another or from one food resource to another.

Improved Reproductive Success It has been proposed that bipedalism enabled adult males to carry food manually to their females and offspring who could remain sequestered in a single locality, the *home base*. This mode of provisioning reduced the need for females to be continuously mobile in foraging for themselves and their clinging offspring and accorded three important advantages:

- A relatively stable home base that provided for more constant social relationships and perhaps closer mother-infant relationships that improved infant survival;
- Reduced infant injuries because infants were no longer bring dragged around by their continuously mobile mother.
- Reduction in the spacing between births by allowing parents to successfully care for more offspring.

Speech

The most advanced of primate vocalizations is human speech. Compared with animal vocalizations that are often limited to a sequence of sounds in single tones, human speech provides a rapid means of communication. For example, we can interpret a sequence of dots and dashes, as in Morse code, at a rate of much less than 50 words per minute, whereas we can often easily understand a sequence of spoken syllables delivered at 150 words per minute.

As in other mammalian vocalizations, the larynx, in the upper part of the tracheal tube, provides the basis for speech. In producing sound the larynx acts like a woodwind reed, controlling vocal pitch by opening and closing rapidly so that expired air from the lungs is interrupted to form puffs; the greater the frequency of puff formation, the higher the pitch. To produce the vowels of human speech, laryngeal puffs must pass through a tube-like airway (the pharynx) whose lengths and shape determine the eventual frequency patterns emitted and thus the quality of the different vowel sounds (**Figure 17.14**).

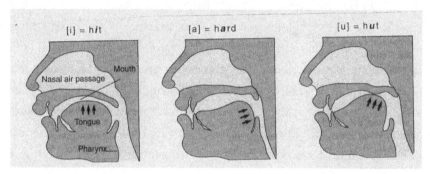

Figure 17.14 A diagrammatic view of how adult humans produce vowel sounds by positioning the tongue (arrows) in different parts of the oral cavity.

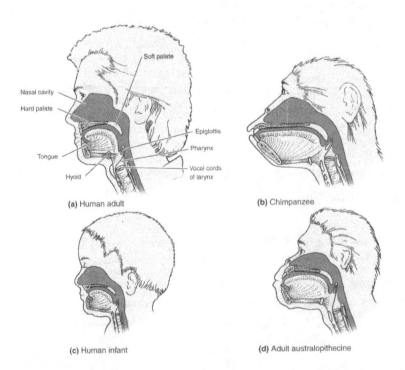

Figure 17.15 The upper respiratory systems of (a) an adult human, (b) a chimpanzee, (c) a human infant, and (d) an adult australopithecine illustrating important structures associated with vocalization.

Figure 17.15 diagrams the upper respiratory and vocalization systems of an adult human, a chimpanzee, an infant human, and a reconstruction of the presumed upper respiratory and vocalization system of an adult australopithecine. The pharynx is much longer in human adults than in chimpanzees because the larynx is displaced downward in the neck, and the bulging tongue formed by shortening of the mandible now forms the anterior wall of the pharynx. As a result, humans can enunciate vowels and syllables more clearly by positioning the tongue in both mouth and pharynx. Newborn human infants show the same overlap between the epiglottis and soft palate as nonhuman primates, but the pharynx lengthens considerably during infancy and childhood, transforming humans from obligate nose breathers as infants, to the adult condition of voluntary mouth breathers.

Because speaking depends so much on soft tissues that do not readily fossilize, researchers find it difficult to uncover the phylogenetic history of speech. We can, however, make important correlations between skull structure and the positioning of the larynx, size of the tongue, and length of the pharynx. According to such studies, the vocal tract of australopithecines was no different from that of nonhuman primates and was probably also true for most, if not all, of the lineage classified as *H. erectus*.

Although an intact *hyoid bone* (used for laryngeal muscle control) has been found with a 60,000-year-old Neandertal skeleton, we still do not know the actual position of the larynx and other soft tissue in this fossil or earlier ones and thus remain uncertain as to the extent of Neanderthal vocalization.

Intelligence

The intellectual capacity of humans or human intelligence is characterized by perception (awareness through senses), consciousness (aware of one's own existence), self-awareness (conscious knowledge of one's own character, feelings, motives, and desires), and volition (the act of making conscious choices or decisions).

As a primate brain, the human brain has a much larger cerebral cortex, in proportion to body size, than most mammals, and a very highly developed visual system. The shape of the brain within the skull is also altered somewhat as a consequence of the upright position in which primates hold their heads. The dominant feature of the human brain is **encephalization**. Increase in the ration of brain mass to body mass during the evolution of a species. Higher degrees of encephalization are correlated with higher degrees of intelligence. Our large cerebral cortex gives the cognitive abilities to learn, form concepts, apply logic and reason, including the capacities to recognize patterns, comprehend ideas, plan, problem solve, make decisions, retain information, and use language to communicate. In the words of Michio Kaku: "The human brain has 100 billion neurons, each neuron connected to 10 thousand other neurons. Sitting on your shoulders is the most complicated object in the known universe."

Rise of the Social Hunting Ape

Humans have been tribal (group) animals since they first walked erect, more than four million years ago. We were forced to gather in groups or tribes out of sheer necessity. Being bipedal, we usually could not out-climb nor out-run either our predators or our prey. Only through tribal cooperation could we hold predators at bay and secure food. For the first two million years of their existence, early hominins were tribal herbivores, but for the last 2 million years they have been social hunters (and in many aspects still are).

Gathering together in groups for the common good is not unique to humans because from ants to birds and orcas to gorillas, other types of animals form tribes and cluster together for survival. With humans, however, tribal (safety in numbers) grouping gave rise to complex social systems; human collectives gradually became ordered and more structured. Some anthropologists believe that one contributing factor in the move from tribal humans to social humans was when males began to assume more of a parental rather than a strictly reproductive role in the collective.

Early human societies would have displayed several basic characteristics:

- *Roles* Each segment of the group—mature men, young men, mature women, young women, small children, and the very old—would have certain duties (roles) they had to perform for the common good.
- *Rules* For the group to function successfully as a coordinated whole, a code of conduct (rules and regulations) of some type would have to have been established.
- *Rulers*. To enforce the rules, rulers (chiefs or tribal elders) would need to be selected.

The first human societies were most likely very similar to hunter-gatherer societies that continue today in places such as Central Africa (*Mbuti Pygmies*), South Africa (*Kalahari Bushmen*), and Australia (*Aborigines*). These groups consist of social communes or bands where males are usually the hunters and females the plant gatherers. Because their omnivorous diet depends on highly variable and often seasonal plant and animal food sources, each band moves several times a year over fairly wide ranges to different home bases or settlements.

Although we do not know when hunting began in human history, it was a significant enterprise for a long enough periods—in many societies, up to the agricultural (Neolithic) revolution about 15,000 to 10,000 years ago—to have seriously influenced human behavior and societal development. First, successful medium and large game hunting requires active cooperation among hunters. Hominin hunters empowered with simple weapons such as wooden spears, clubs, and hand axes, used techniques such as tracking, stalking, and chasing game into swamps, over cliffs, into ambushes, or by continuing the chase until the animal tired. With cooperative hunting, humans could bring down larger animals and feed more people than could single hunters alone.

Second, cooperative hunting and killing of large animals emphasized increased social cohesion during both hunting and the food sharing that followed. Transfer of information in successful hunts became vital, performing a necessary function in many later social interactions of the entire group. Improved communication became especially advantageous as individuals took more complex leadership roles in planning, hunting, helping, food gathering, food sharing, infant care, child training, and other vital activities.

Third, successful hunting emphasized perceiving and retaining information on migratory pathways, watering sites, and home base settlements, whose geographical positions extended over ranges greater than those occupied by most other primates or carnivores. Hominin hunters had to mentally dissect their experiences and observations into component geographical and ecological features, prey behaviors, weather effects, and seasonal changes, and store and synthesize this information into communicable mental maps that enabled prediction, planning, and modification.

Fourth, hominin hunting involved stresses that fostered increased locomotory adaptations such as persistence in the chase (humans can continue jogging for distances that are generally longer than many large animals can continue running), maneuverability in the kill, and long-distance traveling to or between home bases while carrying heavy burdens. Recent anatomical and physiological evidence indicates that adaptations for long-distance endurance running characterized the genus *Homo* from *Homo erectus* onward. Others suggest that the need for increased diffusion of metabolic heat during these pursuits would have selected for the loss of body hair and increased number of sweat glands, features that among primate are restricted to humans.

Finally, hunting placed further social emphasis on the home base to enable food exchange among foraging subgroups, particularly when the food supply was irregular, as it often is in hunting. A home base has value for nursing and pregnant females who could not always or easily cover the long distances necessary for large-scale hunting. The home base would have become a center for food sharing, shelter, hunting preparations, sexual bonding, childcare, and other social interactions in which communication skills tied all members together.

Rise of the Intellectual Tool-Making Ape

Possessing **intelligence** (the capacity to reason, plan, solve problems, and learn) is not unique to humans. In fact, a number of different vertebrate species are regarded as being highly intelligent. To be human is not merely a matter of possessing intelligence, but also a matter of possessing a pliable multifaceted **intellect**. Human intellect is unique in that it developed as a control over our animalistic instincts, allowing us to voluntarily adapt what would otherwise be unchangeable instinctive behavior. Thus, when humans are presented with challenges and situations beyond the scope of our instincts, our refined, multifaceted intellect allow us to adjust, adapt and overcome. To fully understand what it means to be truly and uniquely human from the standpoint of intellect, one only need contrast humans to our closest vertebrate and primate kin—the chimpanzee. A chimpanzee is an ape with raw intelligence driven mainly and usually by instincts, whereas a human is an ape with instincts driven mainly and usually by a refined intellect. (Nothing about animal behavior is absolute. The use of the disclaimers "mainly" and "usually" is meant to account for those times when nonhuman intelligent animals break free of their instincts and, if only for a moment, operate on raw intelligence. However, the phrase was inserted more in consideration of those times when humans lock out their intellect and behave as an instinctive animal.)

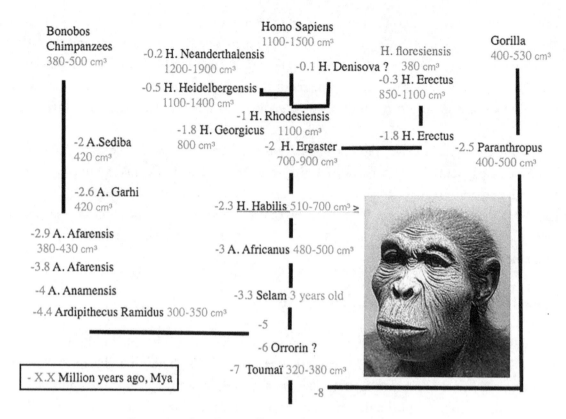

Bonobos
Chimpanzees
380-500 cm³

Homo Sapiens
1100-1500 cm³

Gorilla
400-530 cm³

-0.2 H. Neanderthalensis
1200-1900 cm³

H. floresiensis

-0.1 H. Denisova ?   380 cm³

-0.5 H. Heidelbergensis,
1100-1400 cm³

-0.3 H. Erectus
850-1100 cm³

-1 H. Rhodesiensis

-1.8 H. Georgicus   1100 cm³
800 cm³

-2 A.Sediba
420 cm³

-2 H. Ergaster
700-900 cm³

-1.8 H. Erectus

-2.5 Paranthropus
400-500 cm³

-2.6 A. Garhi
420 cm³

-2.3 H. Habilis 510-700 cm³ ≥

-2.9 A. Afarensis
380-430 cm³

-3 A. Africanus 480-500 cm³

-3.8 A. Afarensis

-4 A. Anamensis

-3.3 Selam 3 years old

-4.4 Ardipithecus Ramidus 300-350 cm³

-5

-6 Orrorin ?

- X.X Million years ago, Mya

-7 Toumaï 320-380 cm³

-8

Figure 17.16 Evolution of brain volume in the primate lineage.

Brain size (specifically the cerebral cortex) and thus the intelligence and intellectual capabilities of humans developed slowly at first and then more rapidly over the 4-million year span of our development (**Figure 17.16**). As time passed, selection pressures and fortunate opportunities allowed humans to develop their basic intelligence to a level achieved by no other animal on the planet—a pliable, multifaceted intellect. Intelligence is but one facet of intellect. Other facets would include the ability to think abstractly and comprehend ideas; the ability to accumulate knowledge or wisdom; the ability to use language; the development of an individual personality and character; and introspection.

We see the first hints of an emerging intellect in the social skills associated with cooperative hunting and other social activities and in the stone implements and tools that date back at least to *Homo habilis*. *Homo erectus* continued making tools, and there is clear evidence of progressive improvement in tool making over time. By 400,000 years ago, *H. erectus* sites commonly had tens of thousands of discarded tools.

With the appearance of the Cro-Magnon culture about 40,000 years ago, tools started becoming markedly more sophisticated. Tool-making skills involve not only manual dexterity, hand-eye coordination, and considerable concentration and focus, but also the ability to plan and visualize an object that is not apparent in the raw material from which it is created. The artisan must conceptualize the final form of a tool in its three dimensions and implement such concepts by mastering a series of techniques. These included finding and recognizing appropriate, workable stones in outcrops that were often widely dispersed, carrying these stones back to a base, and then shaping them into tools by a sequence of precise strokes. The toolmakers also had to supplement the considerable mental abilities they used in tool making with social and communication abilities, in order to transmit such skills to other individuals who could continue the industry (**Figure 17.17**).

Figure 17.17 Hand axes, such as this Acheulean tool, were made as early as 1.7 million years ago and were the Stone Age equivalent of the Swiss army knife.

A wider variety of raw materials such as bone and antler were used, and new implements for making clothing, engraving, and sculpting were developed. Fine artwork, in the form of decorated tools, beads, ivory carving of humans and animals, clay figurines, musical instruments, and spectacular cave paintings appeared over the next 20,000 years.

Rise of the Agricultural Ape

The late arachaic *Homo sapiens* populations were most likely tribal wanderers who exploited a wide range of food sources. This most likely required that they range seasonally over vast territories. Exposed to the elements, hounded by predators, and with reliable and accessible food always in short supply, life for these early hominins must have been a brutal proposition, especially for infants, young children, and the old. The evolution to hunter-gatherer societies improved the lot of *Homo,* but life was still hard. The average Neanderthal male, for example, could look forward to a life-span of about 30 to 35 years.

The next major milestone in the rise of modern humans happened when hunter-gatherers became agriculturists and began to grow rather than seek out food. Greek legend has it that in a burst of goodwill, Demeter, goddess of crops, bestowed wheat seeds on a trusted priest, who then crisscrossed Earth in a dragon-drawn chariot, sowing the dual blessings of agriculture and civilization.

For decades, paleontologists and archaeologists also regarded the birth of agriculture to be such a dramatic and important milestone that they christened it the *Neolithic Revolution.* The contention was that farming was born after the end of the last Ice Age, around 10,000 years ago, when hunter-gatherers settled in small communities in the Fertile Crescent, a narrow band of land arcing across the Near East. Here they quickly learned to grow their own food by sowing cereal grains and breeding better plants. Societies then raised more children to adulthood, enjoyed food surpluses, clustered in villages and developed culture, and headed down the road to civilization. This novel agricultural way of life then diffused across the Old World.

As is the case with many a hypothesis, this scenario did not stand the test of time and has fallen beneath an onslaught of new data. By employing sensitive new techniques from sifting through pollen cones to measuring minute shape changes in ancient cereal grains, researchers are building a new picture of agricultural origins. They are pushing back the dates of both plant domestication and animal husbandry around the world, and many now view the switch to an agrarian lifestyle as a long, complex transition rather than a dramatic revolution. The latest evidence suggests, for example, that hunter-gatherers in the Near East first cultivated rye fields as early as 13,000 years ago. But for centuries thereafter they continued to hunt wild game and gather an ever-increasing range of wild plants, only becoming full-blown farmers living in populous villages around 8500 B.C. And in some cases, villages appear long before intensive agriculture.

A new picture of the origins of agriculture has begun to emerge as evidence continues to be gathered worldwide. In the Near East, not only were some villages born before agriculture but may have even spurred the development of agriculture in some cases. In China, North America, and Mesoamerica, plants appear to have been cultivated and domesticated by nomadic hunter-gathers during the dramatic environmental shifts that accompanied the final phase of the last Ice Age. To many, it no longer makes sense to suppose a strong casual link between farming and settled village life. Indeed, in many regions, settled villages of agriculturists emerged only centuries or millennia after cultivation, if at all. Many ancient peoples simply straddled the middle ground between foraging and farming, creating societies and economies that blended both together. These societies in the middle ground certainly are not failures, but rather societies that found an excellent long-term solution to their environmental challenges.

Rise of the Cultural Ape

Eventually, for reasons still unclear, many of the early domesticators became true agriculturalists by 10,500 years ago in the Near East, 7,000 years ago in China, and later in the Americas and Africa. And during this transition, human populations did indeed soar as hamlets became villages and villages became cities. Archeological sites in the intensively studied Fertile Crescent, for example, show that fields increased more than 10-fold in size from 0.2 hectares to 2.0 to 3.0 hectares during this time of transition. As food supplies became more reliable, the lifestyle became more stable and easier to bear, in turn, a safer, more stable, and longer life permitted humans more opportunities to develop their intellectual side, and as they did so, human societies were transformed into human culture and civilization.

Consider New York City as an example. The physical structure of the city—buildings, roads, electrical and plumbing systems, transportation systems, and so on—are the monuments of present-day *civilization*. The pyramids of Egypt are the remains of a past civilization, whereas the space station may be a preview of future civilizations.

The institutions and systems that build and maintain the physicality of the city as well as those that protect, feed, clothe, house and entertain the human residents of that city-sized chunk of civilization are the *culture*. Even rural areas have civilization and culture, just not on such a grand scale. However, there is no one standard human culture homogeneously layered around the world. Humans in different geographical areas put their own twist on human culture, and so regional differences crop up often within the same country or even within a large city.

Cultural Hallmarks

What are the cultural components and systems that make modern civilization possible?

Communication systems

As the necessary centers in the human brain evolved, simple vocalization gave way to speech. As our intellect evolved, speech gave way to language with the result that today *H. sapiens sapiens* is the only animal with written language.

The ability to communicate swiftly and precisely is so critical to modern culture and civilization that all other cultural systems are totally dependent on it. In fact, humans love to communicate and seem almost addicted to doing so. The evolution of our intellect has produced communication systems that are truly amazing. We now live in an age of electronic communication in which it is possible to sit at a computer and converse with someone on the other side of the planet as if they were in the same room or use a cell phone in Boston to wish your cousin in Seattle happy birthday. And each day billions of people worldwide watch human culture and civilization fly out of their television sets.

One benefit of communication-based culture is the easy access to knowledge and information. The dark side, however, is information overload as so succinctly described by E.O. Wilson when he said: "We are drowning in information while starving for wisdom."

Our compulsion to communicate has driven us to even attempt to communications with other sentient beings on other planets somewhere in the void beyond Earth. The Pioneer 10 and 11 spacecraft launched in 1972 and 1973, were designed for a close flyby of two giant outer planets—Pioneer 10 past Jupiter and Pioneer 11 past Saturn. Bolted to the main frame of each spacecraft was a 6 x 9-inch gold anodized plaque displaying graphic representations of who sent the craft and from what planet they had originated. These plaques were added in the outside chance that sometime in the far distant future these tiny spacecraft might be picked up by a space faring alien race. Launched in 1977, the twin satellites Voyager I and II were designed to take advantage of a rare alignment of the outer planets by flying close to these planets one after the other collecting data and taking pictures. Inside each of these spacecraft is a golden record containing 115 images and variety of natural sounds as well as "Greetings from Earth" in 55 languages as well as music from around the world.

Again, these records were added to the spacecraft in hopes that someday somewhere out among the starts where they might be found and understood. Pioneer 10 and 11 have long since left the solar system and are no longer in contact with us. Voyager I has become the most distant of all human-made objects at a distance of 16 billion km (10 billion miles) it flies another 1.6 million km (1 million miles) farther from the sun each day. Voyager II is about 80 percent as far away as its twin. The power supply of each satellite should last for another 20 years or so. Many thousands of years from now these spacecraft will be closer to other stars than to our sun. We have tossed a spacecraft bottle into the cosmic ocean in hopes that someone or something may eventually open it and read the message inside.

Currently, a number of organized efforts collectively known as SETI (**S**earch for **E**xtra-**T**errestrial **I**ntelligence), survey the sky to detect the existence of transmissions from civilizations on distant planets. Although SETI searches for communication from aliens, METI (**M**essages to **E**xtra-**T**errestrial **I**ntelligence) projects create electronic messages and beam them out into space in hopes they will be received by an alien intelligence which, in turn, might communicate with us.

Legal systems

Humans are the only animals with morals (sense of right and wrong) and a value system of justice (fair play). Ancient human societies must surely have realized the necessity of establishing codes of conduct and rules to account for moral and justice issues. Those ancient rules and codes evolved into the legal systems of modern times. Legal systems vary from culture to culture around the world, but most have three basic components: (1) those that make the regulation or laws (legislators), (2) those who interpret the regulation or laws (judges), and (3) those who enforce the regulation or laws (governments, police, and armies). The modern legal system functions as did the ancient system of rules to prevent the culture from slipping into chaos and anarchy. This fact has been repeated time and again throughout modern history as countries are caught in a civil war or the government of a country is forcibly disrupted and displaced either by forces from outside or within its borders.

Protective systems

Although not exclusively a trait of the human intellect, **altruism** (concern for others not related to us) is perhaps more powerful in humans than in other animals for no other animals cares as much for the welfare of others of its own kind as does the human ape. As such, a number of different protective cultural systems designed to ensure the safety and protection, health, and general well-being of other humans have been developed over the course of our evolution. Protective systems may be public (governmental) or private in nature and funding.

Public Safety and Protection is ensured by police and fire departments at the local level and armies at the state and national level. Some governmental agencies attempt to ensure the safety of everything from the food we eat to the drugs we take while others set and enforce regulations to improve the safety of working conditions for employees in all kinds of industries and jobs.

Public Health is protected by our medical system that ranges from governmental agencies such as the Surgeon General and the Center for Disease Control to your personal physician. Ongoing research into the prevention, improved treatment, and even cure of the host of microbial and degenerative diseases that plague humankind is also an important part of the modern human health cultural system.

Public well-being refers to the medical, domestic, and financial health of an individual. Medical well-being is a function of the health system of our culture. Domestic well-being (being properly fed, clothed, and housed) and financial well-being are the providence of a host of different agencies and groups, both private and governmental. From feeding the hungry in soup kitchens to building housing for the homeless to job training and financial assistance, our modern culture attempts to aid those less fortunate.

Economic systems

An economic system may be defined as a particular subset of cultural institutions that deal with the production, distribution, and consumption of goods and services. Economic systems vary from place to place and from country to country, but they all have money at their root. Be it in the form of paper, coin, or polished pieces of shell, money functions as a medium of exchange for goods and services rendered, a unit of account (the price or value of goods and services), and a store of value (financial capital that can be reliably saved, stored, and retrieved. Simply put, the economic system generates money and the money funds the other cultural systems.

Aesthetic systems

The rise of human intellect allowed humans to develop an aesthetic sense. That is, humans view and value certain aspects of the world around them not for their practicality but for the pleasure they provide. From Niagara Falls to the Grand Canyon to a fiery sunset, humans appreciate the beauty of natural things and are filled with a sense of awe by the majesty these things.

Not content to appreciate only natural beauty, humans have developed their own aesthetically pleasing schemes such as art, music, dance, acting, and literature. What we regard as aesthetically pleasing is one of the most intensely personal parts of our individual being.

Innovation systems

The rise of intellect fostered a literal explosion in the creativity and inventiveness (science and engineering) of humans. Innovation driven by evolving intellect has gone through a number of phases or stages: stone tools, metal tools, machines, electric machines, and technology. The first stage appears to have tool-making. At first tools and weapons were crafted from natural objects such as rocks, bones, and antlers. With the discovery of techniques for processing metals, tools and weapons were forged from copper and bronze at first and later from iron and steel. The next historical leap in innovation saw the construction of machines of all types from vacuum cleaners to steam engines. Modern times have seen the application of electricity to machines to the extent that our modern culture is literally driven by electricity. Often taken for granted, our deep dependence on electricity was driven home to my family personally several years ago when a huge ice storm devastated the area where my family lives. Those without generators were forced to live for days as if the calendar had been turned back to the early 1900s.

The latter part of the 20th century found humankind entering yet another and still current phase of innovation we call the technology stage. Science and engineering taken to the technological level have given us everything from space travel and computers to artificial hearts and genetically altered plants and animals and cellular phones and flat-screen televisions to artificial intelligence and nanobots.

Civilization and all the components and systems of culture that accompany it are the result of the ever-increasing refinement of our multifaceted intellect. Our intellect created and drives human culture and civilization, but this culture and civilization, in turn, challenges our intellect to maintain and improve it. Presumably, as our intellect continues to evolve so will our culture and civilization in unknown ways that **futurists** take great pleasure in trying to predict.

In Summary

- Humans are creatures with an almost alien-sized brain that can write poetry, raise giant cities, travel in space, and even contemplate and explore their own mind; an advanced technological being wrapped in the body of a primate, a naked primate at that.

- Primates, especially humans, are remarkably recent animals on the evolutionary stage. Primate-like mammals (or proto-primates) first arose in the early Paleocene epoch about 65 million years ago.
- The order Primate embraces 230-270 species aligned into two suborders and 13 families:

Suborder Strepsirrhini
> Family Cheirogaleidae—dwarf and mouse lemurs
> Family Lemuridae--Lemurs
> Family Lepilemuridae—Sportive lemurs
> Family Indriidae—Wooly lemurs and allies
> Family Daubentoniidae—Aye-aye
> Family Lorisidae—Lorises, pottos, and allies
> Family Galagidae—Galagos

Suborder Haplorhini
> Family Tarsiidae—Tarsiers
> Family Callitrichidae—Marmosets and tamarins
> Family Cebidae—Capuchins and squirrel monkeys
> Family Cecopithecidae—Mandrils, baboons, and macaques
> Family Hylobatidae—Gibbons
> Family Hominidae
>> Subfamily Ponginae—Orangutan
>> Subfamily Homininae
>>> Tribe Gorillini—Gorillas
>>> Tribe Hominini

- Characteristics of primates:
1. Shortening of the face accompanied by reduction of the snout
2. The general retention of five functional digits on the fore and hind limbs. Enhanced mobility of the digits, especially the thumb and big toes, which are opposable to the other digits.
3. A semi-erect posture that enables hand manipulation and provides an optimal position preparatory to leaping.
4. Claws modified into flattened and compressed nails to support the sensitive digital pads on the last phalanx of each finger and toe.
5. A shoulder joint allowing a high degree of limb movement in all directions and an elbow joint permitting a rotation of the forearm.
6. A reduced snout and olfactory apparatus
7. A reduction in the number of teeth compared to primitive mammals.
8. Teeth and digestive tract adapted to an omnivorous diet
9. A complex visual apparatus with high acuity and a trend toward development of forward-directed binocular eyes and color perception. Well-developed hand-eye motor coordination.
10. A large brain relative to body size, in which the cerebral cortex is particularly enlarged.
11. Only two mammary glands (some exceptions).
12. Typically, only one young per pregnancy associated with prolonged infancy and pre-adulthood.

- Modern humans seem to be defined anatomically by two main characteristics: bipedalism and a large brain with developed sites supporting speech and intelligence.
- Modern humans have gone through a number of different stages and types of evolution: anatomical, agricultural, social, and technological.

Review and Reflect

1. *My Father or My Cousin?* Some people believe that human evolution means we evolved directly from apes. Is this a scientifically accurate interpretation of human evolution? Explain

2. *The Fossils Speak* Fossil evidence suggests humans have evolved over millions of years and that apes and humans shared a common ancestor. Describe several ways scientists can analyze fossils in order to provide evidence for evolution and explain several examples of fossil evidence that gives clues about human evolution.

3. *Compare and Contrast* In what ways are humans similar to apes—chimps, gorillas, and orangutans? How are we different than apes?

4. *A Critical Difference* A lecture on human evolution you attended has just ended. Your friend seated next to you says. "The professor made an issue out of the difference between intelligence and intellect. Aren't they the same thing really?" What would you say?

5. *The Cultural Ape* We have used the term "cultural ape" to describe modern humans at the peak of their evolution (if where we are today actually is the "peak" of our evolution). Do you agree or disagree with our usage of that term to describe modern humans? Defend your answer.

6. *Look into the Crystal Ball* Given what you now know about where we may have come from (evolution) and present day human culture, civilization, and behavior, play the role of a futurist and predict what human culture and civilization will be like 500 years from now.

7. *What's in a Quote?* Arthur Koestler said: "The evolution of the brain not only overshot the needs of prehistoric man, it is the only example of evolution providing a species with an organ which it does not know how to use." What do you think he meant by that remark? Interpret this quote in your own words.

Create and Connect

1. *Humans Arise* Write a position paper on the origin of humans and how they developed into modern *Homo sapiens sapiens*.

   ### Guidelines
   A. Compose a position paper, not an <u>opinion</u> paper. Defend your position with as many facts, figures, quotes, and pertinent information as possible.
   B. Your work will be evaluated not on the "correctness" of your position but the quality of the defense of your position.
   C. Your instructor may provide additional details or further instructions.

2. ***A Day in the Life*** Describe one day in the life of *Homo erectus*. Personalize this by writing it as if it were a daily entry in a diary.

3. ***The Human Condition*** You are the science editor of a large metropolitan newspaper. Your editor has given you the following assignment: Write a column entitled, *The Human Condition—What It Means to be Human.* How would you proceed and what would you say? Format your work as if it were an actual newspaper article.

APPENDIX
SCIENTIFIC WRITING

How well we communicate is determined not by how well we say things but by how well we are understood.

—Andrew Grone

For the scientific enterprise to be successful, scientists must clearly communicate their ideas and work. Scientific findings are never kept secret. Instead, scientists share their ideas and results with other scientists, encouraging critical review and alternative interpretations from colleagues and the entire scientific community. Communication, both verbal and written, occurs at every step as the science cycle turns.

One of the objectives of each chapter in this book is to develop your communications skills through a variety of written review questions and creative challenges. In these assignments, you will be asked to write everything from formal scientific reports to essays to position papers to short stories. The exact format and details will be given with each assignment but before you write and as you write consider these guidelines, hints, and suggestions:

- ☑ ***We can't read your mind***. Your instructors can only evaluate what you have actually written. Do not make the mistake of using the old excuse that, "They will know what I mean" to rationalize incomplete work.

- ☑ ***Mean what you say and then write what you mean***. Refer to a dictionary and thesaurus to ensure clarity and proper word usage. For example, you allude to a book, but you elude a pursuer. Also, use correct composition and grammar.

- ☑ ***Be clear and concise***. Delete unnecessary words such as adjectives and adverbs that have limited use in describing your work. Write clearly in short and logical but not choppy sentences. For example, "The biota exhibited 100% mortality response." is a wordy and pretentious way of saying, "All the organisms died." Keep it simple and straightforward. Do not use vague or ambiguous words or phrases. For example, "An adequate supply of lettuce seeds will be needed" reveals no useful information. Instead employ a specific phrase, "100 lettuce seeds will be the sample size."

- ☑ ***Keep related words together***. The following sentence is taken from an actual scientific publication. "Lying on top of the intestine, you perhaps make out a small transparent thread." Do we really have to personally lie on top of the intestine to see the thread? The author should have written, "A small transparent thread lies atop the intestine."

☑ ***Use active voice whenever possible.*** Using "I" instead of "me" makes the paper easier to read and understand. However, in the Methods and Materials section of a formal scientific paper, passive voice should be used so that the focus of the writing is on the methodology, rather than the investigator.

☑ ***Think metric.*** Use metric units for all measurements and numerals when reporting measurements, percentages, decimals, and magnification. When beginning a sentence, write the number as a word. Numbers of ten or less that are not measurements are written out. Numbers greater than ten are given as numerals. Decimal numbers less than one should have a zero in the one position (e.g., 0.153; not .153).

☑ ***Proofread and edit.*** Carefully proofread your work even though your word processor has checked for grammatical and spelling errors. These programs cannot yet distinguish between "your" and "you're," for example. Also, edit then revise and then do it again.

☑ ***Do your own work and write your own words.*** Plagiarism is a serious offense, and it can result in serious consequences.

☑ ***Regarding references:***

A. No footnotes are used, but each reference should be cited in the body of the paper. For example, Bicak (1999) reported that red-tailed hawks nest in large trees.

B. How to cite:

Journal articles	Watson, J.D. and F. H. Crick. "Molecular Structure of Nucleic Acids: A Structure for Deoxyribose Nucleic Acid." *Nature,* 1953, vol 171, pp. 737-738.
Books	Darwin, C. R. *On the Origin of Species.* London: John Murray, 1859.
Government Publications	Office of Technology Assessment. *Harmful Non-indigenous Species in the United States.* Publication no. OTA-F-565. Washington, D. C. U.S. Government Printing Office, 1993.
Encyclopedia-	Bergman, P. G. editor. (2002). "Annelids." In the *Encyclopedia Britannica.* (Vol 1, pp. 223-225). Chicago: Encyclopedia Britannica
Internet Sources-	Council of Biology Editors. 1999. Oct. 5. home page. <http://www. councilscienceeditors.org> Accessed Oct 7, 2004.

NOTE: Personal or popular websites are normally not legitimate references. Check with your instructor if in doubt about citing an on-line reference.

Interview-	An interview with Dr. Joseph Springer regarding Sandhill cranes. Oct 12, 2004.

NOTE: Direct quotations must be cited. For example:

"Sandhill cranes have been staging along the Platte River for thousands of years." (Springer, 2004)

GLOSSARY

Abiogenesis: The now discredited theory that living organisms can arise spontaneously from inanimate matter; spontaneous generation

Abiotic: Any nonliving factor of the physical environment

Abyssal Plains: The seafloor beyond the continental slope

Abyssopelagic: The open sea beyond the continental shelf below 4,000 meters

Acanthor Larva: Shelled larva of an Acanthocephalan

Accommodation: Adjusting for near and distant objects

Acetabulum: A ventral, hookerless sucker of digenetic flukes

Acid: A molecule or other entity that can donate a proton or accept an electron pair in reactions

Acoelomate: An animal lacking a true coelom

Acontium or Acontial Filament: A free thread attached to the aboral end of the mesentery

Acrodont Teeth: Teeth attached on the surface of the jaw

Acrorhagi: Inflatable tentacles loaded with nematocysts found in sea anemones

Actinopterygii: Ray-finned fishes

Actinula Larvae: One possible larval stage of a hydrozoan that forms directly from the planula larva

Active Transport: Movement of molecules across a cell membrane from an area of low concentration to an area of high concentration

Aculeus: The stinger at the end of the metastoma in scorpions

Adaptive Radiation: The diversification of a group of organisms into forms filling different ecological niches

Adhesion: The action or process of adhering to a surface or object

Adhesive Tubes: Tubes that contain cement glands that secrete adhesive substance used to temporarily attach a gastrotrich to objects in the environment

Advertisement Calls: Calls produced to attract a mate

Aerobic: Requiring oxygen

Aeroplankton or Aerial Plankton: Small invertebrates (insects and spiders) that float in the air.

Aflatoxin: Any of a class of toxic compounds that are produced by certain molds found in food, and can cause liver damage and cancer

Age Structure: The distribution of individuals across different cohorts (ages) in a population

Agriculture: The practical application of raising crops and producing livestock

Agroecosystem: An artificial system of agriculture created by humans that has replaced prairies

Albumin: The white of an egg

Alleles: One of two or more alternative forms of a gene that arise by mutation and are found at the same place on a chromosome

Allelopathy: The chemical inhibition of one plant (or other organism) by another, due to the release into the environment of substances acting as germination or growth inhibitors

Alloparenting: Members of the same species help the breeding pair in raising the young

Allopatric: Existing in different geographical areas

Altricial: Birds born bare with eyes closed

Altruism: Concerns for others not related to us

Altruistic Behavior: A behavior in which an individual appears to act in a way that benefits others at cost to itself

Alula: A small projection on the anterior edge of the wing; it functions as a freely moving first digit and bears three or four feathers; also called a bastard wing or a spurious wing

Alveoli: Air pockets that fill the mammalian lung

Ambulacral Grooves: Grooves located down the length of each arm of sea stars where tube feet protrude

Ametabolous Metamorphosis: The process of development in which the larval body structure is a smaller, immature version of the adult (incomplete metamorphosis)

Amictic: Females produced via parthenogenesis (asexual)

Amino Acids: Small molecules that contain an amine group and a carboxylic acid and make up proteins

Amphids: Chemoreceptors found on the head of nematodes

Amplexus: Mating squeeze by a male frog on a female frog

Ampullae of Lorenzini: Electroreceptors that form a modified lateral line in sharks

Anabolism: Processes by which large macromolecules are formed from smaller subunits.

Anadromous: Hatch and grow in freshwater, migrate to the ocean to mature, and then return to freshwater to spawn

Anaerobic: Not requiring oxygen

Anal Canals: Tubular structures in cnidarians that collect undigestible wastes that are expelled out of the body

Anal Fin: A single fin near the anal opening of fishes

Anal Pores: Small holes that connect the anal glands to the outside

Analogous: Similarity in structure that results from independent evolutionary events

Anaphase: The stage of mitosis when chromosomes are split and the sister chromatids move to opposite poles of the cell

Anapsida: The earliest amniotes that lack openings in the skull

Anatomy: The study of the internal structure of organisms

Ancestral Characteristics: Characters found in the ancestors of a group: basal traits

Aneuploidy: Presence of an abnormal number of chromosomes in a cell

Anion: A negatively charged ion

Annuals: Plants that complete their life cycle in one year

Annual Rings: The layer of wood produced by a single year's growth of a woody plant

Annulations: A form of sexual reproduction by which an organism regularly breaks along preformed lines

Annuli: Rings found on the body segments of annelids

ANOVA: A statistical test that is used to determine if the means of a trait or characteristics are significantly different among more than two groups

Antagonistic Muscles: Opposing muscles

Antennal Glands: Excretory glands that contain a single pair of nephridia that are found on the head in association with the antennae

Anterior: Pertaining to the head end

Anthropoids: Halplorhini: group consisting of tarsiers, tamarins, baboons, monkeys, apes, and humans

Anthropology: Study of humans

Anthropomorphism: Ascribing human characteristics to plants and animals

Anticodon: A sequence of three nucleotides forming a unit of genetic code in a transfer RNA molecule, corresponding to a complementary codon in messenger RNA

Antlers: A pair of bony, branched structures that protrude from the frontals of the skull

Aorta: The largest artery in the body that transports blood from the heart to the body

Apocrine Glands: A type of sweat gland that opens either into a hair follicle or into a space where there once was a hair follicle

Apolysis: Digestion of the old procuticle by hormones secreted from glands in the hypodermis

Aposematic or Warning Coloration: A bright, conspicuous color pattern used to advertise toxicity to predators

Appendicular Skeleton: The limbs and girdles of the skeleton

Aquiferous System: The cells and tissues that carry water through the canals of the body of a sponge

Aboreal: Pertaining to tree-dwelling

Arboreal: The bottom end of polyps that attaches to the substratum

Archaeocytes: Large motile ameboid cells that can change into any other type of sponge cell and, thus, assist with building the skeleton, nourishing eggs, distributing food, and mediating chemical and physiological responses

Aristotle's Lantern: A feeding structure that consists of five protractible calcareous teeth and muscles that control protraction, retraction, and grasping movements of the five teeth

Arthropodization: A suite of highly successful adaptations that evolved in the arthropods

Articular: A small bone at the back of the lower jaw

Ascites: Accumulation of excess fluid in the abdomen

Asconoid Sponges: Simple, small, vase-like sponges that do not contain body folding

Asexual Reproduction: A type of reproduction in which offspring arise from a single organism

Associative Behavior: Behavior that is learned

Associative Learning: Learning that occurs by association with certain events

Assortative Breeding: A mating pattern and a form of sexual selection in which individuals with similar phenotypes mate with one another more frequently than would be expected under a random mating pattern

Asymmetrical: Lacking any definite geometric design in body parts

Atmosphere: Layers of air surrounding the planet

Atoke: the anterior portion of some tube-dwelling and burrowing worms that does not differentiate into specialized structures during the breeding season

Atomic Mass: The mass of an atom of a chemical element expressed in atomic mass units. It is approximately equivalent to the number of protons and neutrons in the atom (the mass number) or to the average number allowing for the relative abundances of different isotopes

Atomic Nucleus (chemistry): The small, dense region consisting of protons and neutrons at the center of an atom

Atomic Number: The number of protons in the nucleus of an atom, which determines the chemical properties of an element and its place in the periodic table

Atoms: Matter composed of a central nucleus with positively charged protons and neutral neutrons and an outer cloud of negatively charged electrons; the smallest particle of an element that still retains the properties of that element

ATP (adenosine triphosphate): A small molecule that transports chemical energy within cells for metabolism.

Atrial (excurrent) Siphon: An outlet for water and waste expelled by tunicates

Atriopore: A ventral pore or opening in the atrium

Atrium: A large, muscular chamber of the heart

Atrium: In the flatworm, a chamber that contains the penis and often the terminal portion of the female tract

Atrium: The interior cavity of a sponge

Auricles: Spike-like projections on the side of the head of turbellarians that contain chemoreceptors

Auricular Feathers: Feathers found around the external ear opening in birds

Autosomes: Body chromosomes

Autotomize: To detach one's own limbs (occurring in some molluscs, echinoderms, and arthropods)

Autozooids: the conspicuous secondary polyps of sea pansies, and sea pens

Auxospore: A reproductive cell in diatoms usually resulting from the union of two smaller cells or their contents and associated with rejuvenescence in cells that have become progressively smaller because of repeated divisions

Axial Skeleton: the cranium, backbone, and ribs of the skeleton

Axial Skeleton: A stiff internal axial rod embedded in the body tissue of many colonial anthozoans

Bacteriology: The study of bacteria

Bacteriophages: A virus that parasitizes a bacterium by infecting it and reproducing inside it

Ballooning: A means of dispersal of small spiders in which they release a loop of parachute silk that is caught by the wind and blows them into the air

Barbels: Short sensitive tentacle-like structures that surround the mouth of hagfish

Barbicels: Tiny hooks that hold together the barbules of feathers

Barrier: Any physical obstacle that prevents a species from occupying an otherwise biologically and geographically acceptable area

Base: Substances that accept protons from any proton donor, and/or contain completely or partially displaceable OH- ions.

Batesian Mimicry: The process of palatable species copying the coloration, structure, and behavior of a poisonous or unpalatable species

Bathypelagic: The open sea beyond the continental shelf from 1,000 meters to 4,000 meters

Behavior: The organized and integrated pattern of activity by which an organism responds to its environment

Bellows Lung: Mammalian lung

Benthic Zone: The seafloor

Bilateral Symmetry: A body design in which one side of the body is a mirror image of the other

Binary Fission: The process in which a parent cell splits into two nearly equal sized daughter cells

Binocular Vision: Vision that produces a sharp image in front of the head due to the eyes being set far to the front of the head

Binomial Nomenclature: A Latin naming system that provides each organism with a unique genus and species combination

Biodiversity: The number of different species of plants and animals

Biennials: Plants that take two years to complete their live cycle

Bioerosion: the process by which invertebrates bore into the shells of molluscs and corals and kill them

Biogenesis: The notion that any organism arises from another organism(s) of the same species

Biogeochemicals: Chemical elements that cycle between biota and the physical environment

Biogeography: The study of the spatial and temporal distribution of organisms

Biology: The study of life and living things

Biological Community: All of the populations of different species living together in the same place at the same time

Biological Magnification: The process by which toxic chemicals become increasingly concentrated as they pass up the food chain

Bioluminescence: Producing light by biological mechanisms

Biomass: The largest geographical biotic unit that consists of various ecosystems and their attendant communities with similar environment requirements

Biome: The largest geographical biotic unit and consists of various ecosystems and their attendant communities with similar requirements of environmental conditions.

Biomineralization: The process of removing minerals from the blood and crystallizing them for another purpose

Biomolecules: Molecules of carbohydrates, lipids, proteins, or nucleic acids

Bioprospect: Search for organisms with medicinal value

Biosensor: An analytical device, used for the detection of an analyte that combines a biological component with a physicochemical detector

Biosphere: A layer 100 feet or so above and within the planet where life occurs

Biotechnology: The manipulation of living organisms for practical purposes and/or to produce useful products

Biotic: Describes anything living (living organisms)

Bipedal Locomotion: Habitually walking upright on the hind legs

Biphasic Life: Aquatic larvae and terrestrial adults

Biradial Symmetry: A combination of radial and bilateral symmetry

Biramous: Branching appendages

Bladder: An organ that stores excess liquid

Blastocyst: An embryo

Blastopore: The first opening that develops in an embryo

Blastula: The fluid-filled ball of cells that forms a developing embryo after multiple cleavage stages

Bond: To join together (as in a chemical bond or union)

Book Gills: The posterior of five pairs of appendages on the opisthosoma

Botany: The study of plants

Bower: A structure built by male bower birds to attract mates where they display a complex set of behaviors

Brachial Heart: A heart that is located at the base of the gills in cephalopods and collects unoxygenated blood, which is then pumped to the gills

Brachiate: Move by swinging through the trees

Branchial System: Gas exchange system of enteropneusts including the gill pores, branchial sacs, and gill slits

Brille: Ocular scale

Brood Parasitism: The process by which an organism lays its eggs along with a host's eggs causing the host to provide for the intruder's young often at the expense and death of the host's young

Brood Patches: Places where the feathers fall out on the underbody of the parent to expose the warm, bare skin that can be pressed directly against the eggs

Brown Fat: Special fat deposits that produce extra heat when metabolized

Buccal Cavity: The external opening/cavity into the digestive system

Buccal Pump: A positive pressure system for ventilating the lungs

Buccal Pumping: Pumping action of the mouth and opecula to create a respiratory current across the gills

Buccal Respiration: Gases exchanged across moist membranes of the mouth

Buccal Tube Feet: A wreath of tube feet that extend from the ring canal (in addition to the five radial canals) around the mouth

Budding: The process by which bumps or protrusions that form on the adult body and break off from the adult body eventually form a new individual (asexual reproduction)

Buds: Bumps or protrusions that form on the adult body and break off from the adult body to eventually form a new individual

Buffer: A solution that resists changes in pH when acid or alkali is added to it

Bulbus Arteriosus: A large tube that conducts blood into the aorta

Calamus: Quill

Calcareous Skeleton: The skeleton of the stony coral, which is made of secreted calcium carbonate

Calciferous Glands: Calcium secreting glands that function to remove excess calcium taken in with food, to remove excess CO_2, and to regulate the acid-base balance of body fluids.

Calyx: A cup-like body form

Calyx: Attaches the arms of a crinoid to the stalk

Cambium: Acellular plant tissue from which phloem, xylem, or cork grows by division, resulting (in woody plants) in secondary thickening

Cameral Fluid: Liquid that a *Nautilus* uses to fill the chambers of its shell to provide buoyancy

Canal Pores: Openings to the outside from the lateral line system

Candidiasis:

Candling: Holding fish fillets up to bright light to check for the presence of worms

Canines: Teeth used for piercing

Capitulum: Anterior portion of a mite that houses the mouthparts

Carapace: Upper, domed shell of a turtle

Carbohydrates: Any of a large group of organic compounds occurring in foods and living tissues and including sugars, starch, and cellulose. They contain hydrogen and oxygen in the same ratio as water (2:1) and typically can be broken down to release energy in the animal body.

Cardiac Stomach: A portion of an echinoderm's stomach that is inverted during feeding

Carnivores: Pertaining to animals that eat other animals

Carotenes: Fat-soluble yellow, orange, and red pigments found in plant chromatophores

Carrier Proteins: Proteins that transport specific substance through intracellular compartments, into the extracellular fluid, or across the cell membrane

Carrying Capacity: The maximum number of individuals of a population that can be supported by the available resources of an area for an indefinite period of time

Cartwheeling: A mating behaivor in birds in which a potential mating pair soars to great heights and then plunge to the ground and goes back up again just before reaching the ground

Catabolism: Process by which large macromolecules are reduced to smaller subunits

Catadromous: Migrating from fresh water to saltwater to breed

Catalyst: A substance that increases the rate of a chemical reaction without itself undergoing any permanent chemical change

Catastrophism: The belief that huge planet-wide geological catastrophes occurred over short time periods

Catch Connective Tissue: A dense, white fibrous and mutable connective tissue found in the body of sea cucumbers

Cation: A positively charged ion

Caudal Autonomy: Easy separation of a tail from the body

Caudal Fin: Tail fin

Cecum: An accessory pouch of the mammalian digestive system

Cells: The smallest unit of life; living building blocks

Cell Division: The process by which replicate

Cell Theory: A theory in biology that includes one or both of the statements that the cell is the fundamental structural and functional unit of living matter and that the organism is composed of autonomous cells with its properties being the sum of those of its cells

Cellulose: A biologically important polysaccharide that composes plant cell walls

Cell Walls: A rigid layer of polysaccharides lying outside the plasma membrane of the cells of plants, fungi, and bacteria. In the algae and higher plants, it consists mainly of cellulose

Central Vacuole: A large vacuole found only in plant cells

Centriole: A cylindrical cell structure composed mainly of a protein called tubulin that is found in most eukaryotic cells

Centromere: The region of a chromosome to which the microtubules of the spindle attach, via the kineto-chore, during cell division

Centrosome: An organelle that serves as the main microtubule organizing center of the animal cell as well as a regulator of cell-cycle progression

Cephalic Shield: A muscular, modified proboscis

Cephalic (head) Tentacles: Tentacles that extend from the head and may bear eyes as well as tactile and chemoreceptor cells

Cephalization: Development of a defined head end

Cereals: A grass producing a cereal grain, grown as an agricultural crop

Cerebellum: A section of the mammalian brain that interprets visual cues and coordinates muscles for movement and equilibrium

Cerebral Ganglion: A rudimentary brain

Cerebrum: A section of the mammalian brain responsible for behaviors such as eating, singing, navigation, mating, and nest building in birds

Cervicals: Vertebrae of the neck

Chaetae or Setae: Bristle-like structures that extend from the body

Chaparral: A shrubland ecosystem found primarily in California

Character: (Genetics) A heritable feature that varies among individuals

Character Displacement: A type of reservoir partitioning in which over time natural selection acts on the plasticity of animal forms and selects different forms and shapes of an organism and changes the morphology of populations (or species) that shared fundamental niches

Chelae: Pinchers

Chelicerae: Anterior appendages of an arachnid often specialized as fangs

Chemical Bonds: Any of several forces by which atoms or ions are bound in a molecule

Chemistry: A branch of the physical sciences that studies the composition, structure, properties and change of matter

Chemoautrotrophs: An organism, typically a bacterium, that derives energy from the oxidation of inorganic compounds

Chemoheterotrophs: Bacteria that use organic compounds as both carbon and energy sources

Chemoreception: Detection of chemical substances such as food chemicals and hormones

Chemoreceptors: Receptors that bind specific chemicals in the environment leading to physiological changes in an organism

Chi Square Test: A statistical test that is used to determine if two or more samples are significantly different from one another (if they are drawn from different populations

Chitin: A fibrous substance consisting of polysaccharides and forming the major constituent in the exoskeleton of arthropods and the cell walls of fungi

Chloragogen Cells: Specialized nitrogen-accumulation cells that line the coelom and extract nitrogenous wastes from the blood

Chloroplasts: A plastid that contains chlorophyll and in which photosynthesis takes place

Choanocytes: Flagellated sponge cells that help move water currents through the chambers and canals of a sponge

Choanoderm: The inner lining of the atrium of a sponge

Chorus: Calls produced by a group

Chromosomes: Coiled and thickened chromatin

Chromatin: Mass of genetic material composed of DNA and proteins that condense to form chromosomes during eukaryotic cell division

Chromatophores: Specialized pigment-containing cells in the dermis

Chromoplasts: Plastids that produce and store pigments

Chromatids: Each of the two threadlike strands into which a chromosome divides longitudinally during cell division

Chrysalis: The last larval exoskeleton of butterflies that surrounds the pupa during metamorphosis into the adult

Cilia: Minute hairlike organelles, identical in structure to flagella, that line the surfaces of certain cells or unicellular organisms and beat in rhythmic waves

Circumoral Nerve Ring: A pentagonal ring of nerves around the mouth of an echinoderm

Cirri: Ciliated, finger-like projections extending from the hood of cephalochordates and function to trap and prevent particles from clogging the hood

Cirri: Long, jointed appendages in barnacles used for feeding

Cirri: Projections that extend from the stalk of a crinoid and are arranged in whorls

Cirri: Sensory tentacles

Civilization: Complex human societies that have developed social structure and technologies

Clade: A group that includes a common ancestor and all the descendants of that ancestor

Cladogram: A branching line diagram that reveals the phylogenetic relationships between select groups

Claspers: Male sex organs used during copulation

Classical Approach: the philosophical approach to science (founded by the ancient Greeks) in which the natural world is explained by reason and logic rather than by experimentation

Classical Conditioning or **Pavlovian Conditioning**: A learning process by which a response that was initially triggered by stimulus is transferred to another stimulus that had no effect

Cleavage: The process by which the zygote divides into increasingly smaller cells

Climax Stage or **Climax Community**: The period of time marked by the presence of the final successors, typically *k*-selected species in an area

Clitellum: An external mucus producing band found on oligochaetes

Cloaca: A single posterior opening that empties the digestive, urinary, and genital tracts

Clone: An organism or cell, or group of organisms or cells, produced asexually from one ancestor or stock, to which they are genetically identical

Closed System: A system in which matter (chemicals, energy, etc) is recycled within the system

Cnidae: Stinging or adhesive structures found in cnidarians

Cnidoblasts: Cells that form from interstitial cells in the epidermis of cnidarians and that produce cnidocytes

Cnidocil: The trigger –hair on the cnidocyte once stimulated by mechanoreception causes the cnidocyte to discharge

Cnidocyte: A fully formed cnidocyte that contains the cnidae

Coaxial Glands or **Malphigian Tubules**: Paired, thin-walled spherical sacs that absorb nitrogenous wastes from the hemolymph

Coccyx: A line of small vertebrae at the base of the spinal column of chordates

Cochlea: Part of the inner ear concerned with hearing

Cocoon: A sealed capsule that forms around the fertilized eggs in the mucus ring of some worm species

Codominance: A heterozygous individual expressing two traits simultaneously without any blending

Codon: A sequence of three nucleotides that together form a unit of genetic code in a DNA or RNA molecule

Coelom: A fluid-filled body cavity that forms within the mesoderm

Coelomic Sinuses: Openings in the coelom

Cofactors: Inorganic ions or nonprotein organic molecules that must be present for an enzyme to work properly

Cognition: Conscious intelligent acts

Cohesion: The sticking together of particles of the same substance

Cohort: A group of individuals of the same age within a population

Collar: Neck-like structure

Collencytes: Ameboid cells found in the mesohyl of sponges that secrete fibrous collagen

Colloblasts: Special adhesive cells of ctenophores

Colonies: Large, ordered groups of the same species

Comb Plates: Rows of ciliated cells on ctenophores used for locomotion

Community: An ecological unit consisting of populations of different species that coexist

Competitive Inhibition: A situation in which an inhibitor and a substrate compete for the active site of an enzyme

Competitive Exclusion: The removal of a species from an area due to negative interactions with another species sharing that same habitat and niche

Complimentary base pairing: Describes the manner in which the nitrogenous bases of the DNA molecules align with each other

Compound: A thing that is composed of two or more separate elements; a mixture

Concertina Movement: A type of locomotion in snakes in which they extend forward while bracing S-shaped loops against the sides of an enclosing passage

Conchin: A protein that composes the cuticle of molluscs

Conditioned Stimulus: A stimulus that initially fails to cause a response without trial and error

Cones: Photoreceptors that receive information about color vision

Conjugation: The process by which one bacterium transfers genetic material to another through direct contact

Conispiral: Coiled in the shape of a cone

Conservation biology: The study of the preservation, protection, or restoration of the natural environment, natural ecosystems, vegetation, and wildlife

Conspecies or **Intraspecies Competition**: Interactions between individuals of the same species

Conspecies: Animals of the same species

Constitutive Exocytosis: Proteins constantly packaged into vesicles by the Golgi complex are carried promptly to the plasma membrane where they released to the extracellular environment or incorporated into the plasma membrane

Constitutive Genes: A gene that is transcribed at a relatively constant level regardless of cell environmental conditions

Consumers or **Heterotrophs**: Organisms that consume other organisms as an energy source

Continental Drift: The gradual movement of the continents across the earth's surface through geological time

Contractile Vacuole: A vacuole in some protozoans that expels excess liquid on contraction

Control Group: The group in an experiment that does not receive the variable under examination and is used to compare the results with the test group

Convergent Evolution: Evolution of similar traits in species that are not closely related due to adaptation to a similar environment or niche

Conversion Efficiency: In ecology the ratio of net production from one trophic level to the next highest level

Coprolites: Fossilized fecal droppings

Copulatory Dart: A sharp, calcareous structure that is driven into the body wall of a partner as part of a mating ritual

Copulatory Spicules: External reproductive structures of male nematodes

Coral Bleaching: The process by which corals lose their color because of the loss of their photosynthetic symbionts, zooanthellae, due to an environmental stressor

Corona: The anterior, ciliated portion of a rotifer

Coronal Muscles: Circular sheets of epidermal muscle fibers around the bell margin and over the subumbrellar surface of medusae

Corpora Allata: Neurosensory cells that store and release PTTH and produce juvenile hormone

Cortex: The middle layer of a hair

Countershading: Dark colors on top (dorsal) and light colors on the belly (ventral)

Courtship Rituals: Elaborate and complex behaviors associated with mating

Covalent Bonds: Chemical bonds that involve the sharing of electron pairs between atoms

Coxal Glands: Excretory glands that contain as many as four pairs of nephridia as the bases of walking legs

Cranial Nerves: Nerves that pass information from the outside directly to the brain

Cranium: Braincase

Cristae: Each of the partial partitions in a mitochondrion formed by infolding of the inner membrane

Crop: A large thin-walled chamber that stores food coming down the pharynx and esophagus

Crossing Over: The exchange of genes between homologous chromosomes, resulting in a mixture of parental characteristics in offspring

Cross-pollination: A situation in which the pollen of one plant fertilizes the eggs of another plant

Crown: Composed of the calyx and arms

Cryptic Coloration or **Camouflage**: A defensive coloration strategy that allows organisms to effectively disappear against certain backgrounds

Cryptic Species: Individuals that are morphologically identical to each other but belong to different species

Ctenidal Vessels: The site of gas exchange of hemolymph before it is drawn back into the heart and pumped to the hemocoel

Ctenidia: Gills of bivalves

Cud: Partially chewed food

Culture: The institution and systems that build and maintain the physicality of an area as well as those that protect, feed, clothe, house, and entertain the human residents of that area

Cupula: A gelatinous structure that surrounds the neuromast cells

Cursorial: Walking

Cutaneous Respiration: Gases exchanged across the skin

Cuticle: The outer layer of a hair

Cuticle: A nonliving thin and transparent outer body layer on some animals and most plants

Cyclomorphosis: Morphological variations in body form due to seasonal or nutritional changes

Cydippid Larvae: Planktonic larval stages of ctenophores

Cynodonts: A type of therapsids that superficially resembles mammals

Cysticercus: A larval stage of a tapeworm that hatches from the oncosphere larva and develops in the muscle of the host where it breaks out of a cyst and develops into the adult worm

Cytokinesis: The process in which the cytoplasm of a single eukaryotic cell is divided to form two daughter cells

Cytology: The study of the structure and function of cells

Cytoplasm: The part of the cell enclosed by the plasma membrane

Cytoskeleton: A scaffolding of protein filaments and tubules found within the cytosol

Cytosol: A translucent gel in which all internal cell components are suspended

Dactylozooids: The defensive polyps of a hydrozoan

Decay: Break down

Decomposers: Organisms that feed on dead organisms

Deductive Reasoning: Drawing specific conclusions from broad generalizations

Definitive Host: Animals that harbor the adult form of a parasite

Dehiscent: The splitting at maturity along a built-in line of weakness in a plant structure in order to release its contents

Dehydration Synthesis: Reactions in which the equivalent of a water molecule—an –OH (hydroxyl group) and an –H (hydrogen atom) is removed as subunits are joined

Density: The number of individuals living in a particular area

Density Dependent: Abiotic factors whose effects on a population changes with the density of the population

Density Independent: Abiotic factors whose effect on a population is not affected by the density of the population

Dentary: The bone that forms each half of the mandible in mammals

Denticulated: Possessing teeth

Deoxyribonucleic Acid (DNA): The double-stranded genetic code, composed of nucleic acids, for the formation of proteins

Dependent Variable: The variable being tested in a scientific experiment. The dependent variable is 'dependent' on the independent variable

Dermal Gills or **Papulae**: Thin folds in the body wall that function in gas exchange

Dermis: Spongy, inner layer of skin

Desiccation: Drying out

Determinate Growth: A pattern in which growth stops in the adult

Detritus: Waste or remains of living organisms

Deuterocereburum: A section of the supraesophageal ganglion

Deuterstomes: Organisms with embryos that develop via radial cleavage of the first cells and from an anus first at the blastopore

Developmental biology: The study of the preservation, protection, or restoration of the natural environment, natural ecosystems, vegetation, and wildlife

Diadromous: Migrating between fresh water and saltwater

Diaphragm: A muscle that separates the thoracic cavity from the abdominal cavity

Diapsida: Amniotes that possess two temporal openings

Diencephalon: Connects the forebrain and midbrain and functions in homeostasis

Differentially Permeable Membrane: A membrane that allows the passage of small molecules but not of large molecules

Diffusion: The process by which atoms and/or molecules move from an area of high concentration to an area of low concentration

Digestive Ceca: The digestive gland formed as an evaginations off the midgut

Digitgrade Locomotion: Walking and running with raised heels

Dioecious: Possessing separate sexes

Diphyodont: Milk or deciduous teeth replaced by a set that lasts the rest of the lifetime

Dipleurula Larva: A basic larval type of an echinoderm with bilateral symmetry, three coelomic sacs, and bands of cilia

Diploblastic: Possessing two cell layers

Diploid: Having a full set of chromosomes (2n)

Direct or **Synchronous Flight**: A mode of flying in which the muscles acting on the bases of the wings contract to produce a downward thrust and other muscles attach dorsally and ventrally on the exoskeleton to produce upward thrust

Disaccharides: A class of sugar in which two monosaccharides are joined

Distal: Farther down the body

Distress Call: A call produced in response to pain or being seized by a predator

DNA (deoxyribose nucleic acid): An extremely long, double-stranded nucleic acid molecule arranged as a double helix that is the main constituent of the chromosome and that carries the genes as segments along its strands

DNA polymerase: A type of enzyme that is responsible for forming new copies of DNA, in the form of nucleic acid molecules

Domain: The largest/highest taxonomic category above kingdom

Dominant Trait: (Genetics) A trait that is phenotypically expressed in heterozygotes

Dormancy: A period in an organism's life cycle when growth, development, and (in animals) physical activity are temporarily stopped

Dorsal: Pertaining to the top or back side

Dorsal Aortas: Arteries that receive blood from the ventral aorta in cephalochordates

Dorsal Fins: A fin located on the back (dorsal) region

Dorsal Vessel: A blood vessel located over the digestive tract on the back side of an organism

Double Covalent Bonds: When two pairs of electrons are shared between atoms

Dyck Texture: Structural texture effect in microscopic portions of the feather that lead to certain blue and green colors and metallic sheens

Dynamic Soaring: Flight that requires continuous winds in which the bird is carried high; a wind-pushed glide

Eccrine Glands: A type of sweat gland that releases a clear, watery fluid onto the surface of the skin

Ecdysis: The process of molting (shedding the outer skeleton layer and forming a larger one

Ecological Niche: All of the resources and physical conditions that a species requires for survival

Ecological Pyramids: Diagrams that represent the relationship between energy and trophic levels within an ecosystem

Ecological Succession: The process by which communities gradually change from more simple to more complex assemblages

Ecology: The branch of biology that studies the interactions between live and the physical environment within the biosphere

Ecosystem: A complex of communities and their interactions with the physical environment in a particular area

Ectoderm: The outer cellular layers of the developing gastrula

Ectoneural (oral): A sensory neural network that coordinates the tube feet and consists of a pentagonal circumoral nerve ring and radial nerves

Ectoparasites: An organism ((parasite) that feeds on the outside of its host

Ectothermic: Derived heat from external sources

Electron: A stable subatomic particle with a charge of negative electricity, found in all atoms and acting as the primary carrier of electricity in solids

Electron Cloud: An informal term in physics used to describe where electrons are when they go around the nucleus of an atom

Electronegativity: A measure of the tendency of an atom to attract a bonding pair of electrons

Electron Transport Chain: A chemical reaction where *electrons* are transferred from a high-energy molecule to a low-energy molecule

Elements: Types of atoms

Embryo: An unborn or unhatched offspring in the process of development

Embryology: The study of the development of embryos starting with the fertilization of the zygote

Embryonic Diapause: A strategy in which the embryo (blastocyst) does not immediately implant in the uterine wall but is maintained in a state of dormancy

Emigration: Process of individuals leaving an area or population

Empiricism: The approach to science (founded in Western culture during the Renaissance) in which the natural world is explained through accurate observation, measurement, and experimentation

Encephalization: An evolutionary increase in the complexity or relative size of the brain, involving a shift of function from noncortical parts of the brain to the cortex

Encystation: The process by which a foreign object that enters a bivalve is covered in many layers of nacre

Endergonic reactions: Reactions that require an input of energy to occur

Endocytosis: A process by which cells engulf large extracellular particles

Endoderm: The inner cellular layers of the developing gastrula

Endoparasites: An organism that feeds on the inside of its host

Endoplasmic Reticulum: A network of tubules and flattened sacs that perform a number of functions within eukaryotic cells

Endosperm: The part of a seed that acts as a food store for the developing plant embryo, usually containing starch with protein and other nutrients

Endostyle: An organ in cephalochordates that binds iodine and produces strings of mucus that trap food from the water as it passes through the gill slits and into the atrium

Endosymbiosis: A symbiosis in which one of the symbiotic organisms lives inside another

Endosymbiotic Theory: An explanation for the rise of eukaryotic cells

Endosymbiont: An organism living inside another organism benefitting both organisms

Endothermic: Heat derived from internal processes

Energy levels: The fixed amount of energy that a system described by quantum mechanics, such as a molecule, atom, electron, or nucleus, can have

Enterocoely: The process of the blastopore developing into the anus

Entomology: The study of insects

Entoneural (aboral): A neural system that consists of a nerve ring around the anus and radial nerves in echinoderms

Enzymes: Biological catalysts responsible for thousands of metabolic reactions that sustain life

Enzymatic–gland Cells: Wedge-shaped ciliated cells of the gastrodermis of cnidarians that produce the enzymes for extracellular digestion

Epicuticle: The outermost layer of the exoskeleton of arthropods that functions as a barrier to water and microorganisms

Epidermis: The outer most layer of cells in animals with a tissue grade of organization

Epidermis: Outer layer of skin

Epifaunal: Bottom dwelling but not a burrower

Epigenetics: A field of genetics that examines the reversible and heritable changes that can occur without a change in DNA sequence

Epigyne: Female genital opening of spiders

Epiparasitism: The process of one parasite feeding on another parasite

Epipelagic: The open sea beyond the continental shelf from the surface down to 200 meters. The sunlit layers

Epiphytes: A plant that grows on another plant upon which it depends for mechanical support but not for nutrients

Epistasis: The interaction of genes that are not alleles, in particular the suppression of the effect of one such gene by another

Epitoke: The posterior portion of some tube-dwelling and burrowing worms that develop gonads during the breeding season

Equilibrium Model: A biological model depicting stable and unchanging communities

Ergotism: Poisoning produced by eating food affected by ergot fungus, typically resulting in headache, vomiting, diarrhea, and gangrene of the fingers and toes

Esophagus: A tube that extends from the buccal cavity (or pharynx) to the stomach and is part of the digestive system

Estivation: A type of hibernation that occurs under stressful conditions in which the metabolic rate approaches zero until conditions are favorable again

Estrous Cycle: Estrus cycle that occurs at regular intervals throughout the year

Estrus: Reproductive receptivity in mammals, heat

Estuary Ecosystems: An ecosystem that is located at the junction of fresh and salt water and contains a mixture of the two (brackish)

Etching Cells: Specialized cells in boring invertebrates that chemically remove fragments or chips of calcareous material

Ethology: The study of animal behavior

Episodic Behavior: A behavior that is adjusted based on recalled social content of a previous event

Ethology: The study of animal behavior

Eukaryotes: Organisms that have cells with membrane-bound organelles

Eukaryotic Cell: Cells that contain membrane-bound organelles

Euprimates: True primates

Euryhaline Fishes: Fishes that can live in both salt and freshwater

Eusociality: Members group of the same species living together in a social group in which there is a division of labor and reproductive and nonreproductive individual

Eutely: Phenomenon in which adults of a cell line always have the same number of cells

Eutrophic Lakes: Lakes with an abundant supply of organic matter and minerals and high biological activity

Eutrophication: The process by which oxygen is depleted from the water due to the rapid/explosive growth of algae, which in turn causes the suffocation of fish and other aquatic fauna

Evolution: Change over time

Evolutionary Biology: The study of the origin and descent of species over time

Evolutionary Theory of Aging: The theory that ageing is a result of natural selection acting on a species' life span

Exchange Pools: Places where chemical are stored for a short period of time (typically within biota)

Exergonic reactions: Spontaneous reactions that release energy

Exocytosis: A process by which a cell directs the contents of secretory vesicles out of the cell membrane and into the extracellular space

Exoskeleton: A chitinous skeleton found primarily in arthropods

Experimental Theory: A theory that is derived from an experiment and has a median level of truth probability

Experimentation: To try or test, especially in order to discover or prove something

Exponential Growth: Rapid growth of a population

External Ectoderm: The outer germ layers of an embryo that will become the skin, skin glands, hair, and nails

External Ear: Anatomical structure from the sounds-conducting canal to the ear drum

Extinction: Cessation of a behavior

Extinction: No longer existing or living; an extinct species

Extracellular: Outside the body

Extrinsic: Relating to something outside of itself

Exumbrella: The convex upper surface of the bell of a cnidarian

Eyespots or **ocelli**: Structures that contain photoreceptors and can detect light but are not able to form images

Facilitated Transport: The process of spontaneous passive transport of molecules or ions across a cell's membrane via specific transmembrane integral proteins

Facultative Anaerobes: An organism that makes ATP by aerobic respiration if oxygen is present, but is capable of switching to fermentation or anaerobic respiration if oxygen is absent

Fangs: Specialized teeth in snakes used to deliver venom into victims

Fat: A biologically important hydrophobic organic molecule

Fatty acids: A long hydrocarbon chain with an even number of carbons and a –COOH (carboxyl) group at one end

Fauna: Animals

Fecundity: Potential reproductive capacity

Fermentation: A process in which sugars are converted to ethyl alcohol and carbon dioxide

Filial Imprinting: Bonding between parent and offspring

Fitness: Chance of reproductive success

Flagellated Canals: Evaginations of sponge body into finger-like projections that are lined by cells possessing flagella and can beat to move water through the canal

Flame Cell: A cell with a tuft of beating cilia that functions to bring water and wastes into the canals of the protonephridia for elimination

Flapping Flight: Flight produced by up and down motion of the wings

Fledge: Become mature enough to fly

Flexors: Muscles that bend the body or limb at the articulation point

Flight Feathers: The longer contour feathers of the wings and tail

Flooded Forests: Forests covered completely or at least occasionally in water.

Flora: Plants

Fluid Mosaic Model: A theory that holds that the plasma membrane is constructed of a bilayer of phospholipids

Flying Phalangers or **Wrist-winged Gliders**: Animals with a fold of skin running from the wrists to the ankles and used in gliding

Food Chain: A linear map of who eats who (or what) in an ecosystem

Food Vacuole: A vacuole with a digestive function in the cytoplasm of a protozoan

Food Web: Overlapping and interlocking food chains

Foot: A muscular structure used for locomotion and burrowing in molluscs, which develops from the ventral body wall

Foot: The attachment organ of a rotifer that possesses pedal glands

Foraging: The process of finding food

Forebrain: Composed of the olfactory lobes and the telencephalon

Foregut: The anterior most section of the arthropod digestive track that serves to ingest, transport, store, and mechanically digest food

Fossil: Any remains, impressions, or traces of living organisms from past geological ages

Fossorial: Pertaining to digging and burrowing

Founder Effect: The reduced genetic diversity that results when a population is descended from a small number of colonizing ancestors

Fovea: A portion of the eye with dense concentrations of receptor cells that perceives images more sharply

Frontal Plane: A plane through the body that cuts the body into dorsal and ventral sections

Functional Group: A group of atoms responsible for the characteristic reactions of a particular compound

Fundamental Niche: The total range of environmental conditions suitable for survival of a species that discounts the effects of interspecific competition and predation

Furcula or **Wishbone**: A forked bone found in theropod dinosaurs and birds that functions to strengthen the thoracic skeleton for the rigors of flight

Fusiform: Funnel-shaped

Futurists: Those who try and predict the future

Gametes: The sex cells; cells that will combine with another cell to form a new organism; sperm and egg in animals and pollen and egg in plants

Gametophyte Generation: Multicellular haploid generation

Gape: Resting for long periods of time on land with the mouth open

Gas Gland: A gland that is the site where gases from the bloodstream enter the swim bladder

Gastric Filaments: A series of filaments in scyphozoans and cubozoans that possess cnidocytes

Gastric Pouches: Extensions of the stomach of a scyphozoan and cubozoans that aid in the circulation of water

Gastrocoel: The inner body cavity that forms due to the invagination of the endoderm of the gastrula

Gastrodermis: The inner most layers of cells of diploblastic organisms

Gastrovascular Cavity or **Coelenteron**: A sac-like or branched cavity in a cnidarian that contains a single opening that serves as both mouth and anus

Gastrozooid: The feeding polyp of a hydrozoan

Gastrula: The embryonic structure that forms after the invagination of the outer cells to form an inner body cavity called the gastrocoel

Geckos: Lizards that are nocturnal, small in size, agile, lack eyelids, and possess large eyes and toe pads

Gemmules: Small spherical structures produced by sponges at the onset of winter or during drought that are resistant of freezing or drying

Genes: a distinct sequence of nucleotides forming part of a chromosome, the order of which determines the order of monomers in a polypeptide or nucleic acid molecule which a cell (or virus) may synthesize

Genetics: The study of genes and heredity

Genetic Linkage: The tendency of alleles that are close together on a chromosome to be inherited together during the meiosis phase of sexual reproduction

Genetic Mass: The amount of DNA in an organism

Gene Pool: The genetic makeup of a population

Genetically Modified Organism (GMO): An organism that has had it DNA manipulated in some way; also called transgenic

Genetic Disorders: A genetic problem caused by one or more abnormalities in the genome, especially a condition that is present from birth (congenital)

Genital Opercula: The first pair of appendages on the opisthosoma that covers the genital pores

Genome: The complete set of genes or genetic material present in a cell or organism

Genophagy: A theory that dominant genes push recessive genes out of existence

Genotype: (Genetics) The genetic makeup of an individual

Geologic Time: The long period of *time* occupied by the earth's *geologic* history

Geological Uniformitarianism: The proposal that the Earth is shaped by slow-acting natural forces over very long periods of time

Germ Layers: Cellular layers that arise during embryonic development

Germinal Cell: Any cell that gives rise to gametes

Germination: The sprouting of a seedling from a seed of an angiosperm or gymnosperm

Germinative Zone: The area just behind the scolex in a tapeworm where new proglottids arise

Germline Mutation: Mutations that occur in the DNA of reproductive cells and can be passed to offspring

Gestation Period: Period of time spent developing and growing inside the mother's uterus

Gestation: The period of intrauterine development

Gill Rakers: A fan-like projection that filters plankton and particles for feeding

Gill Slits: Opening along the pharynx for gas exchange in enteropneusts

Gill Slits: Slits in the body of an animal that cover the gills and allow water to flow out

Girdle: An extension of the body that encircles the outer edge of a chiton shell and can be used to clamp down on rocks and trap prey items

Gizzard: A muscular chamber that leads from the crop and is where food particles are ground up

Gladius or **Pen:** A small, thin internal shell of some molluscs

Gliding: The simplest form of flying in which a bird produces several strong wing strokes and then glides

Glochidia: The larvae of certain species of freshwater mussels which are released into the water and attach to the gills or fins of a fish and develop onto juvenile mussels

Glomerulus: An excretory organ of hemichordates with finger-like outpockets that extract metabolic wastes and release them to the outside

Glottis: Tube-like tracheal opening that allows snakes to breathe while swallowing

Glucose: A biologically significant six-carbon sugar

Glycerol: A 3-carbon compound with three polar –OH groups

Glycogen: A biologically important polysaccharide used to store energy by animals

Gnathochilarium: The lower plate-like jaw of a millipede

Golgi Complex: An organelle consisting of layers of flattened sacs that take up and processes secretory products

Gonopods: Modified trunk appendages used to transfer sperm to females in some arthropods

Gossamer: Spider silk that floats on the wind

Ground State: The lowest energy state of an atom or other particle

Graviportal: Locomotion produced from animals possessing short hind limbs in which the diameter of the bones is disproportionately large

Guard Hair: Coarse, long protective hairs that produce coloration

Gut: A tube that runs down the length of the body from the mouth to the anus

Habitat: The place where an organism normally lives or where the individuals of a population live within their range

Habituation: A form of nonassociative learning in which the responder stops responding to a stimulus over time

Half-life: The amount of time it takes for an isotope to break down by fifty percent

Haploid: Having a half set of chromosomes (n)

Haplorhini: Anthropoids; group consisting of tarsiers, tamarins, baboons, monkeys, apes, and humans

Hardy-Weinburg Equilibrium: A probability model that shows that gene frequencies are inherently stable, but that evolution is expected in all populations all of the time

Heart Vesicle: A contractile circulatory organ located within the proboscis of enteropneusts

Hectocotylus: The modification of a cephalopod arm for transferring sperm

Hedonic Glands: Specialized courtship glands in salamanders

Hemal: A blood-vascular system derived from coelomic sinuses

Hemimetabolous: Having larval stages that undergo gradual changes to become adults

Hemocoel: A body cavity that contains blood or hemolymph

Hemocoel: The body cavity in some invertebrates that is bathed in hemolymph

Hemocyanin: A copper-containing protein found in the hemolymph of molluscs that transports oxygen throughout the circulatory system.

Hemolymph: The watery substance that makes up the blood of invertebrates

Hemotoxins: Anticoagulants that destroy the clotting ability of blood resulting in heavy bleeding

Hepatic Cecum: A structure associated with the gut that functions in lipid and glycogen storage and protein synthesis in cephalochordates

Herbivores: Animals that consume plants

Hermaphroditic: Possessing both male and female reproductive capabilities

Herpetology: The study of amphibians and reptiles

Heterocercal Tail: Caudal fin of sharks in which the vertebral column turns upward at the tail

Heterodont: Teeth varying in size and shape based on function

Heterosis: Tendency of a crossbred individual to show qualities superior to those of both parents

Heterosporous: Producing two different kinds of spores

Heterozygous: Two different alleles for a specific gene

Hibernation: A period of inactivity in which metabolic and respiratory rates slow and body temperature drops

Hindbrain or **Metencephalon**: Posterior part of the brain that controls respiration and osmoregulation as well as controlling some muscles and organs

Hindgut: The posterior most section of the arthropod digestive tract that serves to absorb water and prepare fecal matter

Hirudotherapy: The use of leeches to restore blood circulation to grafted or severely damaged tissue

Histology: The study of the microscopic anatomy of cells and tissues

Histones: Proteins complexed with DNA strands

Holistic Concept: The notion that as the climate and other factors change species respond to these changes in body structure

Holometabolous: Having immature that are very different in body form to the adult and require a drastic change in body structure to become an adult

Home Base: A single location where individuals reside

Homeostasis: The proposition that all organisms maintain stable internal conditions that do not change as conditions in the external environment do change

Homocercal Tail: Caudal fin of bony fish that is suspended from the tip of the vertebral column

Homologous: Similarity of structure that results from descent from a common ancestor

Homologous Chromosomes: A chromosome as the same gene sequence as another

Homosporous: Producing spores of one kind only that are not differentiated by sex

Homozygosity: State of possessing two identical forms of a particular gene, one inherited from each parent

Homozygous: Having the same alleles at a one or more gene loci on homologous chromosome segments

Home Range: The area an individual covers or patrols in its normal routine

Horns: Paired, permanent structures that protrude from the frontals

Host: An organism fed upon by a parasitic organism

Hybridization: The process of genetically crossing different variants to create a hybrid organism

Hydrocarbons: Compounds containing hydrogen and carbon

Hydrolysis: The chemical breakdown of a compound due to reaction with water

Hydrophilic: Having a tendency to mix with, dissolve in, or be wetted by water

Hydrophobic: Tending to repel or fail to mix with water

Hydrosphere: Layers of water found on and in the planet

Hydrothermal Vents: An opening in the sea floor out of which heated mineral-rich water flows

Hydrostatic (turgor) Pressure: The pressure exerted by a fluid at equilibrium at a given point within the fluid, due to the force of gravity

Hydrostatic Skeleton: A support structure that consists of a water-filled interior constrained by the muscular walls of the body

Hydrothermal Vents: Fissures in the deep sea where magma rises close to the seafloor and heats water that then rises out of these fissures

Hyperosmotic: body fluids contain more salt and less water than in the external environment surrounding them

Hypertonic: A situation in which the concentration of water is greater outside a cell than in

Hypertrophy: Excessive enlargement

Hypodermic Impregnation: A form of reproduction whereby the copulating partners stab each other with the penis and inject sperm through the body wall of the other

Hypodermis: The body wall layer that secretes the cuticle in arthropods

Hyponeural: A deep system of parallel nerves that controls motor function

Hypothesis: A tentative explanation or answer to a problem or question

Hypotonic: A situation in which the concentration of water is greater inside a cell than out

Ichthyology: The study of fish

Iguanas: The most familiar New World lizards including marine iguanas and flying dragons

Ilium: Bone of the pelvis

Imago: The adult that emerges from the cocoon, chrysalis, or puparium

Immigration: The process of individuals entering an area

Immunocompetence: The ability to distinguish between self and nonself

Imprinting: Species recognition and bonding

Incisors: Teeth with sharp edges for snipping and biting

Inclusions: Small organisms or pieces of organisms that become trapped and are preserved in amber

Incomplete Dominance: A form of intermediate inheritance in which one allele for a specific trait is not completely expressed over its paired allele

Incurrent Canals: Canals/passageways in a sponge formed by finger-like invaginations of a sponge body

Incurrent Pores: Opening in a sponge's body through which water flows into the sponge

Incus: A bone of the middle ear in mammals

Indehiscent: A pod or fruit that does not split open to release its seeds

Independent (control) Variables: Factors that are kept constant among different experimental groups over the course of an experiment

Indeterminate Growth: A pattern of growth in which the skeleton continues to grow throughout most of the lifetime

Index Fossils: A fossil that is useful for dating and correlating the strata in which it is found

Indirect or **Asynchronous Flight**: A mode of flying in which muscles act to change the shape of the exoskeleton to produce upward and downward strokes

Individualistic Concept: The notion that as the climate and other factors change species respond to these changes independently

Inductive Reasoning: Making broad generalizations from specific observations

Infaunal: Sediment dwelling burrower

Infundibulum: The expanded end of the oviduct

Innate: Behaviors based on preset neural pathways

Inner Ear: Location of the hearing organ

Inorganic Compound: Any compound that is not organic compound. Some simple compounds which contain carbon are usually considered inorganic. These include carbon monoxide, carbon dioxide, carbonates, cyanides, cyanates, carbides, and thiocyanates

Inquilinism: The use of a second organism for housing

Insectivorous: Animals that eat insects, worms, grubs, and other small invertebrates

Instars: Molt stages in between the egg and adult

Inorganic Compounds: Compounds that do not contain carbon, and do not consist of nor are derived from living material

Ion Channel Proteins: A protein that acts as a pore in a cell membrane and permits the selective passage of ions

Isotonic: A situation in which the concentration of water is the same outside and inside a cell

Instinctive: Something that is not learned but possessed at birth

Integrational Theory: A theory that is supported by a large body of evidence and has a high level of truth probability

Intellect, Intelligence: Ability to think, reason, and learn

Interglacial Periods: Periods of warming between glacial periods

Interphase: The phase of the cell cycle in which a typical cell spends most of its life

Interspecies Competition: Between individuals of different species

Interspecific: Describes interactions between members of different species

Interstitial Cells: Totipotent cells with a large nucleus that give rise to sperm and egg cells and are located between the epidermal layer and epitheliomuscular cells in cnidarians

Interstitial Fauna: Animals that inhabit the water-filled spaces between grains of sand

Intertidal or Littoral Region: A zone in the marine ecosystem where the ocean meets the land

Intestine: A tube that leads from the stomach to the posterior end of the digestive system

Intracellular: Within the body

Intraspecies Competition: Between individuals of the same species

Intraspecific: Describes interactions between members of different species

Intrauterine Cannibalism: Eating of siblings by the largest embryos in the uterus

Intrinsic Rate of Increase: The exponential growth rate of a population

Introduced Species: Species that are new residents to a particular area where they are not normally found

Ion: An atom or molecule with a net electric charge due to the loss or gain of one or more electrons

Ionic Bonds: the complete transfer of valence electron(s) between atoms

Iridophores: A type of pigment cell, or chromatophore, containing a silvery pigment that reflects light

Island Gigantism: The observation that some animals on islands grow to be considerably larger than their continental counterparts

Insoluble: Incapable of being dissolved in water

Isomers: Each of two or more compounds with the same formula but a different arrangement of atoms in the molecule and different properties

Isotopes: Elements that possess the same number of protons but different numbers of neutrons

Iteroparity: The ability to reproduce continually throughout the lifetime

Jacobsen's Organ: An organ responsible for the sense of smell in reptiles that contains a region of chemically sensitive nerve endings

Johnston's Organs: Long setae located at the base of antennae in most insects that vibrate when hit with certain frequencies

Jump Dispersal: A means of dispersal (way to spread out in space) in which an actively reproducing population becomes established beyond the normal range of the population

Juvenile Hormone: A hormone produced in the corpora alta and secreted during molting of the larval stages

Karyotype: The number and visual appearance of the chromosomes in the cell nuclei of an organism or species

k-selected: Species or populations with lower reproductive rates that survive best when the population is close to its carrying capacity

Keeled: possessing a sternum

Kentrogens: A developmental stage of a parasitic barnacle that injects parasitic cells into the hemocoel of their crab host

Keratin: Tough, fibrous protein that provides protection against abrasion and dehydration in outer epidermal cells

Keystone Species: An influential species that has an unusually strong direct effect on the composition of a community

Kin Selection: Changes in gene frequency across generations that are driven by interactions between related individuals

Kinetic Skull: Adapted to enable animals to deliver a faster and more powerful bite and open its mouth wider to facilitate the capture and handling of prey

Kleptocnidae: The process of incorporating cnidae (stingers) from another animal by digesting them

Kleptoparasitism: The process of a parasite stealing food that belongs to the host

Kreb's Cycle (or Citric Acid Cycle): The sequence of reactions by which most living cells generate energy during the process of aerobic respiration

Labrum: The flap-like mouth structure formed by a fusion of the second maxillae

Labyrinth Organ: A simple lung-like structure found in some freshwater fish that takes in oxygen from the air

Lachrymal Glands: Tear glands

Lacunae: Thin walled spaces that sit among blood vessels in a closed circulatory system

Lacunar System: Complex circulatory channels found in the epidermis

Lamellae: Plate-like folding of gills

Lamellae: Thin, flat plates or discs

Larynx: Structure that encloses the vocal cords; the voice box

Lateral Canals: Many smaller canals that are part of the water vascular system of a sea star, which extend from the radial canals and end in a bulbous ampullae

Lateral Diverticula: Side pouches of an organ

Lateral Line Canal: A variation of the lateral line system of fish in which receptor cells (neuromasts) lie beneath the skin rather than be directly exposed to the environment

Lateral Line System: Mechanoreception that occurs along the sides of a fish

Lateral Undulation: Locomotion patterns in snakes in which an S-shaped path is left behind the moving snake

Law of Superposition: A geologic principle that states that the lower a layer of rocks is in the Earth, the older the rocks

Learned or **Associative Behavior**: A behavior that develops from modification of a behavior through experience

Lek: An area where males display (dance and make noises) to attract females

Leucoplasts: Organelles in plant cells that store starch, lipids or proteins, or have biosynthetic functions creating a variety of organic compounds

Leuconoid Sponges: Sponges that typically lack an atrium but possess many small flagellated chambers

Leucophores: Pigment cells that reflect light

Life Span: The average length of time from birth to death of an organism

Limnetic Zone: Open freshwater from shore to shore and down as far as light penetrates

Lipids: Any of a class of organic compounds that are fatty acids or their derivatives and are insoluble in water but soluble in organic solvents.

Lipochromes: Red, orange, and yellow pigments found only in the feathers of parrots

Lithosphere: The rocky outer part of the Earth consisting of the crust and uppermost mantel

Littoral Zone: The shallow edge around the shore where freshwater meets land

Loop of Henle: A specialized structure in the nephron that allow kidneys to concentrate urine

Lophophore: A fringe of hollow tentacles used for feeding

Lorica: The plate-like encasement of the body of a Loriciferan

Loricate Larva: The larval form of a priapulid that lives in mud and morphologically resembles a loriciferan

Lunate Wrist Bone: A bone found in theropod dinosaurs and birds that allows for twisting motions required for flight

Lung: The respiratory organ found in most terrestrial organisms

Lymphatic Hearts: Contractile vessels that move around the lymphatic fluid

Lymphatic System: A network of vessels that filter fluids, ions, and proteins from capillary beds in tissue spaces and returns them to the circulatory system

Lysosome: An organelle in the cytoplasm of eukaryotic cells containing degradative enzymes enclosed in a membrane

Macroecosystems: A large area containing living biota and abiotic media in which life exists, such as forests, prairies, deserts, lakes, ponds, streams, and rivers

Macroevolution: Evolution on the large scale

Macrofossils: Large fossils

Macromolecules: Large and complex assemblages of atoms joined together

Macropores: Pores or openings in sediment produced by burrowing animals

Madreporite or **Sieve Plate**: The opening to the water vascular system of a sea star

Maggot Debridement Therapy: The placement of maggots into a wound dressing allowing maggots to eat the dead, decaying tissue leaving the wound clean

Male Pore: Openings on the body where sperm is released during mating

Malleus: A bone of the inner ear of mammals

Malpighian Tubules: Paired, thin-walled spherical sacs that absorb nitrogenous wastes from the hemolymph

Mammary Glands: Modified sweat glands that secrete milk for the initial development period of young mammals

Mammalogy: The study of mammals

Mandible: The lower jaw

Mandibles: Invertebrate jaws

Mantle: Fleshy lobes that form from the dorsal body wall and secrete calcareous spicules, shell plates, or shell in molluscs

Mantle: The inner membrane that lines the tunic of tunicates

Mantle Cavity: A body cavity that houses the visceral mass and opens to the outside to allow the exchange of gases, food, wastes, and reproductive cells with the outside

Manubrium: A fold of the body wall that hangs from the center of the subumbrella and surrounds the mouth of a hydromedusa

Marine Biology: The study of ocean ecosystems and the living things that inhabit those ecosystems

Mark and Recapture: A method of calculating the density of a species in which a small portion of the population is captured, tagged, released, and ideally recaptured

Marsupium: A special pouch where the young of marsupials complete their development

Mastax: The pharynx of a rotifer

Materialism: A theory that physical matter is the only or fundamental reality and that all being and processes and phenomena can be explained as manifestations or results of matter

Maxilla: The upper jaw

Maxillae: Either of a pair of irregularly shaped bones of the skull fused in the midline supporting the upper teeth and forming part of the eye sockets, hard palate, and nasal cavity

Maxillary Glands: Excretory glands that contain a single pair of nephridia that are found on the head in association with the maxillae

Maxillary Teeth: Small cone teeth around the edge of the upper jaw in frogs

Maxillipedes: Paired venomous poison claws in centipedes

Maxillules: The first maxillae in crustaceans

Means of Dispersal: A mechanism for species to spread out in space and time

Medulla: The inner layer of a hair

Megaphylls: A leaf with several or many large veins branching apart or running parallel and connected by a network of smaller veins

Megasporangia: A plant structure in which megaspores are formed

Megaspores: The larger of the two kinds of spores produced by some ferns

Meiosis: A type of cell division that results in four daughter cells each with half the number of chromosomes of the parent cell, as in the production of gametes and plant spores

Melanin: Brown, black, gray, and red-brown pigments

Melanophores: A type of chromatophore containing brown or black melanin

Mesencephalon: Midbrain

Meristems: A region of plant tissue, found chiefly at the growing tips of roots and shoots and in the cambium, consisting of actively dividing cells forming new tissue

Mesenchyme or **Mesoglea**: A layer of partially cellular gelatinous material derived from the ectoderm and located between the epidermis and gastrodermis of cnidarians

Mesenteries: Membranous sheets

Mesoderm: The middle germ layer that develops in some animals from the migration of endodermal cells in the developing gastrula

Mesoecosystem: A medium-sized ecosystem within a macroecosystem

Mesohyl: A middle layer of a sponge between the pinacoderm and choaoderm that contains the ameboid cells, collenocytes, spongocytes, sclerocytes, and archaeocytes

Mesopelagic: The open sea beyond the continental shelf from 200 meters to around 1,000 meters down

Mesosoma: The first six segments of the abdomen in scorpions

Mesothorax: The middle-most segment of an insect body

Messenger RNA (mRNA): The form of RNA in which genetic information transcribed from DNA as a sequence of bases is transferred to a ribosome

Metabiosis: A mode of living in which one organism is indirectly dependent on another for the preparation of the environment in which it can live

Metabolism: The chemical processes that occur within a living organism in order to maintain life

Metaphase: A stage in mitosis in which the chromosomes align in the equator of the cell before being equally separated into each of the two daughter cells

Metamerism: Possessing a segment body

Metamorphosis: The process of development in which the body of the larvae changes drastically to become an adult

Metanephridia: A type of nephridia that has an open ciliated funnel on one end and opens to the outside on the other end and functions as a kidney in invertebrates

Metasoma: A long tail composed of six segments that bears the stinger in scorpions

Metastasize: The process by which cancers spread through the body

Metathorax: The posterior-most segment of an insect body

Metazoans: Complex, multicellular animals

Microbiology: The study of microscopic organism

Microecosystem: A small-sized ecosystem within a mesoecosystems

Microevolution: Evolution on a small-scale

Microfilaments: Contractile fibers found within the cytosol

Microfossils: Tiny fossils

Microhabitat: An extremely localized small-scale environment

Microphylls: A type of plant leaf with one single, unbranched leaf vein

Microtriches: Tiny folds within the tegument of flukes and tapeworms that function to increase surface area

Mictic: Amictic females that produce ova by meiosis

Microtubules: A component of the microskeleton

Midbrain or **Mesencephalon**: A portion of the brain associated with vision, hearing, motor control, sleep/wake, alertness, and temperature regulation

Midden: An ancient mound of shells that were discarded by early humans

Middle Ear: The portion of the ear internal to the eardrum and external to the oval window of the cochlea

Midgut: The middle section of the arthropod digestive tract that serves to produce enzymes, and to digest and absorb nutrients

Migration: Mass movement of animals

Milk: Nutritious secretion of the mammary glands

Mimicry: The process of copying the color patterns, behavior, or structural adaptations of another species

Mitochondria: An organelle found in large numbers in most cells, in which the biochemical processes of respiration and energy production occur

Mitosis: The process, in the cell cycle, by which the chromosomes in the cell nucleus are separated into two identical sets of chromosomes, each in its own nucleus

Mitotic Spindle: The subcellular structure that segregates chromosomes between daughter cells during cell division

Mixotrophs: An organism capable of existing as an autotroph or a heterotroph at the same time

Moist Deciduous and Semi-evergreen Forests: Forests in areas with a cool dry winter during which time many of the trees drop some or all their leaves

Molars: Teeth for crushing and grinding

Molecules: a group of atoms bonded together, representing the smallest fundamental unit of a chemical compound that can take part in a chemical reaction

Molecular Biology: The study of biological functions at the molecular level

Molecular Clock: The average rate at which a particular kind of gene or protein evolves

Molecular Formula: A formula giving the number of atoms of each of the elements present in one molecule of a specific compound

Molting: The replacement of old feathers as they become worn and drop out

Molting Gel: A fluid that enters the space between the hypodermis and the old exoskeleton during molting

Monocular Vision: Vision in which an animal can see things on each side of their head at the same time as well as in front because the eyes are on the sides of the head

Monoecious: Possessing both male and female gonads in the same body

Monoesterous: Possessing one estrus period per breeding cycle

Monogamous: Mated pair remains together for at least one breeding season

Monohybrid Cross: A genetic cross between parents that differ in the alleles they possess for one particular gene, one parent having two dominant alleles and the other two recessives

Monogamy: A reproductive strategy in which one male and one female mate exclusively

Monomers: A molecule that can be bonded to other identical molecules to form a polymer

Monophyletic: A group of species whose members are related to one another through a unique history of common descent

Monosaccharides: Simple sugars such as glucose, galactose, and fructose

Monounsaturated fats: Fat molecules that have one unsaturated carbon bond in the molecule

Montane Rain Forests: Forests that experience at least 40 inches of rain and are found in mountainous areas with cooler climates

Morphology: The exterior form and structure of an organism

Mortality: Death

Motility: Possessing the ability to move

Mouth Brooding: Fertilized eggs are brooded in the mouth until hatching

Mullerian Mimicry: The process of several unrelated species resembling one another

Mutagens: Something that induces mutations in DNA

Mutation: permanent altering of a nucleotide sequence in the genome of an organism

Medulla or **Myelencephalon**: Posterior part of the brain that controls muscles and organs

Mycelium: The vegetative part of a fungus, consisting of a network of fine white filaments (hyphae)

Mycology: The study of fungi

Mycotoxins: Any toxic substance produced by a fungus

Myomeres: Zigzag band arrangement of musculature

Myoneme: Contractile myofibril in epitheliomuscular cells in cnidarians

Nacre: A smooth iridescent, aragonite-based secretion deposited on the inner surface of some bivalve shells

Naiads: The aquatic, immature Hemimetabolous stages of terrestrial insects

Nares: External openings of the nasal cavity

Nasohypophyseal Opening: An opening on the dorsal side of the head of a lamprey

Natality: Birth

Natatorial: Swimming

Nauplius Larva: An arthropod larval stage characterized by the presence of three head appendages

Nebulae: Cloud of gas and dust in outer space, visible in the night sky either as an indistinct bright patch or as a dark silhouette against other luminous matter.

Neck: The region located immediately behind the scolex in a tapeworm

Necrosis: Death of tissue

Necrotic: Dead

Nematocysts: Harpoon-like tubes that are connected to toxin sacs and are used to impale and paralyze prey

Neo-Darwinism or **Modern Evolutionary Synthesis:** The combination of Darwinian natural selection and Mendelian inheritance that explains both micro- and macroevolutionary changes

Neolithic Revolution: The birth of agriculture

Neonates: Undeveloped young

Neoteny: A form of paedomorphosis in which an organism displays physiological maturity but not sexual maturity

Nephridia: Tube or funnel-like structures that filter wastes from the body

Nephridioducts: Excretory pores that lead from the protonephridia to the external environment

Nephridiopore: An opening to the outside (or to the mantle cavity) that connects to the nephridia

Nephrostome: An open ciliated funnel of a metanephridia

Neritic Zone: A zone in the marine ecosystem from the low-tide line to the edge of the continental shelf

Nerve Net: Connection between the between the hyponeural, ectoneural, and entoneural systems of echinoderms

Nerve Net: A diffuse network of nerve cells forming a primitive nervous system in cnidarians, ctenophores, and certain other organisms

Nerve Rings: Bundles of nerves concentrated around the bell margin of a hydromedusae

Nested Hierarchy: Clades nested within one another

Neurobiology: The study of nervous systems

Neural Crest: An embryonic ectodermal structure that becomes facial cartilage, skin pigment cells, ganglia of the autonomic nervous system, dentin of teeth, spiral septum of the heart, and ciliary body of the eye

Neural Tube: An embryonic ectodermal structure that becomes the brain, spinal cord, retina, and posterior pituitary of the adult

Neuromast Cells: A cluster of hair cells that are joined together in a gelatinous cupula

Neuropathy: Degenerative changes to nerves

Neurotoxins: Toxins that act on the nervous system

Neutral Atom: An atom in which the number of proton charges balance the number of electron charges

Neutron: A subatomic particle of about the same mass as a proton but without an electric charge, present in all atomic nuclei except those of ordinary hydrogen

Niche Differentiation: A type of resource partitioning in which populations (or species) utilize different parts of a shared resource

Nictitating Membrane: A transparent or translucent third eyelid in some animals that can be drawn across the eye for protection

Nitrogenous Base: A molecule that acts as a base and contains nitrogen

Nonassociative Learning: The simplest form of learning in which a behavioral change occurs due to a repeated presentation of a stimulus

Noncompetitive inhibition: An occurrence in which an inhibitor binds to an enzyme at a location other than the active site

Nonequilibrium Models: Biological models depicting unstable and changing communities

Nonobligate Brood Parasites: Birds that are capable of raising their own young but lay eggs in the nest of conspecies to increase reproductive output

Nonpolar Molecules: Molecules in which the electrons are shared equally by the atoms resulting in no electrical charge

Notochord: A hard tube-shaped cluster of cells that are wrapped in a fibrous sheath and extends along the central nervous system and provides a stiff structure of the attachment of muscles

Nuchal Organs: Sensory pits or grooves that function as chemoreceptors located in the head region

Nucleus: A dense organelle present in most eukaryotic cells, typically a single rounded structure bounded by a double membrane, containing the genetic material

Nucleic Acid: A complex organic substance present in living cells, especially DNA or RNA, whose molecules consist of many nucleotides linked in a long chain

Nucleolus: An organelle found within the nucleus of eukaryotic cells

Nucleons: A proton or neutron

Nucleosome: Structural unit of a eukaryotic chromosome, consisting of a length of DNA coiled around a core of histones

Nucleotides: Molecules that contain a sugar, nitrogenous base, and a phosphate group

Null Hypothesis: A type of hypothesis used in statistics that proposes that no statistical significance exists in a set of given observations

Nutritive-muscle Cells: Large, elongated cells of the gastrodermis of cnidarians that engulf small food particles by phagocytosis

Nymphs: Immature hemimetabolous individuals

Obligate Anaerobes: An organism that can grow only in the complete absence of molecular oxygen

Obligate Brood Parasites: Birds that are incapable of raising their own young and must lay eggs in the nest of other species

Obligate Intracellular Parasites: Parasites that cannot reproduce outside their host cell

Obligate Mutualism: The process of two species being so tightly associated that they cannot live separately

Obligate: Necessary for survival

Occipital Condyle: Bony protrusions from the base of the skull that seat the skull on top the first vertebrae

Ocelli: Discs or pits of photoreceptor cells that allow the animal to respond to light

Ocelli: Small, simple eyes of some invertebrates

Octet Rule: A chemical rule that reflects the observation that atoms of main-group elements tend to combine in such a way that each atom has eight electrons in its valence shell, giving it the same electronic configuration as a noble gas

Oils: Triglycerides produced by plants in liquid form

Oldowan Tools: Regularly patterned stone tools associated with Homo habilis

Olfactory Nerves: Nerves that transport sensory information from the olfactory lobes to the brain

Olfactory Rosette: Sensory pads that detect chemicals and line the chamber that extends from the nares

Oligotrophic Lakes: Lakes with a small amount of organic matters and minerals and low biological activity

Ommatidia: The fused receptor units of a compound eye

Omnivores: Organisms that eat both plants and other animals

Oncosphere Larva: A larval stage produced by some tapeworm species that hatches and lives within an intermediate host

Oogenesis: The process by which ova (eggs) are formed

Oophagy: Eating of unfertilized eggs by developing embryos

Ooze: Thick blanket of mud and decaying matter that has accumulated over millions of years on the deep sea floor

Open System: A system in which matter or energy flows through and can be lost from the system

Operant Conditioning: Learning through trial and error

Operculum: A middle-ear bone that transmits vibrations from the pectoral girdle to the inner ear in amphibians

Operculum: Gill cover in fishes

Operon Model: A model depicting the coordinated expression of a group of genes

Opisthaptor: The posterior attachment organ of monogenetic flukes

Opisthosoma: The posterior tagmata of arthropods that consists of the abdomen

Optic Lobe: A region of the brain attached to the midbrain and is responsible for vision, flight, and food acquisition in birds

Optimal Foraging Theory: The most successful animals tend to feed on prey that maximizes their net energy intake per amount of foraging time

Oral (incurrent) Siphon: An inlet for water and the mouth opening of tunicates

Oral Arms: Extensions of the manubrium of cnidarians that contain cnidocytes and aid in the capture and ingestion of prey

Oral Disc: The flared oral end of a sea anemone that possesses eight to several hundred hollow tentacles

Oral End: The top end of polyps that contains the mouth and tentacles

Oral Hood: An anteriorly projecting structure that surrounds the mouth on the body of a cephalochordate

Oral Sucker: An anterior, hookerless sucker that surrounds the mouth of diagenetic flukes

Orbitals: Each of the actual or potential patterns of electron density that may be formed in an atom or molecule by one or more electrons, and that can be represented as a wave function

Orbits: Eye sockets

Organic Compounds: Compounds that contain carbon or are derived from living things

Organs: Part of a living thing, distinct from the other parts, that is adapted for a specific function

Organelles: Membrane bound structures found in eukaryotic cells

Organic: Containing carbon atoms

Organic Compound: Any member of a large class of gaseous, liquid, or solid chemical compounds whose molecules contain carbon

Organ Systems: A group of *organs* that work with one another in order to achieve or perform a particular task or set of tasks

Ornithology: The study of birds

Orthogenesis: The hypothesis that life has an innate tendency to progress from simple forms to higher, complex forms

Osculum: An opening found at the top of some sponge

Osmoconformers: Animals whose internal osmotic concentrations is the same as the surrounding environment

Osmoregulators: Animals that maintain the osmotic difference between their body fluid and the surrounding environment

Osmosis: The diffusion of water molecules from an area of high concentration to an area of low concentration

Osmotic Pressure: The minimum pressure which needs to be applied to a solution to prevent the inward flow of water across a semipermeable membrane

Osphradia: Chemoreceptors patches on the gill or mantle wall

Ossicles: Inner-ear bones consisting of the malleus, incus, and stapes

Ossicles: Separate plates that arise from mesodermal tissue and form a calcareous skeleton

Osteoderms: Bony plates located beneath the keratinized dermis scales and reinforces them in lizards and crocodiles

Ostia: Pores that perforate the body of a sponge

Ostia: Perforations in the heart wall that allow blood to flow into the heart in some animals

Otoliths: Earbones

Outgroup: An organism that is not contained within a group of related organisms

Ovale: A muscular valve in the posterior dorsal region of the bladder that opens to release the high internal pressure in the bladder

Ovaries: Female reproductive organs where eggs are released

Overfishing: Reduction of fish stocks below an acceptable level

Overturn: A seasonal mixing of the zones in large, temperate lakes

Oviduct Funnels: Tubes that connect the ovaries to the oviducts and allow for the passage of eggs

Oviducts: Openings on the body where eggs are released

Oviparity: Egg-laying in which eggs are externally fertilized

Oviparous: A developmental strategy in which egg development occurs within the female's body until they are ready to hatch

Ovipositors: Female structures used to place eggs on or in a substrate

Ovoviparous: A developmental strategy in which egg development occurs within the female's body until they are ready to hatch

Ovum: The single-celled egg

Ozone: A form of atmospheric oxygen that blocks ultraviolet radiation from the sun; the Earth's "sunscreen"

Paedomorphosis: Adults retain some but not all juvenile characteristics and are sexually mature

Palate: Roof of the mammalian mouth

Paleontology: The study of fossils

Pallial Oviduct: A tubular structure of the female reproductive system of molluscs in which eggs are fertilized and sperm is temporarily stored

Pallial Tentacles: Sensory tentacles located in the margin of the mantle of a bivalve and contain tactile and chemoreception cells

Papillae: Horny, keratinous outgrowths from specialized areas of the skin

Paraphyletic: As assemblage in which member species are all descendants of a common ancestor but which does not contain all the species descended from that ancestor

Parapodia: Fleshy lobes that extended from the body

Parasitoids: Parasites that lay eggs of living hosts that then feed on the host's tissues until the host dies

Parasitology: The study of parasites and parasitism

Parenchyma: Tightly packed cells that are derived from the mesodermal layers

Parietal Eye: An eye buried beneath opaque skin that detects changes in light intensity

Pars Superior: Upper section of the inner ear of a fish that determines balance

Parthenogenesis: A mode of reproduction involving the development of an unfertilized egg into a function individual

Patagium: The fold of skin that runs from the wrist to the ankles in flying phalangers

Pearl: Formed by layers of nacre that were laid down by a bivalve in response to the presence of a foreign object between the shells

Pectines: A feathery response organ in scorpions

Pectoral Fins: Anterior pair of fins

Pectorals: Breast muscles

Pedal Cords: Ventral nerve cords that innervate the muscle of the foot in molluscs

Pedal Disc: The flattened aboral end of a sea anemone used for attaching to the substrata

Pedal Gland: An organization of cells that produces large amounts of mucus to lubricate the sole of a snail's foot during locomotion

Pedal Glands: Located on the foot of a rotifer and produce an adhesive secretion

Pedal Laceration: A method of asexual reproduction in sea anemones in which parts of the pedal disc are left behind while moving; these remains form new adults

Pedicel: Connection between the antler and their bony attachment

Pedicellariae: Pincher-like structures that protrude from the aboral surface of some sea stars and function to clean the body of debris and protect the body

Pedigree: The recorded ancestry of a family or individual

Pedipalps: Specialized sensory appendages located near the mouth of an arachnid

Peer Review: A process in which scientific work is evaluated and reviewed by experts in the field

Pelagic Zone: The open sea beyond the continental shelf

Pellicle: A thin layer supporting the cell membrane in various protozoa

Pelvic Fins: Posterior pair of fins

Pelycosaurs: The collective term for the first herbivorous and carnivorous synapsids

Penis Fencing: A reproductive behavior in hermaphroditic flatworms in which the first male to successfully pierce the skin of another flatworm becomes the male of the mating pair and delivers its sperm to its mate (the loser in the fight assumes the female role)

Pentaradial: A body symmetry in which the body parts are arranged in fives or a multiple of five around an oral-aboral axis

Peptide: Two or more amino acids bonded together

Peptide Bonds: The bonds that join amino acids together

Perennials: Plants that live for more than two years

Pericardial Cavity or **Pericardial Coelom**: The cavity containing the heart

Pericardial Sinus: A circulatory structure in some invertebrates that collects the blood as it returns to the heart

Perihemal Sinuses: Coelomic channels that enclose the hemal system of echinoderms

Periodic Table: A table of the chemical elements arranged in order of atomic number, usually in rows, so that elements with similar atomic structure (and hence similar chemical properties) appear in vertical columns

Periostracum: A thin, organic surface coat on the inside surface of a molluscan shell

Peristomium: The second anterior segment that surrounds the mouth and bears sensory tentacles

Peritoneum: A slick epithelial lining of the coelom

Permafrost: Permanently frozen ground

Permian-Triassic Extinction Event: A mass extinction event informally known as the "Great Dying" spread out over a few million years when 96% of marine species and 70% of terrestrial species went extinct

Phagocytosis: Engulfment of a particle or organism by a cell

Phagosome or **Food Vacuole**: A vesicle formed around a particle (usually food) absorbed by phagocytosis

Pharyngeal Gill Slits: Slits in the body of an animal that cover the gills and allow water to flow out of the body of chordates

Pharynx: A tube that leads from the mouth into the gastrovascular cavity of an anthozoans

Pharynx: A muscular structure in the anterior portion of the digestive system superior to the esophagus

Phasmids: Unicellular sensilla in the lateral tail region of certain species of nematodes

Phenotype: (Genetics) The physical appearance of an individual

Pheromone: A body secretion that functions as a mode of communication

Phloem: The living tissue that carries organic nutrients to all parts of the plant where needed

Phoresy: The use of a second organism for transportation

Phosphate Group: A molecule that contains phosphate

Phospholipids: Triglycerides in which the third fatty acid attached to glycerol has been replaced by a polar phosphate group

Photoautotrophs: Organisms that carry out photosynthesis

Photoreceptors: Receptors that detect the presence of light

Photosynthesis: A process carried out by plants and photosynthetic bacteria that uses solar energy to convert carbon, hydrogen, hydrogen, and water molecules into simple sugars and carbon dioxide

Phototaxis: Ability in animals to respond to light

Phyllopidia: The expanded and flattened swimming limbs of some crustaceans

Phyllotaxy: The arrangement of leaves on an axis or stem

pH scale: A measure of acidity or alkalinity of water soluble substances

Phylogenetic Trees: Depiction of evolutionary relationships among more than two groups of taxa

Phylogenetics: A biological classification scheme that groups organisms based on their evolutionary relationships

Phylogeny: The history of organismal lineages over time

Physiology: The study of the normal functions of organisms and their parts

Physoclistous: Lacking a pneumatic duct

Physostomous: Possessing a pneumatic duct

Phytoplankton: The microscopic primary producers in aquatic ecosystems

Pinacocytes: A layer of flattened cells on the outer surface of a sponge

Pinacoderm: The outer surface of a sponge

Pineal Body: Located above the diencephalon and functions to detect light, maintain circadian rhythms, and control color changes

Pinna (pl.Pinnae): Funnel-shaped flap of skin at the external ear opening

Pinnules: Smaller branches with a feathery appearance

Pinocytosis: The ingestion of liquid into a cell by the budding of small vesicles from the cell membrane

Pioneer Species: The first species that inhabit newly disturbed land, typically *r*-select species

Pioneer Stage: The period of time in an area marked by the presence of the pioneer species

Placenta: A spongy mass embedded in the wall of the mother's uterus that nourishes the embryo

Planospiral: A coiled, spiral shaped that is flattened

Plantigrade Locomotion: Whaling and running on the soles of the feet

Planula Larva: A free-swimming cnidarian larva that turns into either a polyp or medusa

Plasma Membrane: A bilayer membrane that surrounds and contains a cell

Plasmids: A circular DNA molecule in bacteria that is separate from chromosomal DNA and can replicate independently

Plasmolysis: Contraction of the cytoplasm of a cell due to water loss

Plastids: Any of a class of small organelles, such as chloroplasts, in the cytoplasm of plant cells, containing pigment or food

Plastron: The flat lower shell of a turtle

Pleiotropy: The production by a single gene of two or more apparently unrelated effects

Plesiadapiformes: Proto-primate mammals

Pleurites: The lateral sclerite (exoskeletal plate)

Pleurodont Teeth: Teeth attached to the inner side of the jaw

Pleuroperitoneal Cavity: The cavity containing the lungs

Plumage: The arrangement and appearance of all feathers on a bird's body

Pneumatic Duct: A connection between the gut and swim bladder in basal teleost fish

Pneumatized Bone: Hollow bones with air spaces in them and stiffened by internal struts

Podia: Hollow, muscular sucker that connects to the bulbous ampullae as part of the water vascular system of an echinoderm

Podites: Limb segments or pieces of arthropods

Point Mutations: A small-scale mutation in the DNA usually affecting only a few nucleotides in a single gene

Polar Molecules: A molecule that has a partial positive charge in one part and a complementary negative charge in another part

Polian Vesicles: Blind pouches that arise from the ring canal, which are thought to help regulate internal pressure of the water vascular system of a sea star

Pollination: The process by which pollen is transferred to the female reproductive organs of a plant, thereby enabling fertilization to take place.

Polyandry: A reproductive strategy in which females mate with more than one male

Polyestrous: Estrus cycle that occurs at regular intervals though out the year

Polygamy: A reproductive strategy in which males or females have more than one mate

Polymers: Large macromolecules composed of many subunits of the same type linked together

Polynomial: Possessing several scientific names

Polypeptide: A linear organic polymer consisting of a large number of amino-acid residues bonded together in a chain

Polyphyletic: A group comprising species that arose from two or more different ancestors

Polyploidy: A situation in which a cell contains more than two paired (homologous) sets of chromosomes

Polysaccharides: Complex sugars; many simple sugars units bonded together (starch and cellulose)

Polyunsaturated Fat: A fat that contains more than one double bond in the carbon chain

Population: An ecological unit consisting of a group or groups of individuals of the same species living in the same place at the same time

Population Bottleneck: A decrease in the genetic variation of populations due to genetic drift

Population Ecology: The study of the dynamics of populations

Population Equilibrium: The state of a population when emigration and immigration are more or less in balance

Population Explosions: Times of rapid, exponential growth of a population

Population Genetics: The study of changes in the gene frequencies within populations

Porocyte: A ring-shaped cell of a sponge that forms the incurrent pore

Posterior: Pertaining to the rear or tail end

Precocial: Birds born with eyes open, down feathers, and the ability to move and collect food

Predator: An organism that kills and consumes another organism

Preening: The process of zipping together barbs of feathers that have become separated by running the feathers through the beak

Prehensile Tail: A tail adapted for grasping

Premolars: Teeth for shearing, slicing, crushing, and grinding

Prey: An organism that is consumed by another organism

Primary Consumers (Herbivores): Organisms that feed directly on the primary producers (plants)

Primary Producers (Autotrophs): Organisms that produce their own food through photosynthesis (plants)

Primary Productivity: The amount of photosynthetic activity

Primary Succession: The first stage in ecological succession in which new colonizers inhabit an open space

Primers: Something that elicits slow, long lasting behavioral responses

Prions (proteinaceous infection particles): A protein particle that is the agent of infection in a variety of neurodegenerative diseases

Proboscis: A piercing anterior structure that extends from the mouth in some insects and nemerteans

Proboscis: The flexible conspicuously long snout of some mammals (tapirs, shrews); especially the trunk of an elephant

Programmed Apoptosis: Genetically predetermined cell death

Prokaryotic Cell: A cell in which the nuclear material is not bound by a membrane

Procuticle: The hard inner layer of an insect exoskeleton

Products: A substance formed as the result of a chemical reaction

Profundal Zone: The area below the limits of light penetration in freshwater

Progenesis: A form of paedomorphosis in which sexual maturity is sped up in juveniles

Proglottids: Individual sections of tapeworms

Prohaptor: A posterior adhesive organ of monogenetic flukes

Prokaryotes: Organisms that have cells lacking membrane bound organelles

Prokaryotic Cell: A type of cell that lacks membrane bound organelles

Proton: A stable subatomic particle occurring in all atomic nuclei, with a positive electric charge equal in magnitude to that of an electron, but of opposite sign

Prophase: A stage of mitosis in which the chromatin condenses into double rod-shaped structures called chromosomes in which the chromatin becomes visible

Prosimians: Strepsirrhini; group consisting of tarsiers, lemurs, aye-aye, bush babies, potto, and lorises

Prosoma (Cephalothorax): The anterior tagmata (section) of arthropods that is a fusion of the head and thorax

Prostaglandins: A group of 20 lipids that act as local chemical messengers in vertebrate tissues

Prostomium: The anterior most segment of annelids that contain antennae in sea worms

Protandric: Gonads first produce sperm and then later produce eggs

Protandry: Changing sex from male to female

Proteins: Any of a class of nitrogenous organic compounds that consist of large molecules composed of one or more long chains of amino acids and are an essential part of all living organisms

Prothoracic Gland: A hormone producing organ in the head of insects that releases ecdysone

Prothoracicotropic Hormone (PTTH): A hormone produced by cells in the brain and ganglia that stimulate the release of ecdysone

Prothorax: The anterior-most segment of an insect body

Protocerebrum: A section of the supraesophageal ganglion

Protogyny: Changing sex from female to male

Protonephridium: A tube-like nephridia with a closed bulb at one end and an opening to outside at the other end

Protostomes: Organisms with embryos that develop via spiral cleavage of the first cells and form a mouth first at the blastopore

Proventriculus: First section of a bird's stomach where food is mixed with digestive juices

Proximal: Close to the body core

Proximate Causation: The immediate causes of an animal's behavior

Pseudocoel: A "false" body cavity that develops between the mesoderm and endoderm

Pseudopodia: A temporary protrusion of the surface of an amoeboid cell for movement and feeding

Psuedogenes: Gene remnants that no longer function but continue to be part of an organism's genome

Pterylae: Tracts on a bird's skin from which feathers grow

Ptychocysts: Specialized cnidocytes that produce the tubes of burrowing anthozoans

Puggle: A young echidna

Pulmonary Respiration: Gases exchanged across lungs

Pupa: The last larval molt of a homometabolous species

Puparium: The last larval exoskeleton of flies that surrounds the pup during metamorphosis into the adult

Purines: Nucleotides that have a base with a double ring

Pygidium: The terminal segment of a worm

Pygostyle: The remains of the reptilian tail in birds

Pyloric Ceca: Structures in ray-finned fish that function to absorb fat

Pyloric Ducts: Extensions of the pyloric stomach that stretch into each arm of a starfish

Pyloric Stomach: A portion of an echinoderm's stomach that is not inverted for feeding

Pyramid of Biomass: A diagram that represents the biomass present at each trophic level within an ecosystem

Pyramid of Energy: A diagram that represents the number of calories per trophic level

Pyramid of Numbers: A diagram that represents the relative numbers of organisms at each trophic level

Pyrimidines: Nucleotides that have a base with a single ring

Quadrate: A small bone at the back of the upper jaw

Quadrate: The skull bone that connects to the lower jaw

Qualitative Data: Any data that contains numbers; "hard data"

Quantitative Data: Any data that are observational or anecdotal; "soft data"

r-selected: Species or populations with high reproductive rates that survive best when the population is far below the carrying capacity

Radial Canals: Part of the water vascular system of a sea star that extends from the circumoral canal into each of the five arms of a sea star

Radial Nerves: Large sensory nerves that extend into each arm of an echinoderm

Radial Symmetry: A body design in which the body parts radiate outward from a central axis

Radicle: The first part of a seedling (a growing plant embryo) to emerge from the seed during the process of germination

Radioactive Decay: The process by which the nucleus of an unstable atom loses energy by emitting radiation, including alpha particles, beta particles, gamma rays, and conversion electrons

Radioactive Isotope: Natural or artificially created isotope of a chemical element having an unstable nucleus that decays, emitting alpha, beta, or gamma rays until stability is reached

Radula: A file-like set of hooked teeth used for rasping and tearing

Ram Ventilation: Swimming with an open mouth to create a temporary current across the gills

Ramus: A single, branching appendage

Range: A geographical area that provides the necessary requirements for a population to exist

Range of Tolerance: The environmental and physical requirements for a population to exist

Raptorial Claws: Specialized forelimbs modified for piercing or smashing prey in mantis shrimp

Rate-of-living Theory of Aging: The theory that ageing (and deterioration) is an inevitable outcome of the harmful act of living and cell metabolism

Reactants: Substances initially present in a chemical reaction that are consumed during the reaction to make products

Reaggregation: the process of separated cells fusing together again

Realized Niche: The actual habitat that an organism or population occupies after accounting for competition and predation

Recessive Traits: (Genetics) A situation in which a trait may be phenotypically produced if both genes of an allele pair are identical

Reciprocal Altruism: A behavior in which one individual benefits at the immediate cost to another; however, this individual is guaranteed future benefit

Reciprocal Crosses: A pair of crosses between a male of one strain and a female of another, and then reversing the gender roles and crossing again

Recombinant DNA: A form of biotechnology in which DNA is isolated from one organism, manipulated in some way, and then reintroduced into the cells of another organism

Rectal Glands or Rectal Sacs: Protrusions from the short intestine of an echinoderm

Rectilinear Movement: A type of caterpillar crawl locomotion in some snakes

Reciprocity: A hypothesis suggesting that individuals may form "partnerships" in which mutual exchanges occur because they benefit both participants

Red Tide: A bloom of toxic red algae

Regeneration: The process of producing viable adults from small fragments

Regulated Exocytosis, A situation in which proteins and small molecules stored in vesicles fuse with the plasma membrane after an extracellular signal triggers the cell

Regulated Genes: A *gene* that *regulates* the expression of one or more structural *genes* by controlling the production of a protein (as a *genetic* repressor) which *regulates* their rate of transcription

Release Calls: Produced by either male amphibians that are mistaken as females by other males or by unresponsive female amphibians

Releaser: A trigger that initiates a behavior

Renopericardial Canals: Blood vessels that extend from the heart to the kidneys

Reproductive Caste: The queens and drones in a eusocial species

Repugnatorial Glands: Cells that produce defensive secretions that are foul tasting and can be poisonous or sedative in nature

Reservoir: An area of the environment that serves as a storage place for chemicals over a long period of time

Residence Time: The amount of time an element or chemical remains in a reservoir or exchange pool

Resin: A syrupy material excreted from species of evergreen trees

Resource Partitioning: A subdivision of a shared resource by populations or species with overlapping fundamental niches

Respiration: A process in living organisms involving the production of energy, typically with the intake of oxygen and the release of carbon dioxide from the oxidation of complex organic substances

Rete Mirabile: A mesh of arteries that is a site of high acidity from the release of lactic acid by the gas gland

Retia Mirabilis. A mesh of blood vessels that allows for counter-current heat exchange.

Retinula Cells: Specialized cells of the ommatidium that possess special light-collecting areas called rhabdom, which convert light energy into nerve impulses

Retroviruses: Any of a group of RNA viruses that insert a DNA copy of their genome into the host cell in order to replicate

Rhabdite: A type of rhabdoid that secretes mucus to prevent desiccation in a turbellarian

Rhabdom: Special light-collecting areas of retinula cells that convert light energy into nerve impulses

Rheoreceptors: Receptors that are stimulated by water currents and allow an organism to orient its body to various water movements

Rhizoids: Short, thin filaments found in fungi and in certain plants and sponges that anchor the growing (vegetative) body of the organism to a substratum

Rhizome: A continuously growing horizontal underground stem that puts out lateral shoots and adventitious roots at intervals

Rhombencephalon: Hindbrain

Rhopalia: Specialized projections between the scalloped margins of a hydromedusan's bell that contain statocysts and ocelli

Rhynchocoel: The hollow "nasal cavity" of nemerteans, which houses an eversible proboscis

Ribonucleic Acid (RNA): The single-stranded genetic code that serves as messenger and organized in the construction of those proteins

Ribosomes: A minute particle consisting of RNA and associated proteins, found in large numbers in the cytoplasm of living cells

Ribosomal RNA (rRNA): A type of RNA, distinguished by its length and abundance, functioning in protein synthesis as a component of ribosomes

Ribozymes: Enzymes made of RNA instead of protein

Ricochetal: Using just the hind limbs to hop in fast succession

Ring or Circumoral Canal: Part of the water vascular system of a sea star that forms a ring in the body of the sea star and has five separate radial canals branching off it

Ritualized Threat Displays: Threatening displays that rarely result in injury or death because they have been used time and again to send information to opponents

Riverine: Best adapted to calmer areas in deep fast-flowing rivers

RNA (ribonucleic acid): A polymeric molecule essential in various biological roles in coding, decoding, regulation, and expression of genes

Rods: Photoreceptors that function in less intense light than other kinds of photoreceptors

Root Hairs: Each of a large number of elongated microscopic outgrowths from the outer layer of cells in a root, absorbing moisture and nutrients from the soil

Root Pressure: Osmotic pressure within the cells of a root system that causes sap to rise through a plant stem to the leaves

Rostral Organ: A large gel-filled cavity in the snout of a coelacanth

Rostrum: Snout of a fish

Rough Endoplasmic Reticulum: ER with ribosomes embedded

Ruminants: Mammals that grind their food with broad, flat teeth, swallow it, and later bring it back up for further chewing

Salt-absorbing Cells: Cells in the gill epithelium that transfer salt ions from the water to the blood

Salt-secreting Cells: Special cells that excrete salt from the blood to the surrounding water

Saltatorial: Hopping

Saltatory Locomotion: Leaping movement

Saprotrophs, Saprophytes: Organisms that feed on nonliving organic matter

Saturated Fats: Fats that contain only single bonds between carbon atoms

Savanna: Relatively dry grasslands and bush lands with occasional trees

Scabies: An infestation of itch mites that tunnel through the epidermis of the skin and excrete irritating chemicals

Scalids: Thin pointed structures; stylets

Scansorial: Climbing

Scarified: Weakening, opening, or otherwise altering the coat of a seed to encourage germination

Scavenger (or Detritivores): Organisms that feed on dead and decaying matter

Schistosomiasis or Bilharzia: A debilitating disease in humans caused by the fluke *Schistosoma* living within the circulatory system

Schizocoel: The embryonic gut of an annelid

Schizocoely: The process of the blastopore developing into the mouth

Science: The search for natural truths by exacting individuals (scientists) using precise and reliable methods

Scientific Law: A theory with a considerable amount of evidence that has proven to be correct and true over a considerable amount of time; the highest probability level of truth in science

Scientific Method: A method of procedure consisting in systematic observation, measurement, and experiment, and the formulation, testing, and modification of hypotheses

Sclerites: Four exoskeletal plates of arthropods

Sclerocytes: Amoeboid cells found in the mesohyl of sponges that produce calcareous and siliceous spicules

Sclerosepta: Thin, radiating calcareous septa within the skeletal cup of a stony coral

Sclerotization: The process of chemically cross-linking the layers of proteins in the protocuticle to harden and darken the exoskeleton

Scolex: The anterior head region of a tapeworm that contains suckers and hooks

Scutes: Broad belly scales in snakes

Scyphistoma: A polyploidy larval stage of scyphozoans and cubozoans that lives attached to hard surfaces

Sebum: The fatty, greasy secretion from sebaceous (oil) glands that provides insulation and waterproofing for the skin of mammals

Secondary Consumers or Carnivores: Organisms that feed on the primary consumers

Secondary Palate: The partitioning that separates the oral and nasal cavities

Secondary Productivity: The activity (energy processing) or amount of herbivorous animals

Secondary Succession: The second stage in ecological succession in which the first colonizers are replaced by different plants and animals

Seed Coat: The thin, protective outer layer of a seed

Selective Degradation: Situation in which proteins are systematically broken down after a period of time

Self-pollination: A situation in which the pollen of a plant fertilizes its own eggs.

Semelparity: Reproduce once and then die shortly thereafter

Seminal Vesicles: Part of the male reproductive tract where sperm cells complete differentiation

Senescence: The process of aging

Sensilla: A collective term for several mechanoreceptors and chemoreceptors

Sensory Cells: Cells within the epidermis that detect information from the surroundings

Sensory Palps: A sensory appendage located in the prostomium

Septa: Vertical structures between the gastric pouches of scyphozoans and cubozoans that possess an opening to help circulate water

Septal Filament: A glandular, ciliated band located on the free edge of the mesentery of anthozoans

Septum: A muscular structure that divides the heart into chambers

Sequential Hermaphrodites: Possess male and female sex organs but not at the same time; they change sex at some point during their life

Seral Stage: The period of time in an area that is between the pioneer and climax stages and marked by the presence of both pioneer and climax species

Serendipity: Something good or beneficial that happens by chance

Serous Glands: Large glandular cells embedded in the dermis of amphibian skin and produce a complex chemical mixture of biologically active compounds

Sessile: Not able to move freely

Sex-Linked Trait: A trait that is associated with a gene carried only on the sex chromosomes

Sexual Imprinting: The process by which a young animal learns the characteristics of a desirable mate

Sexual Reproduction: The production of new living organisms by combining genetic information from two individuals of different types (sexes)

Sexual Selection: The process by which interactions between members of the same gender and interactions between males and females (mate choice) causes changes in a population over time

Shell Glands: Cells that secrete calcareous spicules or shells in molluscs

Side-winding: A form of locomotion in snakes in which they move side-to-side quickly minimizing surface contact

Sign Stimulus: A releaser that an animal responds to

Silk: A proteinaceous liquid that hardens as it is drawn out of the spinneret of an arachnid

Simultaneous Hermaphrodites: Possess both male and female sex organs at the same time

Single Covalent Bonds: When only one pair of electrons is shared between atoms

Sinus Venosus: A thin-walled sac where blood from the fish's veins collect prior to entering the atrium

Sinuses: Open spaces around the organs of the body

Siphon or Funnel: A rolled extension of the mantle that functions as an inhalant tube in terrestrial gastropods and allows water in and out of the body of aquatic mollusc's visceral mass

Siphonoglyph: A ciliated groove in the center of the oral disc responsible for pulling water into the gastrovascular cavity of a sea anemone

Siphonozooids: Highly modified polyps used to pump water into the interconnected gastrovascular cavities of corals

Siphuncle: A small tube that is used to control buoyancy in the *Nautilus*

Skinks: A group of lizards with elongated bodies and reduced limbs

Smooth Endoplasmic Reticulum: ER with no embedded ribosomes

Soaring: Flight in which birds rise on thermal currents and then glide down until carried up by another thermal current

Social Insects: Insects that live together in large, ordered groups

Sociobiology: A field of scientific study which is based on the assumption that social behavior has resulted from evolution

Sole: The flat ventral foot of a gastropod

Solute: The minor component in a solution, dissolved in the solvent

Solution: A liquid mixture in which the minor component (the solute) is uniformly distributed within the major component (the solvent

Somatic Cells: Body cells

Somatic Cell Mutations: Mutations that occur in the DNA of body cells that cannot be passed to offspring

Somatic Tube: The outer portion of a eucoelomate that contains the sense organs and muscles

Somites: Sections

Somitogenesis: The formation of somites or body divisions

Sonation: Nonvocal sounds

Spatial Memory: A type of memory in which an animal can recall the location of an object based on its comparative position to other objects in the environment

Speciation: Rise of a new species

Species: The smallest cluster of organisms that possess at least one diagnostic character and that are reproductively isolated from all other species

Species Diversity: A measurement of the different number of species in a community

Sperm Funnel: A ciliated structure part of the male reproductive tract that connects to the vas deferens in annelids

Spermatogenesis: The process of sperm production

Spermatophore: A packet of sperm

Spherical Symmetry: A body design that is round with its parts concentrically arranged around a central point

Spicules: Either calcareous or siliceous skeletal elements of a sponge

Spinal Nerves: Nerves that pass information from the extremities through the spinal cord to the brain

Spinnerets: Conical telescoping organs that produce silk in arachnids

Spiracles: Holes that lead from the tracheae to the outside

Spiral Valve: A valve in the intestine of chondrichthyes that slows the passage of food and increases absorptive area

Spiral Valve: A valve that directs blood into the pulmonary and systemic circuits in some amphibians and lungfish

Spirocysts: Sticky tubes or loops that wrap around and stick to prey

Sponges: The simplest multicellular invertebrate animals that lack symmetry, have a cellular grade of organization, and are sessile

Spongin: Modified collagen fibers found in a sponge

Spongocoel: The internal cavity of a sponge

Spongocytes: Ameboid cells found in the mesohyl of sponges that produce supportive collagen called spongin

Spontaneous Generation: The supposed production of living organisms from nonliving matter

Sporangia: A receptacle in which asexual spores are formed

Sporangiophores: A specialized hyphae bearing sporangia

Sporophyte Generation: Multicellular diploid generation

Stapes or Columella: Middle ear bone that transmits high-frequency vibrations from the tympanic membrane to the inner ear

Starch: A biologically important polysaccharide used to store energy by plants

Statocysts: Pits or closed vesicles containing tiny calcareous statoliths that when stimulated inhibit muscular contraction on that side

Statocysts: Sensory structures that allow an organism to orient to gravity

Statoliths: Calcareous bits found in statocysts that aid in the detection of body orientation

Stenopodia: The long, thin uniramous walking appendages of arthropods

Stereogastrula: A gastrula with no cavity

Stereoisomers: Each of two or more compounds differing only in the spatial arrangement of their atoms

Stereom: An open meshwork structure with living tissue that fills the interstices that develop from the calcite crystals of an echinoderm exoskeleton

Sternite: The ventral sclerite plate of an exoskeleton

Steroids: A class of lipids that can be found in animal cell membranes or that function as a hormone

Stimuli: That which influences or causes a temporary increase of physiological activity or response in the whole organism or in any of its parts

Stolons: Connections between the zooids of pterobranchs or horizontal filaments (hyphae) found in fungi

Stomach: The digestive organ that processes food

Stomata: One of the tiny openings in the epidermis of a plant, through which gases and water vapor pass in and out

Stone Canal: Part of the water vascular system of a sea star that connects the madreporite to the circular ring

Strata: Layers; pertaining to rock

Stratigraphy: The branch of geology concerned with the order and relative position of strata and their relationship to the geological time scale

Strepsirrhini: Prosimians, lemurs, aye-aye, bush babies, potto, and lorises

Sterptoneury: The twisted internal anatomy of some molluscs in which the nervous system takes the shape of a figure eight due to torsion

Streptostyly: Hinge-like configuration of the skull that enables the back of the jaw to move freely

Stereoisomers: Each of two or more compounds differing only in the spatial arrangement of their atoms

Strip Census: A method to determine the population size of birds by identifying and counting the birds in a strip of measured width throughout an entire area

Strobila: The region of a tapeworm just behind the scolex that contains reproductive structures called proglottids

Strong Nuclear Force: One of the four basic forces in nature (the others being gravity, the electromagnetic force, and the weak nuclear force

Structural Formula: A formula that shows the arrangement of atoms in the molecule of a compound

Structural Genes: A gene that codes for any RNA or protein product other than a regulatory factor (i.e. regulatory protein)

Structural Isomers: Any of two or more compounds with identical chemical formulas that differ structurally in the sequence in which the atoms are linked

Stylets: Thin pointed structures

Synthetic Biology: The design and construction of new biological parts, structures or systems and/or the re-design of existing natural biological systems for useful purposes

Subesophageal Ganglion: The posterior portion of the tritocerebrum that loops around the esophagus and connects to the ventral nerve cord

Subumbrella: Undersurface of the bell of a medusa

Sugar: A type of carbohydrate

Summer Eggs: A type of turbellarian egg that possesses a thin shell and hatches rapidly

Supracoracoideus: The muscles that raise the wing of birds

Supraesophageal Ganglion: A bundle of fused ganglia in the head of arthropods

Survivorship: The percentage of the population that survives to a given age

Survivorship Curve: A graph of the percentage of the population that survives to a given age

Swim Bladder: A thin sac located between the peritoneal cavity and the ventral column that functions to maintain buoyancy in fish

Symbiosis: Interactions between different species or populations (interspecies)

Sympatric: Species existing in the same geographical area

Sympatric Speciation: The process through which new species evolve from a single ancestral species while inhabiting the same geographical area

Symplesiomorphies: Common ancestral characteristics

Synapomorphies: Unique derived characteristics

Synapsida: Amniotes that possess one temporal opening

Synconoid Sponges: Sponges with evaginated flagellated canals that create one layer of folding

Syncytial: Multinucleated mass of fused cells

Syrinx: The voice box of birds located near the junction of the trachea and bronchi

Synthetic Biology: An interdisciplinary branch of biology and engineering that combines various disciplines from within these domains, such as biotechnology, evolutionary biology, genetic engineering, molecular biology, molecular engineering, systems biology, biophysics, and computer engineering

t-**Test**: A statistical test that is used to determine if the means of a trait or characteristic are significantly different between two groups

Tadpole Larva: The free-swimming larval forms of a tunicate which metamorphs into an adult

Tagmata: Specialized body regions

Tagmatization: The modification of segments for specialized functions

Taxonomic Categories: Hierarchical rankings of organisms into domain, kingdom, phyla, classes, orders, families, genera, and species

Taxonomy: The scientific grouping (classification) and naming (nomenclature) of living things

Taxons: Taxonomic groups

Tegument: The nonciliated external covering of flukes and tapeworms, which functions to protect the body

Telencephalon: Forebrain equivalent to the cerebrum and functions in olfaction

Telomeres: A region of repetitive nucleotide sequences at each end of a chromatid, which protects the end of the chromosome from deterioration or from fusion with neighboring chromosomes

Telophase: The final stage of mitosis in which the nucleus reforms and the cell membrane or cell wall separates the two daughter cells

Telson: The tail section of certain crustaceans

Temperate Phages: The ability of some bacteriophages (notably coliphage ☒) to display a lysogenic life cycle

Tensile Strength: The amount of force a material can withstand without breaking

Tergite: The dorsal sclerite plate of the exoskeleton of an arthropod

Terpenes: A class of lipids that comprise many biologically important pigments

Territory: Any space an animal defends against intruders

Tertiary Productivity: The activity (energy processed) or amount of carnivores

Test: The shell of echinoderms formed by closely packed dermal ossicles

Test Group: The group that is being tested in an experiment

Testis: Male reproductive organs where sperm cells form and mature

Tetrachromatic: Possessing four different types of color-sensing cone cells in the eye

Tetrapods: Four-limbed animals; all vertebrates above fish

Thallus: A plant body that is not differentiated into stems and leaves and lacks true roots and a vascular system. Thalli are typical of algae, fungi, lichens, and some liverworts

Thecodont Teeth: Teeth set in sockets in the jawbone

Theory: An idea or proposition that is suggested or presented as possibly true but that is not known or proven to be true

Therapsids: A branch of the pelycosaurs that arose in the Permian

Thermogenesis: Generation of heat

Tiedermann's Body: Blind pouches that arise from the ring canal as part of the water vascular system of a sea star

Tissue: A large mass of similar cells that make up a part of an organism and perform a specific function

Toes: Protrusions from the foot of rotifer

Tool: Anything not part of the body that is used to accomplish a given task

Tornaria Larva: A planktonic larval stage of hemichordates with nonyolk eggs that possess ciliary bands

Torsion: Twisting

Totipotent: Possessing cellular plasticity in which a cell is capable of changing form and function

Tracheae: A series of branched, tubes that conduct gases to and from body tissues in insects

Tracheae: A thin-walled cartilaginous tube descending from the larynx to the bronchi and carrying air to the lungs

Tracheids: A type of water-conducting cell in the xylem that lacks perforations in the cell wall

Trait: (Genetics) The genetic variant for each genetic character

Trans fat: A triglyceride that has at least one bond in the trans configuration

Transgenic Animal: Animals that have had a foreign gene deliberately inserted into their genome

Transfer RNA (tRNA): RNA consisting of folded molecules that transport amino acids from the cytoplasm of a cell to a ribosome

Transpiration: The process where plants absorb water through the roots and then give off water vapor through pores in their leaves

Transpirational Pull: The proposed mechanism by which trees draw water through their roots

Transverse Fission: A form of asexual reproduction in which an organism splits in half separating head from tail

Transverse Plane: A plant through the body that cuts the body into anterior and posterior sections

Trichogyne: A hairlike terminal process forming the receptive part of the female reproductive structure in red algae and certain ascomycete and basidiomycete fungi

Triglyceride: A lipid composed of three fatty acid chains each attached to a glycerol molecule

Trinomial: Possessing three scientific names

Triple Covalent Bonds: When three pairs of electrons are shared between atoms

Triploblastic: Possessing three tissue layers

Tritocerebrum: A section of the supraesophageal ganglion

Trochophore Larva: The unique larval stage of the Trochozoa

Trochophore Larvae: Larval stage of a polychaete

Trophi: Hard jaws of some rotifers that are located on the mastax and used to grind food

Trophic Cascades: The indirect interactions between species at different levels of the trophic web

Trophic Levels: A measure of how far an organism is removed from its original source of energy

Trophic Structures: The feeding relationships within communities and ecosystems

Tropical Rain Forests: Forests with diverse biota that are constantly warm between 68° and 77° F and receive at least 40 inches of rain per year

Tube Foot: The ampullae and podia of the water vascular system of an echinoderm that is used for locomotion

Tunic: The body wall of most tunicates, which is composed of a tough connective tissue secreted by the epidermis

Turbinate Bones: Bones of the nasal cavity

Turgid: The state of being turgid or swollen, especially due to high fluid content

Tympanic Membrane: Eardrum, a piece of skin stretched over a cartilaginous ring at the end of the ear canal

Tympanic Organs: Organs composed of a thin, cuticular membrane that covers a large air sac, which, when vibrated by sound waves, leads to the stimulation of sensory nerves in insects

Type I Error: The incorrect rejection of a true null hypothesis ("false positive")

Type II Error: Incorrectly retaining a false null hypothesis ("false negative")

Typhlosole: A ridge or fold that increases the internal surface area of the intestine

Typolite: A fossil that forms as a mold of the external features of an organism

Ultimate Causation: The origin of a behavior

Unconditioned Response: A response that initially occurs due to a stimulus without any conditioning

Unconditioned Stimulus: A stimulus that initially initiates a response

Underhair: Dense, soft and highly insulative hair

Understory: Layers of plant life beneath the taller trees/plants

Ungulates: Hoofed mammals

Unguligrade Locomotion: Walking and running on a single toe capped by a hoof and with the heel raised

Unicellular Glands: Mucus producing glands embedded in the epidermis that connect to the outside through pores

Uniramous Appendages: Single, branched appendages

Unsaturated Fats: Fats that contain at least one double bond between the carbons

Ureters: Tubes that carry urine to the cloaca or urinary bladder

Uropygial Gland: An oil gland located at the base of the tail in birds

Utriculus: Lower section of the inner ear of a fish

Vacuoles: A space or vesicle within the cytoplasm of a cell, enclosed by a membrane and typically containing fluid

Valence Electron: A valence electron is an electron that is associated with an atom, and that can participate in the formation of a chemical bond

Vas Deferens: Part of the male reproductive tract that connects to the sperm funnel and to the male pore in annelid worms

Vas Deferens: The coiled duct that connects the testes with the urethra

Vectors: Carriers

Vegetative Propagation: A method of plant *propagation* not through pollination or via seeds or spores but by way of separating new plant individuals that emerge from *vegetative* parts, such as specialized stems, leaves and roots and allow them to take root and grow

Vegetative Reproduction: A form of asexual reproduction in plants, in which multicellular structures become detached from the parent plant and develop into new individuals that are genetically identical to the parent plant

Veliger: A larval stage unique to molluscs that possess a foot, shell, operculum and some other adult feature

Velum: A shelf-like inward projecting fold of the bell of a hydro medusa that helps increase the force of the water jet for locomotion

Velum: The swimming and feeding structure of the veliger larva

Velvet: Skin and soft hair that cover antlers as they grow

Ventilation: The process of actively moving air or water over the respiratory surfaces for gas exchange

Ventral: Pertaining to the bottom or belly side

Ventral Aorta: An artery that pumps blood to the dorsal arteries and then to body tissues in cephalochordates

Ventral Vessel: A blood vessel that is located under the digestive tract on the ventral side of an organism

Ventricle: A thick-walled muscular chamber of the heart that pumps blood from the heart

Vermiculture: A farming/gardening practice that uses earthworms to process soil

Vermiform: Resembling a worm in form or movement

Vessel Elements: One of the cell types found in xylem, the water conducting tissue of plants

Viral Envelopes: The outer structure that encloses the nucleocapsids of some viruses

Viroids: An infectious entity affecting plants, smaller than a virus and consisting only of nucleic acid without a protein coat

Virology: The study of viruses

Virulent Phage: A bacteriophage that causes the destruction of the host bacterium by lysis

Visceral Tube: The inner portion of a eucoelomate that contains the gut

Vitamins: Small, organic molecules required in trace amounts for the synthesis of some coenzymes and become part of a coenzyme's molecular structure

Viviparous: A developmental strategy in which the female provides nutrients directly to the embryos

Vocal Cords: Cartilaginous membranous folds located in the larynx, which vibrate and produce sound when air passes over them

Vocal Sacs: Membranous pouches under the throat or on the corner of the mouth that distend and act as resonating structures that amplify the call of a frog or toad

Vocalization: Sound production

Volant: Flying

Vomerine Teeth: A ridge of small cone teeth on the roof of the mouth in frogs

Vomeronasal Organ or Jacobson's Organ: An organ in the roof of the mouth of many animals that detects scent molecules

Waggle Dance: A coordinated dance conducted by honey bees in which a bee informs its hive mates of the direction and distance to nectar and pollen by the direction, angle, intensity, and rate of the dance

Wampum: An intricate Native American belt woven out of white and colored beads made from the shell of the channeled whelk and hard-shelled clam

Waxes: A substance that is secreted by bees and is used by them for constructing the honeycomb, that is a dull yellow solid plastic when warm, and that is composed of a mixture of esters, cerotic acid, and hydrocarbons

Weaned: Period when young no longer feed off the milk of their mother

Wheel Organ: The lateral walls of the vestibule bearing complex ciliary bands that drives food particles into the mouth of cephalochordates

Winter or Dormant Eggs: A type of turbellarian egg that possesses thick, resilient shells, which are capable of withstanding cold and desiccation

Worker Caste: Sterile members of a eusocial species

Worm Casts: The excrement that passes out of worms

Xanthophores: A type of chromatophore containing yellow, orange, and red pigments (**xanthophylls**)

Xylem: A compound tissue in vascular plants that helps provide support and that conducts water and nutrients upward from the roots, consisting of tracheids, vessels, parenchyma cells, and woody fibers

Zoea: A larval stage that typically follows the nauplius stages in crustaceans

Zonites: Divided segments of kinorhynchs that bear spines but no cilia

Zooanthellae: Photosynthetic yellow-brown unicellular algae

Zoogeography: A field of biology that studies the patterns of animal distribution in time and space

Zooids: Asexual buds that differentiate along the length of an adult turbellarian's body

Zoology: The study of animals

Zoonotic: Ability of a parasite to infect more than one species

Zoospore: A spore of certain algae, fungi, and protozoans, capable of swimming by means of a flagellum

Zygapophyses: Supportive processes on each vertebrae of an amphibian skeleton that prevent twisting when stress is applied

Zygote: The single, large cell that develops immediately after fertilization

Zygotes: Fertilized eggs

Zygospore: The thick-walled resting cell of certain fungi and algae, arising from the fusion of two similar gametes

Zygosporangium: A sporangium in which zygospores are produced

PHOTO CREDITS

Chapter 1
    1.1 ©Midosemsem/Shutterstock

Chapter 2
    2.8 ©Leremy/Shutterstock

Chapter 3
    3.1 StefanoRR; **3.9** Dirk Hünniger; **3.18** Jacek FH

Chapter 4
    4.13 Mariana Ruiz Villiarreal

Chapter 5
    5.9a Michael Schmid; **5.12a** Ali Zifan; **5.23** Boumpheyfr

Chapter 6
    6.13 Chakazul

Chapter 8
    Opener ©Nobeastsofierce/Shutterstock; **8.3** Ehamberg

Chapter 10
    10.2 Hugo Llits

Chapter 11
    11.22 Thomasione

Chapter 12
    12. 3 Jonsta247; **12.5** BruceBlaus; **12.15** Dyndna; **12.20** ©Blamb/Shutterstock

Chapter 13

13.1 Ellen Levy Finch; **13.3** John Doebly; **13.5** Ali West; **13.8** Ehamberg;

Chapter 14

14.14 Semhur

Chapter 15

15.3 ©Gilles San Martin; **15.4** Ealbert17; **15.5** Elembis; **15.6** Tsaneda; **15.17** Schyleur Shepherd; **15.12** I, Omnitarian; **15.19** Stephano Stabile; **15.20** Dohduhdah; **15.22** Geoff Gallice;

Chapter 16

16.4 Yassine Mrabet; **16.8** Ghedoghedo; **16.31** travelwayoflife;

Chapter 17

Opener ©David Carillet/Shutterstock **17.3a** ©LeonP/Shutterstock; **17.3b** ©Praiseng/Shutterstock; **17.4a** ©Dennis W. Donohue/Shutterstock;**17.4b** ©Matej Hudovernik/Shutterstock; **17.4c** ©Kjersti Joergensen/Shutterstock; **17.8** ©Didier Descouens; **17.12** ©Linda Bucklin/Shutterstock; **17.13** ©Linda Bucklin/Shutterstock; **17.16** P'tit Pierre

INDEX

NOTE: *f* indicates a figure, *t* indicates a table

CPSIA information can be obtained
at www.ICGtesting.com
Printed in the USA
BVHW052221150719
553544BV00002B/4/P